SIX-MEMBERED TRANSITION STATES IN ORGANIC SYNTHESIS

SIX-MEMBERED TRANSITION STATES IN ORGANIC SYNTHESIS

Jaemoon Yang
Montana State University
Department of Chemistry

A JOHN WILEY & SONS, INC., PUBLICATION

Copyright © 2008 by John Wiley & Sons, Inc. All rights reserved.

Published by John Wiley & Sons, Inc., Hoboken, New Jersey.
Published simultaneously in Canada.

No part of this publication may be reproduced, stored in a retrieval system, or transmitted in any form or by any means, electronic, mechanical, photocopying, recording, scanning, or otherwise, except as permitted under Section 107 or 108 of the 1976 United States Copyright Act, without either the prior written permission of the Publisher, or authorization through payment of the appropriate per-copy fee to the Copyright Clearance Center, Inc., 222 Rosewood Drive, Danvers, MA 01923, (978) 750-8400, fax (978) 750-4470, or on the web at www.copyright.com. Requests to the Publisher for permission should be addressed to the Permissions Department, John Wiley & Sons, Inc., 111 River Street, Hoboken, NJ 07030, (201) 748-6011, fax (201) 748-6008, or online at http://www.wiley.com/go/permission.

Limit of Liability/Disclaimer of Warranty: While the publisher and author have used their best efforts in preparing this book, they make no representations or warranties with respect to the accuracy or completeness of the contents of this book and specifically disclaim any implied warranties of merchantability or fitness for a particular purpose. No warranty may be created or extended by sales representatives or written sales materials. The advice and strategies contained herein may not be suitable for your situation. You should consult with a professional where appropriate. Neither the publisher nor author shall be liable for any loss of profit or any other commercial damages, including but not limited to special, incidental, consequential, or other damages.

For general information on our other products and services or for technical support, please contact our Customer Care Department within the United States at (800) 762-2974, outside the United States at (317) 572-3993 or fax (317) 572-4002.

Wiley also publishes its books in a variety of electronic formats. Some content that appears in print may not be available in electronic formats. For more information about Wiley products, visit our web site at www.wiley.com.

Library of Congress Cataloging-in-Publication Data:

Yang, Jaemoon
  Six-membered transition states in organic synthesis / by Jaemoon Yang.
    p. cm.
  Includes index.
  ISBN 978-0-470-17883-6 (cloth)
  1. Reaction mechanisms (Chemistry) 2. Stereochemistry 3. Organic compounds—Synthesis. I. Title.
  QD502.5.Y36 2008
  547'.2—dc22

2007019896

Printed in the United States of America

10 9 8 7 6 5 4 3 2 1

*To my sons,
Walt and Larry*

CONTENTS

Preface — ix

Introduction — 1

1 [3,3]-Sigmatropic Rearrangements — 5

General Considerations, 5
Reactions, 13
1.1 Claisen Rearrangement, 13
1.2 Johnson–Claisen Rearrangement, 20
1.3 Ireland–Claisen Rearrangement, 27
1.4 Cope Rearrangement, 32
1.5 Anionic Oxy-Cope Rearrangement, 36
1.6 Aza-Cope–Mannich Reaction, 43

2 Aldol Reactions — 49

General Considerations, 49
Reactions, 57
2.1 Asymmetric *Syn*-Aldol Reaction, 57
2.2 Asymmetric *Anti*-Aldol Reaction, 78
2.3 Proline-Catalyzed Asymmetric Aldol Reaction, 91

3 Metal Allylation Reactions — 97

General Considerations, 97
Reactions, 102
3.1 Boron Allylation Reaction, 102
3.2 Silicon Allylation Reaction, 127

4 Stereoselective Reductions — 147

General Considerations, 147
Reactions, 151
4.1 Diastereoselective *Syn*-Reduction of β-Hydroxy Ketones, 151

4.2 Diastereoselective *Anti*-Reduction of β-Hydroxy Ketones, 161
4.3 Asymmetric Reduction, 173

List of Copyrighted Materials   **197**

Abbreviations   **199**

Subject Index   **201**

Scheme Index of Natural Products   **209**

PREFACE

When I was a graduate student in the Department of Chemistry at the University of Pittsburgh, I took organic chemistry courses taught by Professors Craig S. Wilcox and Dennis P. Curran. One of the most amazing topics that I learned about from their lectures was stereoselective synthesis in organic chemistry.

Stereochemistry is a concept of paramount importance in chemistry. Stereoselective reactions, be they diastereoselective or enantioselective, are therefore a valuable tool in producing compounds of the desired stereochemistry. Every stereoselective reaction has an energetically preferred transition state that can explain the formation of the major stereoisomer. A reasonable transition state is very important not only in rationalizing the experimental results, but also in further advancing the chemical system that one is studying.

Since the seminal proposal in 1957 by Howard E. Zimmerman and Marjorie D. Traxler regarding the stereoselective Ivanov reaction, six-membered chairlike transition states have been recognized as one of the most convincing methods used in organic chemistry to describe the course of reactions that have a well-organized molecular ensemble geometry. In this book I describe organic reactions that go through well-defined six-membered transition states. The reactions are classified into four categories: [3,3]-sigmatropic rearrangements, aldol reactions, metal allylation reactions, and stereoselective reductions. Each chapter begins with a section on general considerations in which I gather all the computational studies known to me that support the proposal of a six-membered transition state. Each reaction has a brief introduction, a description of the six-membered chairlike transition state, and applications selected from natural product synthesis. In presenting reactions and transition states, I have tried to deliver the arguments and conclusions exactly the way they are outlined in the original references. When questions arise or further information on a transition state is sought, readers are strongly encouraged to study the references listed at the end of each section.

This book will serve as a starting point in learning the amazing features of six-membered chairlike transition states in stereoselective organic reactions. With this book, I hope that students and practitioners alike will be able to propose reasonable transition states for the description of newly discovered stereoselective reactions.

Comments and suggestions from readers are always welcome. I can be reached by email at bizibeaver@yahoo.com.

Acknowledgments

I would like to thank the American Chemical Society and Elsevier for their generous permission to use materials for this book. A list of copyrighted materials is included.

Without the many people who have helped me, this book could not have appeared. I would like to thank Professor Tom Livinghouse of Montana State University for his critical reading of the entire manuscript. Drs. Rohan Beckwith, Xing Dai, Tim Peelen, and Janelle Thompson each read portions of the manuscript and provided a number of helpful comments. Thanks are also due to graduate students Elisa Leonardo and Bryce Sunsdahl at the Livinghouse Laboratory for their invaluable editorial assistance. Thanks also to the anonymous reviewers who read sample chapters and gave me invaluable suggestions. I have one very special person to whom I would like to express my sincere gratitude: Walt Harris, head football coach at the University of Pittsburgh between 1996 and 2005. Not only has coach Harris been a constant source of inspiration to me, he has also been a big part of my family's life, and I truly thank him for that.

Finally, I would like to thank my wife, Wenjing Xu, for her criticisms, encouragement, and suggestions over the course of writing the book.

JAEMOON YANG

Bozeman, Montana

INTRODUCTION

In 1957, Zimmerman and Traxler published their study on the reaction of benzaldehyde with the magnesium enolate of phenylacetic acid: namely, the *Ivanov reaction*[1] (Scheme I). The major product from the reaction is an *anti*- or *threo*-isomer of 2,3-diphenyl-3-hydroxypropionic acid, and the minor product is a *syn*- or *erythro*-isomer.[2] Although in 1957 the Ivanov reaction had been known for quite some time, no reasonable proposal had been put forward to explain the stereochemical outcome observed for the reaction. In explaining the ratio of the two stereoisomers, the authors made a seminal proposal that the condensation reaction would go through a six-membered transition state (Scheme II). The coordination of benzaldehyde carbonyl group with magnesium brings the two reactants in close contact in both chairlike transition state **A** and boatlike transition state **B**. The authors speculated that the particular spatial arrangement of the four substituents in the transition state could determine the stereochemistry of the products. For the Ivanov reaction of benzaldehyde, transition state **A** would be favored over **B** because transition state **A** involves a lower-energy approach to bonding than that of the alternative transition state **B**, which experiences an energetically unfavorable gauche interaction between the two phenyl substituents[1] (Scheme III).

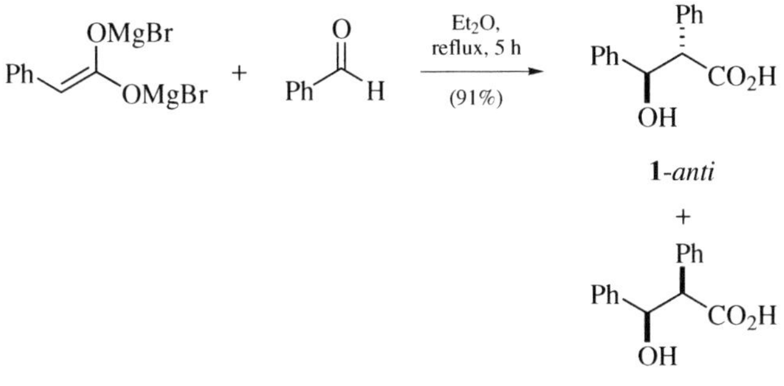

Scheme I

Six-Membered Transition States in Organic Synthesis, By Jaemoon Yang
Copyright © 2008 John Wiley & Sons, Inc.

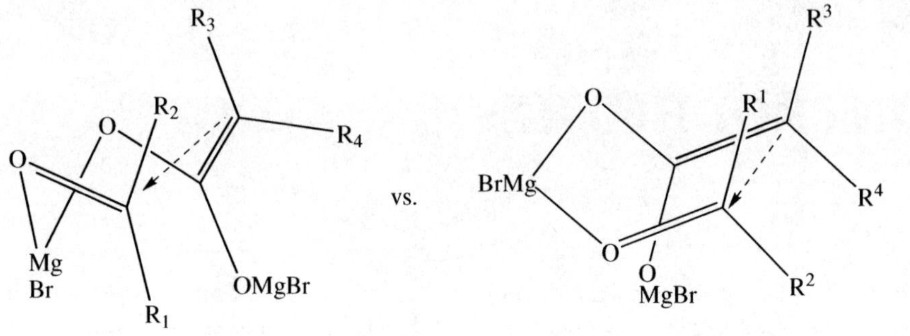

TS A: chairlike transition structure   **TS B**: boatlike transition structure

Scheme II

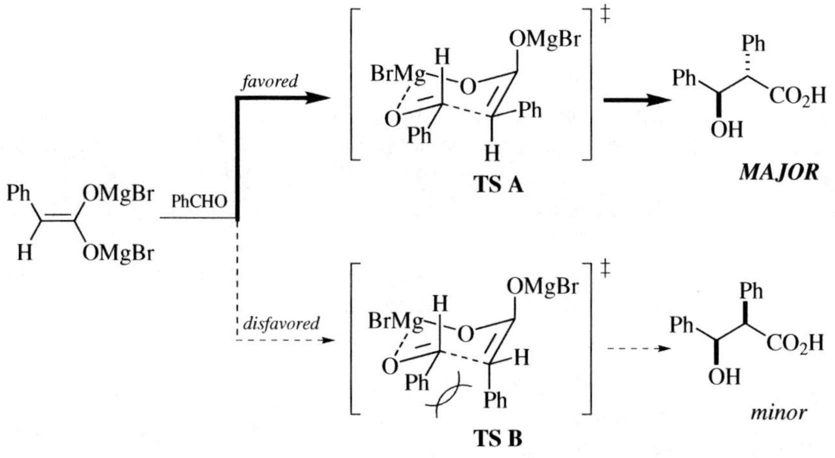

Scheme III

Due to its simplicity and outstanding prediction power, the *Zimmerman–Traxler transition state* has frequently been used in explaining the stereochemical outcome of certain stereoselective reactions. The characteristics of the Zimmerman–Traxler transition state can be summarized as follows:

1. The transition state is for a six-atom system and thus is six-membered.
2. The transition state involves six electrons and thus exhibits the aromatic character of benzene.[3]
3. A chairlike transition state is favored over a boatlike transition state. There are, however, exceptions.

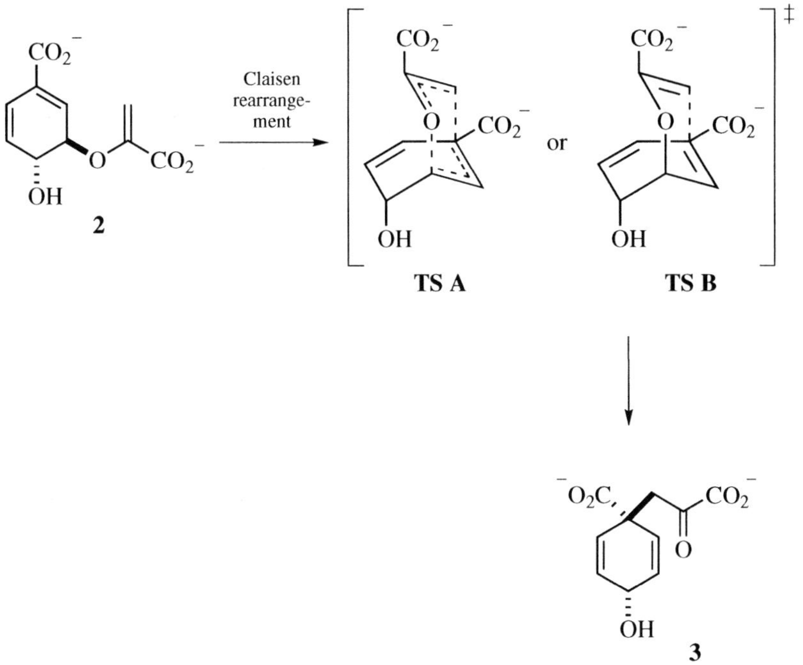

Scheme IV

TABLE 1 Conformational Energies of Monosubstituted Cyclohexanes

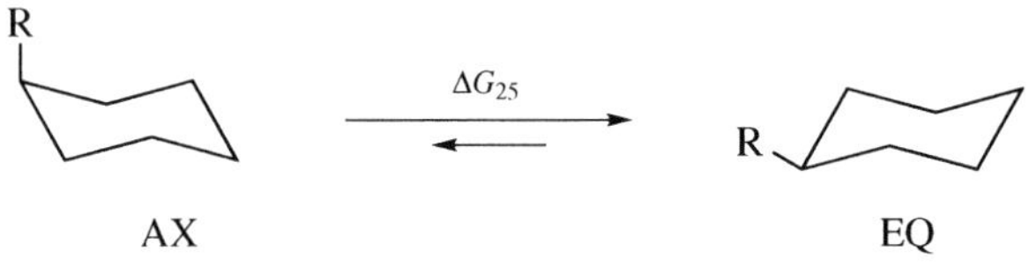

R	$-\Delta G$ (kcal/mol)	EQ/AX
Me	1.74	19:1
Et	1.79	19:1
i-Pr	2.21	42:1
t-Bu	4.7	>99:1
C_6H_5	2.8	>99:1
OMe	0.55	2.5:1

4. When two chairlike transition states compete, the transition state in which a bulky substituent occupies an equatorial position is favored over the state that has the same substituent in an axial position. The free-energy difference between the two potential transition states can be approximated by using the ΔG or A-values of the monosubstituted cyclohexanes[4] (Table 1).

In the following four chapters, readers will find some of the most frequently cited and most synthetically relevant examples of the Zimmerman–Traxler or six-membered transition state. In presenting reactions that go through a six-membered chairlike transition state, I pay special attention to including computational studies, in an effort to prove the existence of a six-membered chairlike transition state. Although not all six-membered transition states have been studied computationally, recent interest in using computers in studies of stereoselective reactions would certainly confirm the legitimacy of Zimmerman–Traxler transition states for many more reactions.[5]

Before we embark on our journey into the world of six-membered transition states, I would like to speak briefly about one reaction, to illustrate how a transition state is drawn throughout the book. The enzyme-catalyzed transformation of chorsimate (**2**) to prephenate (**3**) is a classic example of a [3,3]-sigmatropic *Claisen rearrangement*[6] (Scheme IV). As an old bond is being broken and at the same time a new bond is formed in the transition state, the transition state for the Claisen rearrangement of chorsimate to prephenate would look more like transistion state **A** than like **B**. Still, for the convenience of following the bond connection event clearly, I prefer to draw the transition state like **B**.

REFERENCES

1. Zimmerman, H. E.; Traxler, M. D. *J. Am. Chem. Soc.* **1957**, *79*, 1920.
2. For definitions of *syn* and *anti*, see Masamune, S.; Ali, S. A.; Snitman, D. L.; Garvey, D. S. *Angew. Chem. Int. Ed.* **1980**, *19*, 557.
3. (a) Day, A. C. *J. Am. Chem. Soc.* **1975**, *97*, 2431; (b) Carey, F. A.; Sundberg, R. J., *Advanced Organic Chemistry*, Part A, 3rd ed.; Plenum Press: New York, **1990**; Chap. 11.
4. Eliel, E. L.; Wilen, S. H.; Mander, L. N. *Stereochemistry of Organic Compounds*; Wiley: New York, **1994**; Chap. 11.
5. Lipkowitz, K. B.; Kozlowski, M. C. *Synlett* **2003**, 1547.
6. (a) Andrews, P. R.; Haddon, R. C. *Aust. J. Chem.* **1979**, *32*, 1921; (b) Copley, S. D.; Knowles, J. R. *J. Am. Chem. Soc.* **1985**, *107*, 5306.

1 [3,3]-Sigmatropic Rearrangements

GENERAL CONSIDERATIONS

The Claisen and Cope rearrangements are two of the best known sigmatropic rearrangements in organic chemistry[1] (Scheme 1.I). As the rearrangement involves six electrons in a six-atom system, these two reactions serve as excellent examples of the ubiquitous existence of a six-membered transition state in organic chemistry.

In 1912, Ludwig Claisen discovered that the allyl ether **1** of ethyl acetoacetate underwent a reaction to afford **2** upon heating in the presence of ammonium chloride[2] (Scheme 1.II). Similarly, the allyl naphthyl ether **3** transformed into 1-allyl-2-naphthol (**4**) in 82% yield at 210 °C. The reaction, now known as the *Claisen rearrangement*, is general for a variety of aliphatic and aromatic ethers and is recognized as one of the most synthetically useful reactions in organic chemistry.[3]

The Claisen rearrangement is a thermally induced [3,3]-sigmatropic rearrangement of allyl vinyl ethers to form γ,δ-unsaturated carbonyl compounds.[4] Due to the concerted nature and synthetic utilities of the Claisen rearrangement, much effort has been devoted to understanding the mechanism of the reaction.[5] Although the extent of delocalization of the six electrons involved in the transition state may depend on the nature of the substrates, it is believed that the rearrangement goes through a six-membered aromatic transition state[6] (Scheme 1.III).

To uncover the transition-state structures for Claisen rearrangement of the parent allyl vinyl ether,[7] Vance et al. performed ab initio quantum mechanical calculations[8] (Scheme 1.IV). When the transition structures were calculated using the 6-31G* basis set, the partially formed C_1-C_6 bond length is 2.26 Å and the partially broken C_4-O bond length is 1.92 Å in chairlike transition structure **A**. These two bond lengths were confirmed by Meyer et al. in a later study employing different-level calculations.[9] Another important finding in Vance et al.'s study is that chairlike transition structure **A** is more stable than boatlike structure **B**, by 6.6 kcal/mol. The conclusion thus supports the proposals of chairlike transition structures for the stereoselectivities observed for the Claisen rearrangement reactions of substituted molecules.

Six-Membered Transition States in Organic Synthesis, By Jaemoon Yang
Copyright © 2008 John Wiley & Sons, Inc.

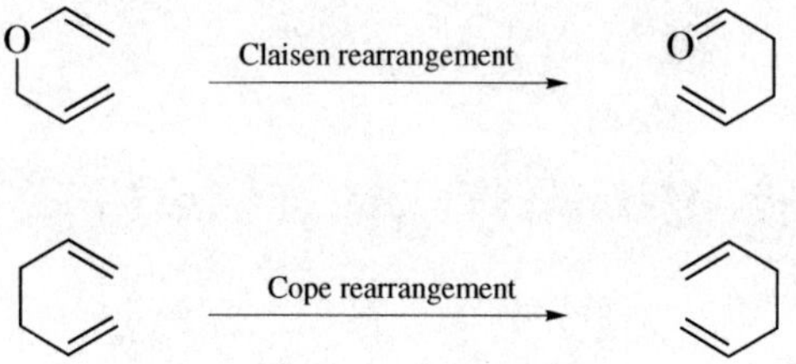

6 [3,3]-SIGMATROPIC REARRANGEMENTS

Scheme 1.I

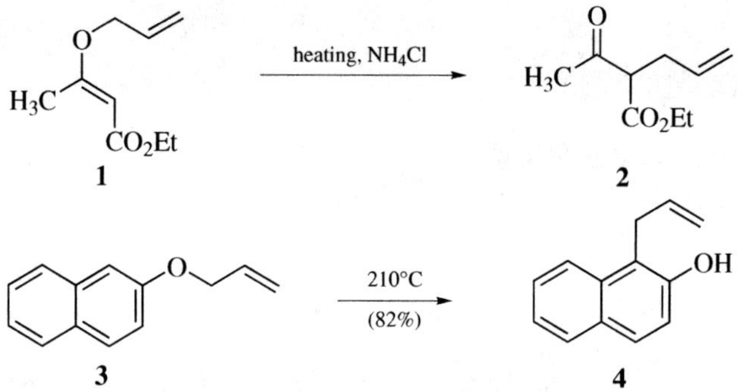

Scheme 1.II

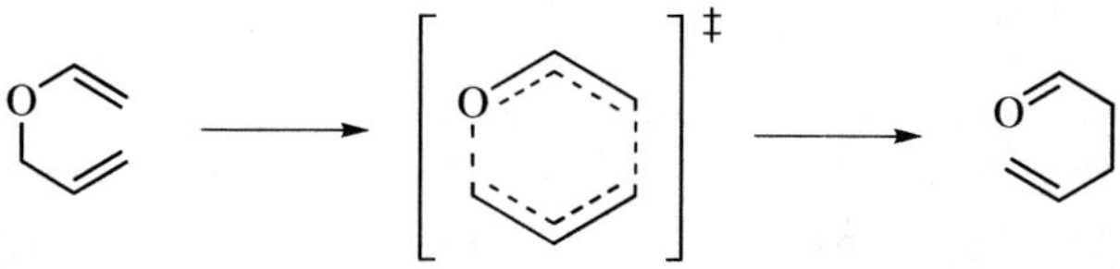

Scheme 1.III

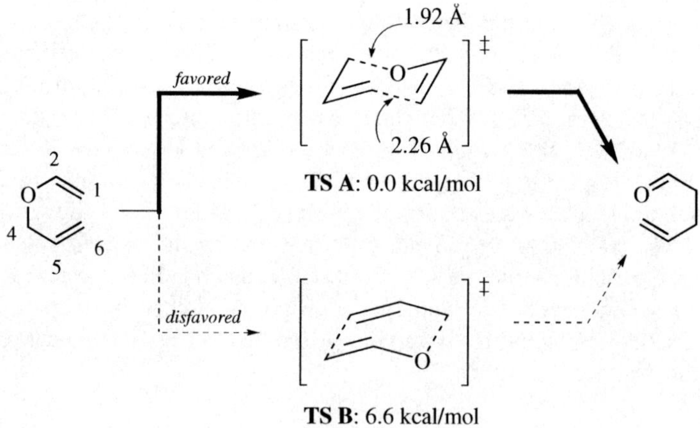

TS A: 0.0 kcal/mol

TS B: 6.6 kcal/mol

Scheme 1.IV

GENERAL CONSIDERATIONS 7

One classic example that confirms the preference of Claisen rearrangement for a chairlike transition state was provided by Hansen and others. In 1968, they investigated the Claisen rearrangement of the crotyl propenyl ethers **5a** and **5b** to examine the stereochemistry of the rearrangement in the gas phase at 160 °C[10] (Scheme 1.V). Both the *E,E*- and *Z,Z*-isomers rearrange to afford a *syn*-isomer as the major product. The stereochemical outcome of the reaction can be explained

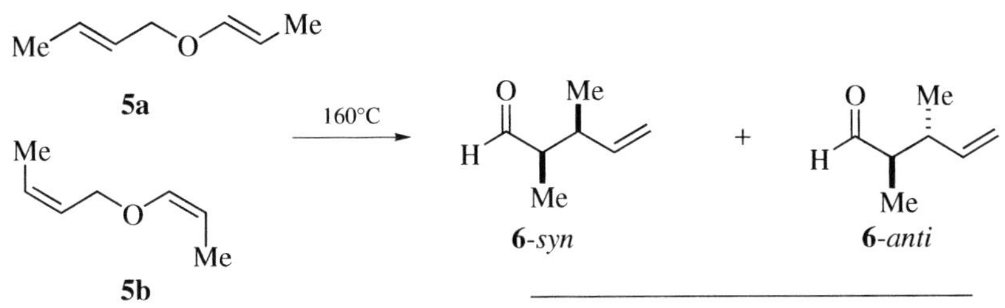

Ether	syn/anti	Rel. Rate
5a	95.9:4.1	9
5b	94.7:5.3	1

Scheme 1.V

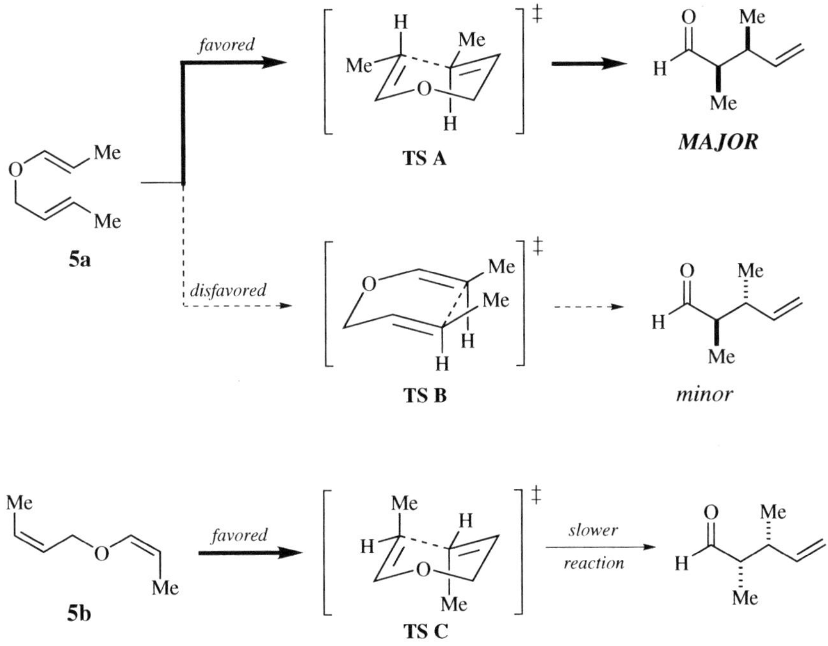

Scheme 1.VI

[3,3]-SIGMATROPIC REARRANGEMENTS

in terms of a six-membered transition state[10] (Scheme 1.VI). Between the two transition states for **5a**, chairlike transition state **A** is favored over boatlike state **B** to afford a *syn*-isomer as the major product. Chairlike transition state **C** can explain the formation of the *syn*-isomer that is enantiomeric to **6-*syn***. Other indirect evidence for the existence of a chairlike transition state is the fact that the *E*,*E*-isomer **5a** reacts nine times faster than the *Z*,*Z*-isomer **5b**. This difference in the reaction rate can be understood by examining transition states **A** and **C**: Transition state **C** for **5b** is of higher energy than transition state **A** for **5a**, due presumably to the 1,3-diaxial interactions arising from the axial methyl groups in transition state **C**.

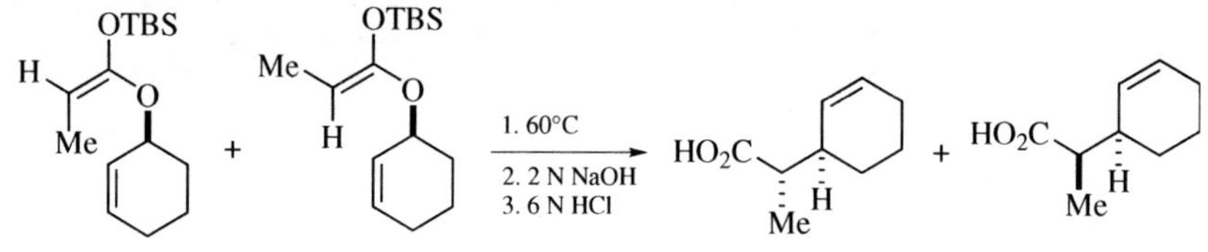

7E/7Z	8a/8b	Yield, %
83:17	84:16	79
4:96	72:28	91

Scheme 1.VII

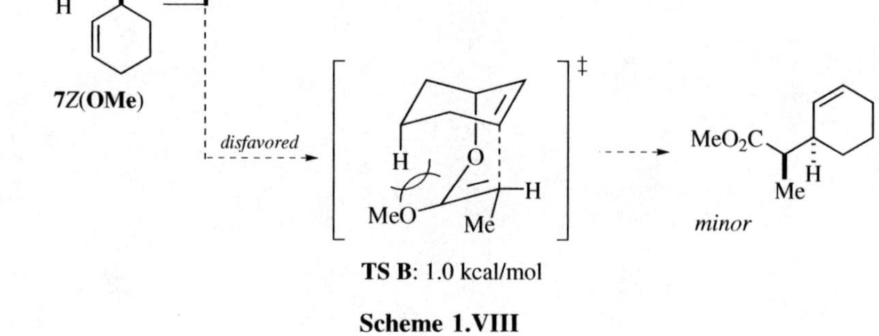

Scheme 1.VIII

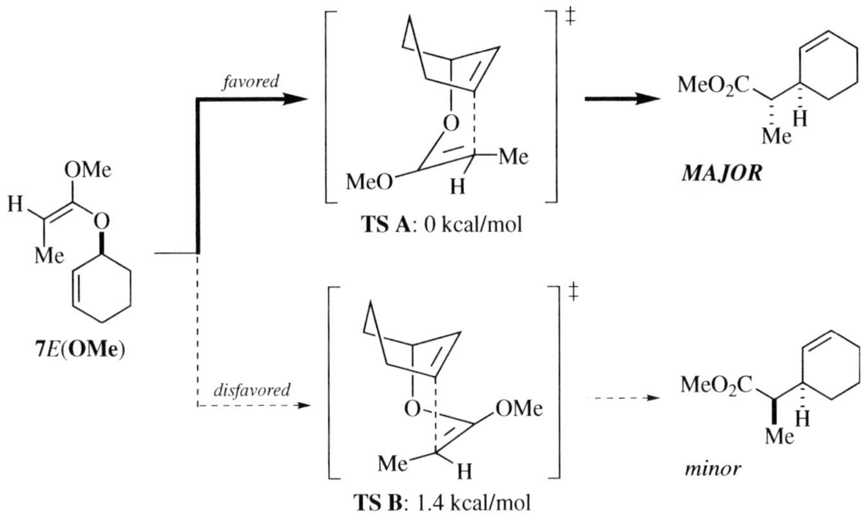

Scheme 1.IX

TS A: 0 kcal/mol (favored) → **MAJOR**

TS B: 1.4 kcal/mol (disfavored) → minor

Starting material: 7E(OMe)

Although a chairlike transition state is favored for the Claisen rearrangement reactions of acyclic substrates, this is not always the case with cyclic systems. For example, Bartlett and Ireland independently studied the rearrangement reactions of cyclohexenyl silylketeneacetals and found that there was competition between the chairlike and boatlike transition states[11] (Scheme 1.VII). Clearly, the *E*-isomer **7E** gives **8a** via a chairlike transition state, whereas the *Z*-isomer **7Z** affords the same product (**8a**) via a boatlike transition state.

To quantitatively understand the preference for the chairlike and boatlike transition states of the Claisen rearrangement, Houk et al. carried out a computational study[12] (Scheme 1.VIII). In the theoretical treatment two methyl acetals, **7Z(OMe)** and **7E(OMe)**, were used as a model system instead of the *tert*-butyldimethylsilyl (TBS) ketene acetal. Calculations locate four transition states for the rearrangement of **7Z(OMe)**, among which boatlike transition state **A** is of the lowest energy that leads to the formation of the major isomer observed experimentally. Chairlike transition state **B** is disfavored, due to steric repulsion between the axial hydrogen of the cyclohexenyl unit and the methoxy substituent of the alkene.

For the reaction of **7E(OMe)**, chairlike transition state **A** is favored over boatlike transition state **B**[12] (Scheme 1.IX). These computational results provide a solid theoretical rationalization of the original proposal by Bartlett and Ireland that the boatlike transition state is favored for the Claisen rearrangement of **7Z**, and the chairlike transition state is preferred for **7E**.

Another important [3,3]-sigmatropic rearrangement is the *Cope rearrangement*, a carbon analog of the Claisen rearrangement. At the eighth National Organic Chemistry Symposium in 1939, Arthur C. Cope and Elizabeth M. Hardy presented their exciting discovery of this new reaction in which an allyl group

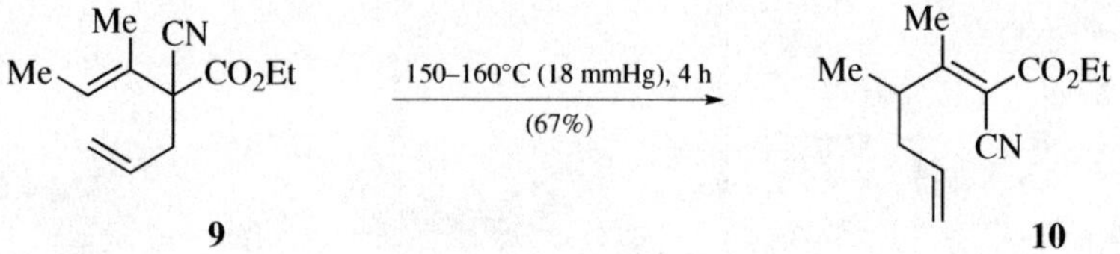

Scheme 1.X

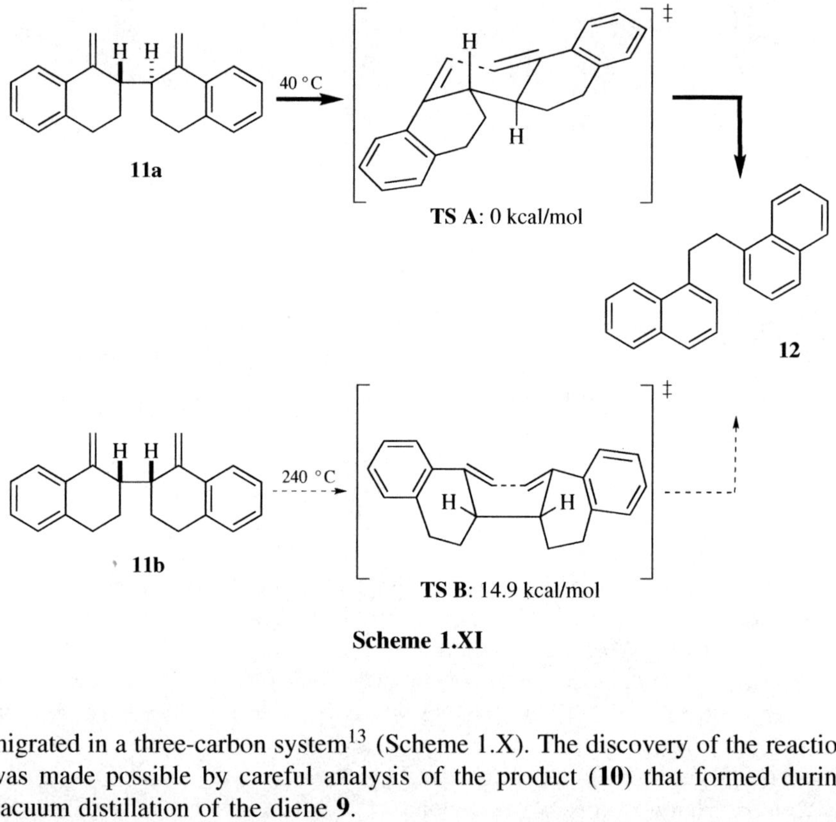

Scheme 1.XI

migrated in a three-carbon system[13] (Scheme 1.X). The discovery of the reaction was made possible by careful analysis of the product (**10**) that formed during vacuum distillation of the diene **9**.

The Cope rearrangement, which is the conversion of a 1,5-hexadiene derivative to an isomeric 1,5-hexadiene by the [3,3]-sigmatropic mechanism, has been studied extensively.[14] As is the case for the Claisen rearrangement, the Cope rearrangement prefers to go through a six-membered chairlike transition state. Shea et al. demonstrated elegantly the preference for the chairlike over the boatlike transition state by carrying out Cope rearrangements of racemic (**11a**) and meso (**11b**) naphthalenes[15] (Scheme 1.XI). It was determined that the racemic 1,5-diene **11a** underwent Cope rearrangement 7 million times faster than the meso diene **11b**. The energy difference between transition states **A** and **B** is calculated to be 14.9 kcal/mol.

GENERAL CONSIDERATIONS   11

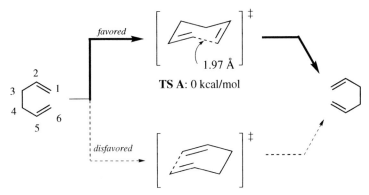

Scheme 1.XII

Scheme 1.XIII

14EE/14ZZ/14EZ = 90:9:<1

Scheme 1.XIV

A number of theoretical studies have been conducted to understand the mechanism of the Cope rearrangement.[16] According to calculations by Houk and co-workers, the chairlike transition state is more stable than the boatlike transition state by 7.8 kcal/mol (Scheme 1.XII). When Schleyer and colleagues performed calculations to compute the magnetic properties of the transition-state structures, transition states **A** and **B** had a magnetic susceptibility of −55.0 and −56.6, respectively. These values are comparable to that of benzene (−62.9), confirming the existence of an aromatic transition state in the Cope rearrangement.

One classic example is an experiment reported by Doering and Roth in 1962[17] (Scheme 1.XIII). Upon heating, racemic 3,4-dimethylhexa-1,5-diene (**13**) rearranged to a mixture of (2E,6E)-octa-2,6-diene (90%), (2Z,6Z)-octa-2,6-diene (9%), and a trace amount of (2E,6Z)-isomer. The experimental results are explained in terms of a six-membered transition state[17] (Scheme 1.XIV). Chairlike transition state **A** is favored over transition state **B** based on the conformational analysis of 1,2-dimethylcyclohexane, in which the methyl substituents prefer to be in an equatorial position. The observation that **14***EZ* was formed in only trace amounts indicates that boatlike transition state **C** is of significantly higher energy than transition state **A** or **B**.

REFERENCES

1. Hoffmann, R.; Woodward, R. B. *Acc. Chem. Res.* **1968**, *1*, 17.
2. Claisen, L. *Chem. Ber.* **1912**, *45*, 3157.
3. (a) Tarbell, D. S. *Org. React.* **1942**, *2*, 1; (b) Rhoads, S. J.; Raulins, N. R. *Org. React.* **1975**, *22*, 1. (c) Hiersemann, M.; Nubbemeyer, U. Eds. *The Claisen Rearrangement: Methods and Applications*; Wiley-VCH: New York **2007**.
4. For reviews, see (a) Ziegler, F. E. *Acc. Chem. Res.* **1977**, *10*, 227; (b) Ziegler, F. E. *Chem. Rev.* **1988**, *88*, 1423; (c) Castro, A. M. M. *Chem. Rev.* **2004**, *104*, 2939.
5. (a) Houk, K. N.; González, J.; Li, Y. *Acc. Chem. Res.* **1995**, *28*, 81; (b) Gajewski, J. J. *Acc. Chem. Res.* **1997**, *30*, 219.
6. Dewar, M. J. S. *Angew. Chem. Int. Ed.* **1971**, *10*, 761.
7. Hurd, C. D.; Pollack, M. A. *J. Am. Chem. Soc.* **1938**, *60*, 1905.
8. Vance, R. L.; Rondan, N. G.; Houk, K. N.; Jensen, F.; Borden, W. T.; Komornicki, A.; Wimmer, E. *J. Am. Chem. Soc.* **1988**, *110*, 2314.
9. Meyer, M. P.; DelMonte, A. J.; Singleton, D. A. *J. Am. Chem. Soc.* **1999**, *121*, 10865.
10. (a) Vittorelli, P.; Winkler, T.; Hansen, H.-J.; Schmid, H. *Helv. Chim. Acta* **1968**, *51*, 1457; (b) Hansen, H.-J.; Schmid, H. *Tetrahedron* **1974**, *30*, 1959.
11. (a) Bartlett, P. A.; Pizzo, C. F. *J. Org. Chem.* **1981**, *46*, 3896; (b) Ireland, R. E.; Wipf, P.; Xiang, J. *J. Org. Chem.* **1991**, *56*, 3572.
12. Khaledy, M. M.; Kalani, M. Y. S.; Khong, K. S.; Houk, K. N.; Aviyente, V.; Neier, R.; Soldermann, N.; Velker, J. *J. Org. Chem.* **2003**, *68*, 572.
13. Cope, A. C.; Hardy, E. M. *J. Am. Chem. Soc.* **1940**, *62*, 441.
14. Carey, F. A.; Sundberg, R. J. *Advanced Organic Chemistry*, 2nd ed., Part B; p. 316.
15. Shea, K. J.; Stoddard, G. J.; England, W. P.; Haffner, C. D. *J. Am. Chem. Soc.* **1992**, *114*, 2635.

16. For theoretical studies, see (a) Wiest, O.; Black, K. A.; Houk, K. N. *J. Am. Chem. Soc.* **1994**, *116*, 10336; (b) Jiao, H.; Schleyer, P. v. R. *Angew. Chem. Int. Ed.* **1995**, *34*, 334; (c) Hrovat, D. A.; Beno, B. R.; Lange, H.; Yoo, H.-Y.; Houk, K. N.; Borden, W. T. *J. Am. Chem. Soc.* **1999**, *121*, 10529; (d) Staroverov, V. N.; Davidson, E. R. *J. Am. Chem. Soc.* **2000**, *122*, 186; (e) Hrovat, D. A.; Borden, W. T. *J. Am. Chem. Soc.* **2001**, *123*, 4069.

17. Doering, W. v. E.; Roth, W. R. *Tetrahedron* **1962**, *18*, 67.

REACTIONS

1.1. Claisen Rearrangement

A very interesting stereochemical outcome is noted for the Claisen rearrangement of substituted allyl vinyl ethers. For example, the allyl vinyl ethers **1** underwent Claisen rearrangement to afford the unsaturated aldehydes **2** in quantitative yields and with high levels of stereoselectivity, which depend largely on the steric bulkiness of the R group[1] (Scheme 1.1a). To explain the stereochemical outcome of the rearrangement, Perrin and Faulkner proposed a six-membered chairlike transition state[1] (Scheme 1.1b). Transition state **A**, leading to formation of the major product, is favored over **B**, in which the bulky alkyl group (R) occupies an axial position, resulting in energetically unfavorable 1,3-diaxial interactions.

The methyl substituent in 2-methyltetrahydropyran prefers to be in an equatorial position[2] (Scheme 1.1c). The formation of **2E** (R = Et) as a major isomer in the rearrangement can be understood qualitatively when 2-methyltetrahydropyran is employed as a model system to estimate the energy difference between transition states **A** and **B**.[3] As the Claisen rearrangement is a concerted reaction, the chirality in the starting material is translated directly to the product without loss of optical purity. For example, upon heating the allyl vinyl ethers **3R** and **3S** both gave the γ,δ-unsaturated aldehyde **4**[4] (Scheme 1.1d). The degree of chirality transfer was calculated to be 98% after correcting the optical purities of the starting materials. The high level of chirality transfer in the foregoing reactions supports the notion that the reaction goes through a six-membered transition

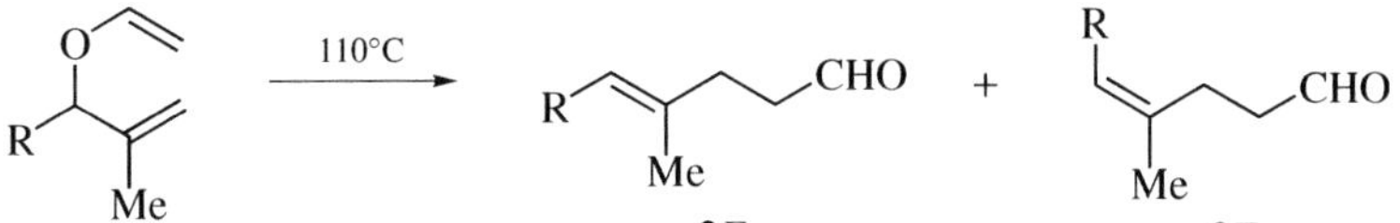

R	E/Z	Yield, %
Et	90:10	100
i-Pr	93:7	100

Scheme 1.1a

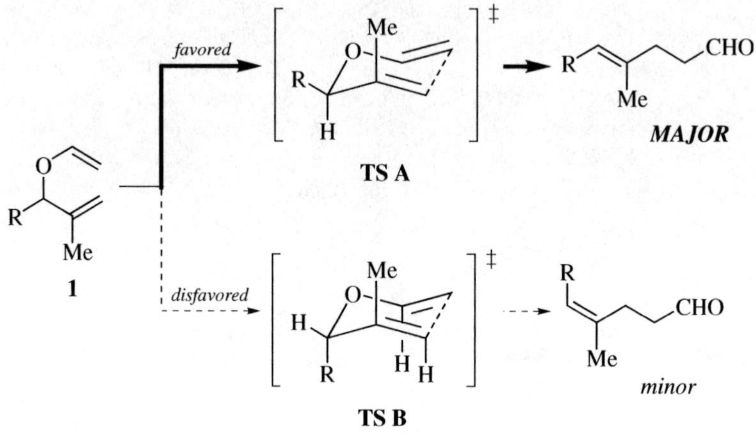

Scheme 1.1b

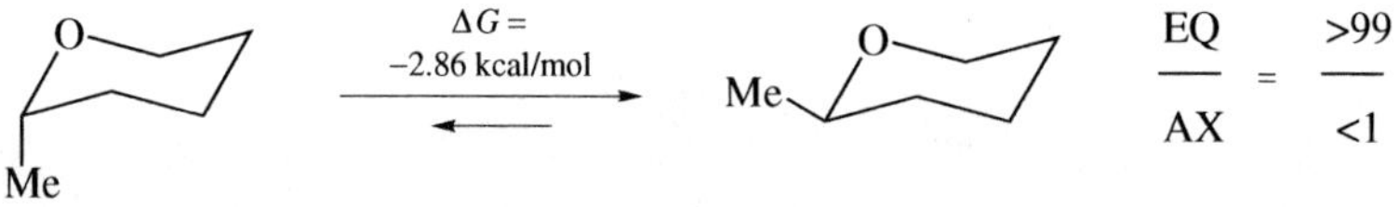

Scheme 1.1c

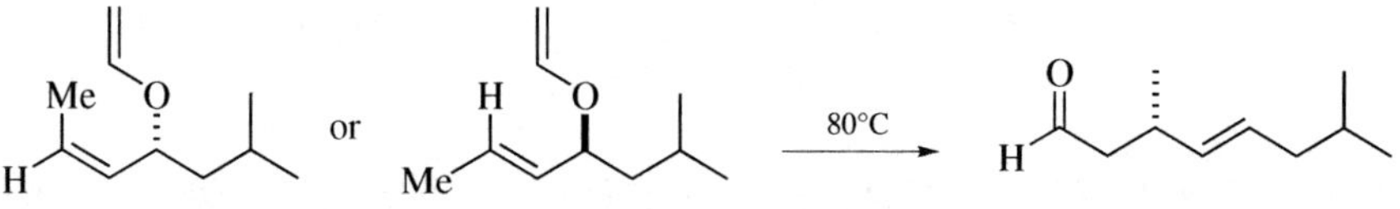

3*R* (93.0% ee) **3*S*** (95.6% ee) **4** (89.8% ee from **3*R***; 91.6% ee from **3*S***)

Scheme 1.1d

state[4] (Scheme 1.1e). The isobutyl group in transition states **A** and **B** occupies an equatorial position in the six-membered chairlike conformation.

The Claisen rearrangement was used in the asymmetric total synthesis of (+)-9(11)-dehydroestrone methyl ether (**5**), a versatile intermediate in the synthesis of estrogens[5] (Scheme 1.1f). The key feature of the synthesis is the successful development of the asymmetric tandem Claisen-ene sequence. Thus, a solution of the cyclic enol ether **6** in toluene was heated in a sealed tube at 180°C for 60 hours to afford the product **9** in 76% isolated yield after deprotection of the silyl enol ether. The Claisen rearrangement of the allyl vinyl ether **6** occurred stereoselectively to give an intermediate (**7**), in which the 8,14-configuration was 90% *syn*. The stereoselectivity in the Claisen rearrangement can be explained

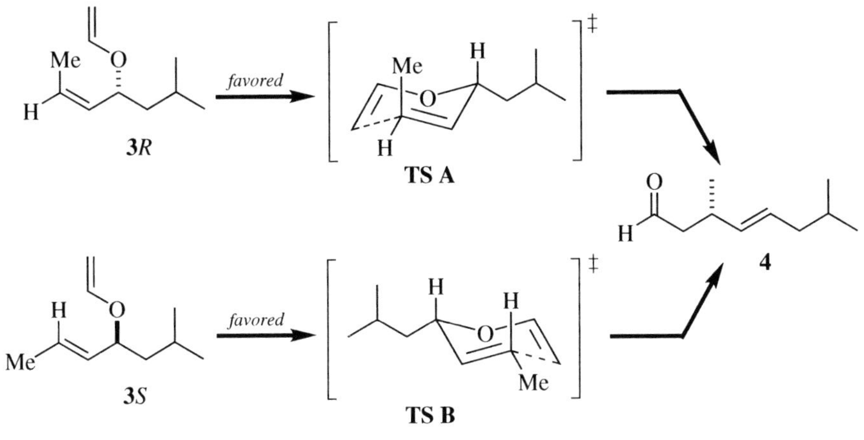

Scheme 1.1e

by the chairlike transition state **6TS**, which has minimal 1,3-diaxial interactions. Therefore, the S-Z chirality of the enol ether **6** is transmitted completely to the 14S chirality in the Claisen product **7**, along with a high 8,14-*syn* selectivity.

Another application of the Claisen rearrangement is given in Boeckman et al.'s synthesis of (+)-saudin (**10**), a natural product that has been shown to possess in vivo non-insulin-dependent hypoglycemic activity[6] (Scheme 1.1g). The allyl vinyl ether **13** was synthesized by O-alkylation of the thermodynamic enolate of **11** with the allylic triflate **12**. The Claisen rearrangement of **13** occurred at −65 °C with excess TiCl$_4$ in the presence of Me$_3$Al as a proton scavenger to afford **14** as the major product. The facial selectivity was rationalized by invoking a six-membered chairlike transition state **13TS**, in which titanium(IV) metal coordinates with oxygens of both the vinyl ether and the ester.

In the total synthesis of the tetrodotoxin **15**, Isobe et al. used the Claisen rearrangement to obtain the highly functionalized intermediate **18**[7] (Scheme 1.1h). The alcohol **16** was treated with 2-methoxypropene and a catalytic amount of pyridinium *p*-toluenesulfonate (PPTS) in tetrahydrofuran (THF) to afford the allyl vinyl ether **17**. Heating **17** in 1,2-dichlorobenzene in the presence of base affected a smooth Claisen rearrangement to provide the ketone **18** in high yield. The Claisen rearrangement was used to prepare an α-allyl carbonyl compound in the total synthesis of garsubellin A (**19**), a polyprenylated phloroglucin natural product with highly potent neurotrophic activity[8] (Scheme 1.1i). O-Allylation of the 1,3-diketone **20** with allyl iodide in the presence of 4-Å molecular sieves gave the enol ether **21**. The Claisen rearrangement of **21** with the use of sodium acetate went smoothly to afford the key intermediate (**22**) in excellent yield.

Metal-catalyzed isomerization of unsymmetrical diallyl ethers is a unique route to synthesize allyl vinyl ethers for the Claisen rearrangement. In 1977, Reuter and Salomon reported that heating the diallyl ether **23** in the presence of a catalytic amount of tris(triphenylphosphine)ruthenium(II) dichloride resulted in the

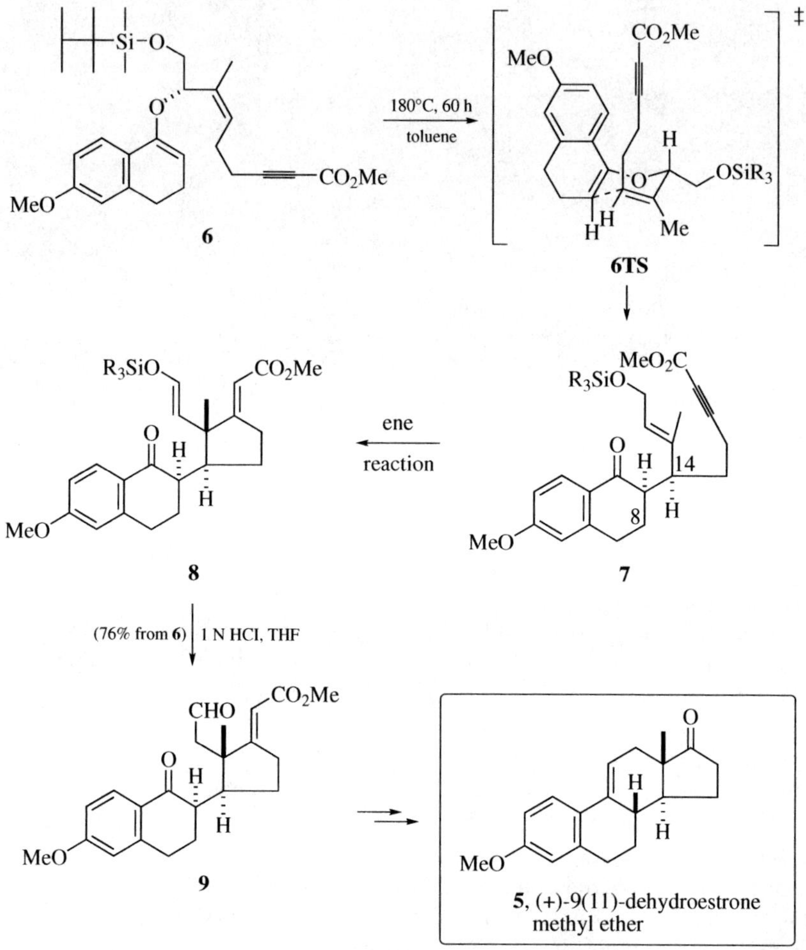

16 [3,3]-SIGMATROPIC REARRANGEMENTS

Scheme 1.1f

formation of the γ,δ-unsaturated aldehyde **25**[9] (Scheme 1.1j). The Claisen rearrangement presumably occurred through the intermediate **24**, which was produced via a highly regioselective isomerization of the monosubstituted alkene.

In accessing chiral allyl vinyl ethers for Claisen rearrangement reactions, Nelson et al. employed the iridium-mediated isomerization strategy. Thus, the requisite enantioenriched diallyl ether substrate **28** was synthesized via a highly enantioselective diethylzinc–aldehyde addition protocol[10] (Scheme 1.1k). The enantioselective addition of Et_2Zn to cinnamaldehyde catalyzed by (−)-3-*exo*-morpholinoisoborneol (MIB; **26**)[11] provided an intermediate zinc alkoxide (**27**). Treatment of **27** with acetic acid followed by *O*-allylation in the presence of palladium acetate delivered the **28** in 73% yield and 93% ee. Isomerization of **28** with a catalytic amount of the iridium complex afforded the allyl vinyl ether

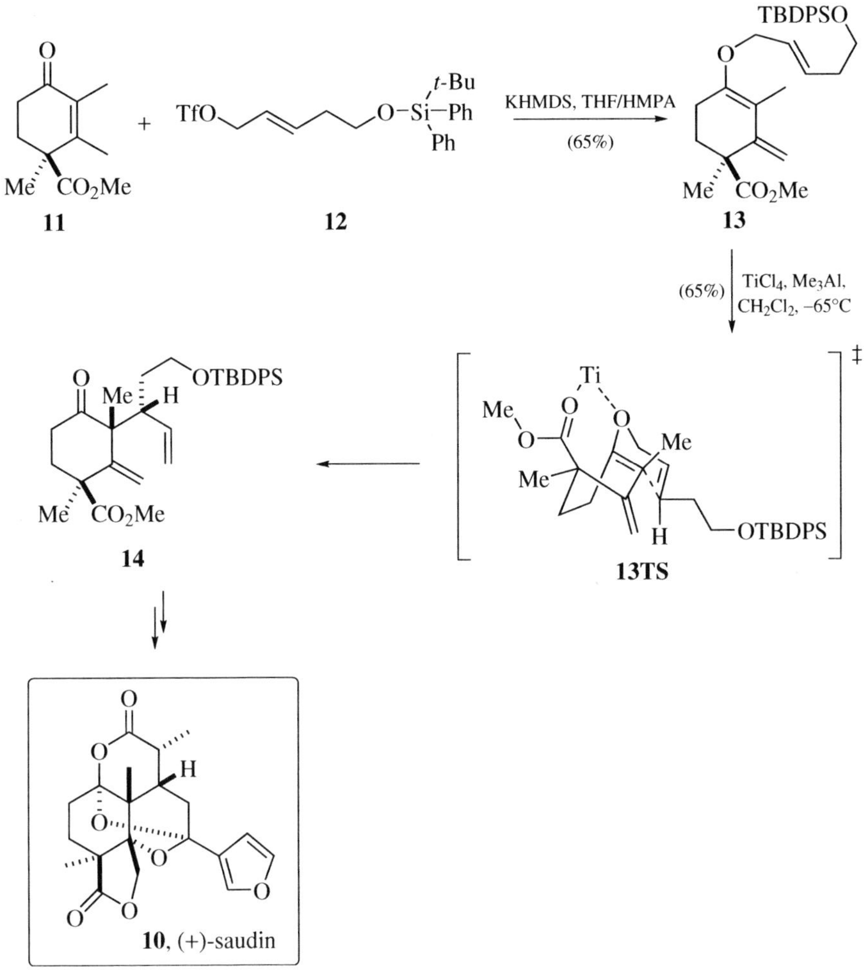

Scheme 1.1g

29, which then underwent Claisen rearrangement to produce the γ,δ-unsaturated aldehyde 30 (*syn/anti* = 95:5) without loss of optical purity.

Nelson and Wang completed an enantioselective synthesis of (+)-calopin dimethyl ether (31), highlighting the potential utility of the olefin isomerization–Claisen rearrangement strategy in asymmetric synthesis[12] (Scheme 1.1l). Reaction of 2,3-dimethoxy-4-methylbenzaldehyde (32) with Meyers's lithio enaminophosphonate reagent gave the enal 33, which was then subjected to the (−)-MIB (26)-catalyzed addition of Et_2Zn, acetic acid treatment, and palladium-catalyzed *O*-allylation to afford the diallyl ether 34 in high enantioselectivity. Isomerization–Claisen rearrangement of 34 and the in situ reduction of the intermediate aldehyde generated the 2,3-*syn*-disubstituted 4-heptenol (35) (90% ee, *syn/anti* = 94:6).

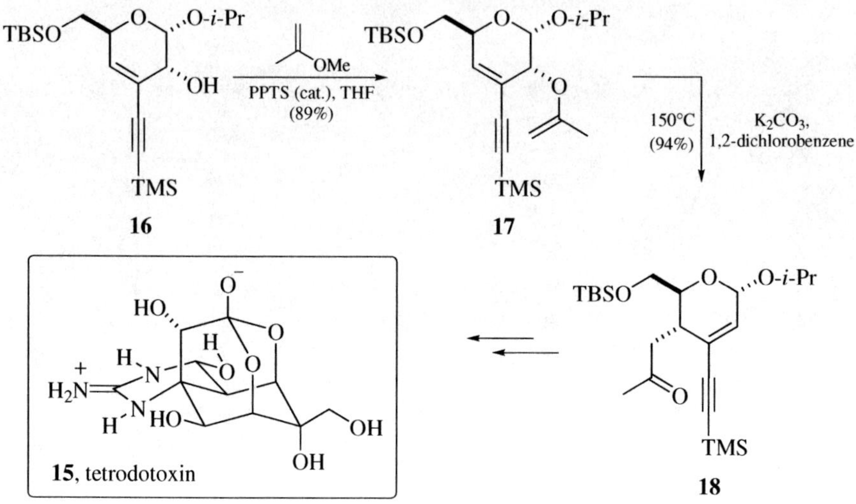

[3,3]-SIGMATROPIC REARRANGEMENTS

Scheme 1.1h

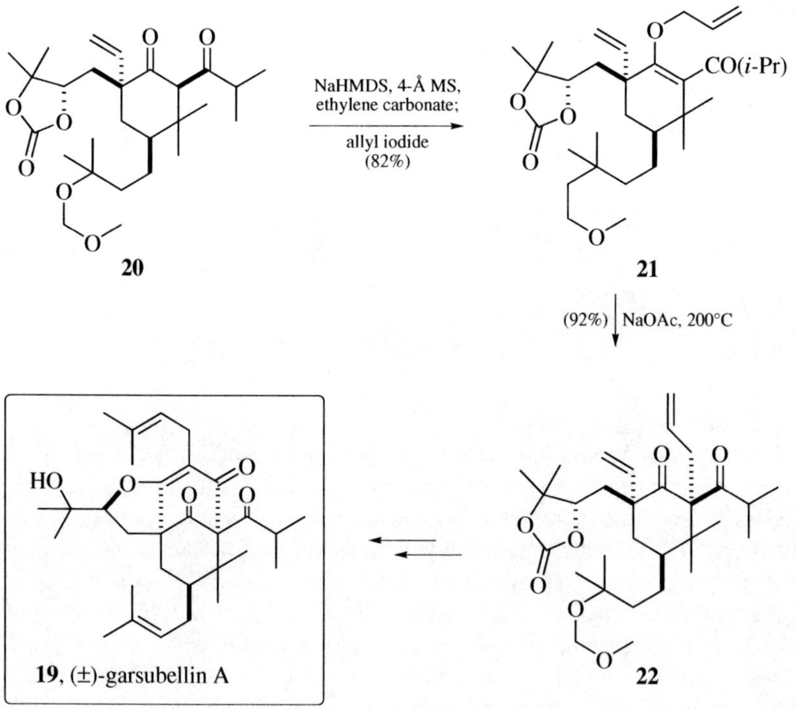

Scheme 1.1i

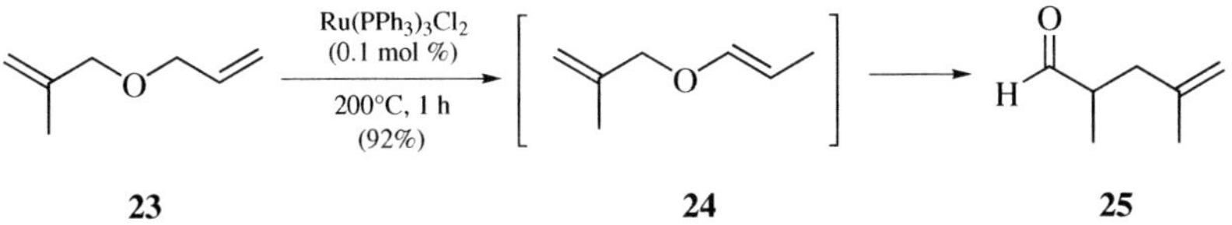

REACTIONS

Scheme 1.1j

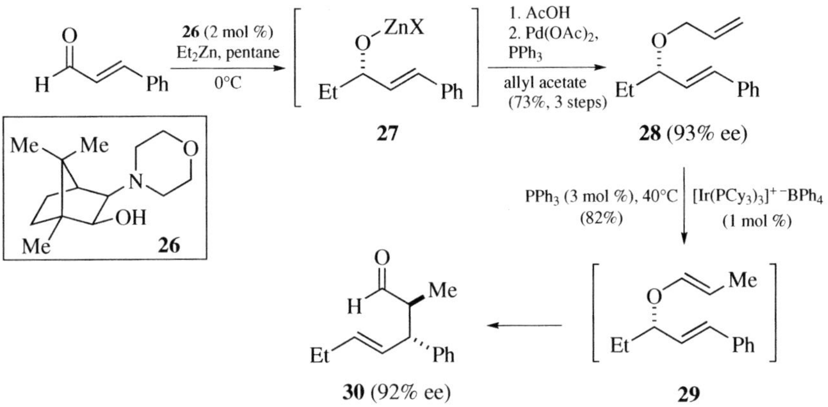

Scheme 1.1k

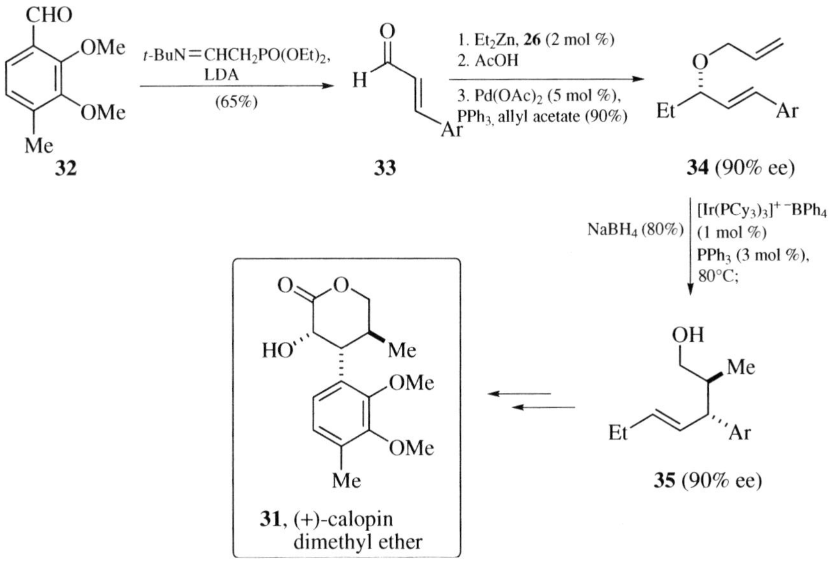

Scheme 1.1l

REFERENCES

1. Perrin, C. L.; Faulkner, D. J. *Tetrahedron Lett.* **1969**, *10*, 2783.
2. Eliel, E. L.; Hargrave, K. D.; Pietrusiewicz, K. M.; Manoharan, M. *J. Am. Chem. Soc.* **1982**, *104*, 3635.
3. Wilcox, C. S.; Babston, R. E. *J. Am. Chem. Soc.* **1986**, *108*, 6636.
4. Chan, K.-K.; Cohen, N.; De Noble, J. P.; Specian, A. C., Jr.; Saucy, G. *J. Org. Chem.* **1976**, *41*, 3497.
5. Mikami, K.; Takahashi, K.; Nakai, T. *J. Am. Chem. Soc.* **1990**, *112*, 4035.
6. Boeckman, R. K., Jr.; Ferreira, M. R. R.; Mitchell, L. H.; Shao, P. *J. Am. Chem. Soc.* **2002**, *124*, 190.
7. Ohyabu, N.; Nishikawa, T.; Isobe, M. *J. Am. Chem. Soc.* **2003**, *125*, 8798.
8. Kuramochi, A.; Usuda, H.; Yamatsugu, K.; Kanai, M.; Shibasaki, M. *J. Am. Chem. Soc.* **2005**, *127*, 14200.
9. Reuter, J. M.; Salomon, R. G. *J. Org. Chem.* **1977**, *42*, 3360.
10. Nelson, S. G.; Bungard, C. J.; Wang, K. *J. Am. Chem. Soc.* **2003**, *125*, 13000.
11. Nugent, W. A. *J. Chem. Soc. Chem. Commun.* **1999**, 1369.
12. Nelson, S. G.; Wang, K. *J. Am. Chem. Soc.* **2006**, *128*, 4232.

1.2. Johnson–Claisen Rearrangement

In 1970, Johnson and others reported a highly stereoselective synthesis of *trans*-trisubstituted olefinic bonds via the Claisen rearrangement. The alcohol **1**, on heating with 7 equivalents of ethyl orthoacetate and a catalytic amount of propionic acid at 138 °C for 1 hour while distilling ethanol, was converted to the diene ester **2** in 92% yield and with more than 98% (*E*)-isomer[1] (Scheme 1.2a). Heating a mixture of allylic alcohol and ethyl orthoacetate in the presence of a small amount of propionic acid gives a mixed orthoacetate that loses ethanol to form the ketene acetal, which then undergoes a [3,3]-sigmatropic rearrangement[1] (Scheme 1.2b). The outstanding stereoselectivity observed in the Johnson–Claisen rearrangement can be explained by a six-membered transition state where the nonbonded interaction between the ethoxy and R groups is a determining factor[2] (Scheme 1.2b).

Other indirect evidence for the existence of a six-membered transition state in the Johnson–Claisen rearrangement is found in Daub et al.'s experiments[3] (Scheme 1.2c). When cinnamyl alcohol was heated with triethylorthopropionate in the presence of an acid catalyst, the products were obtained

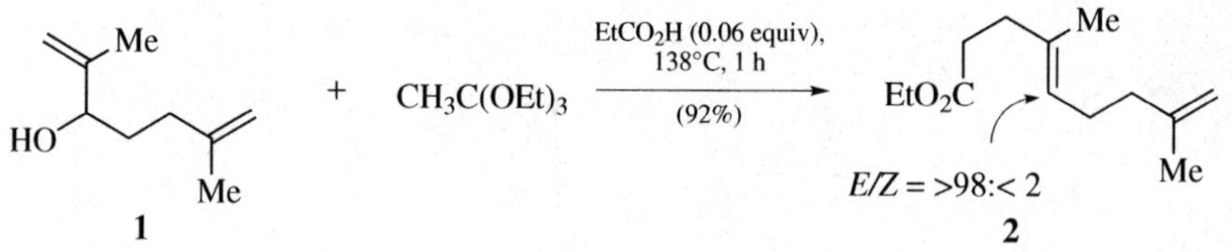

Scheme 1.2a

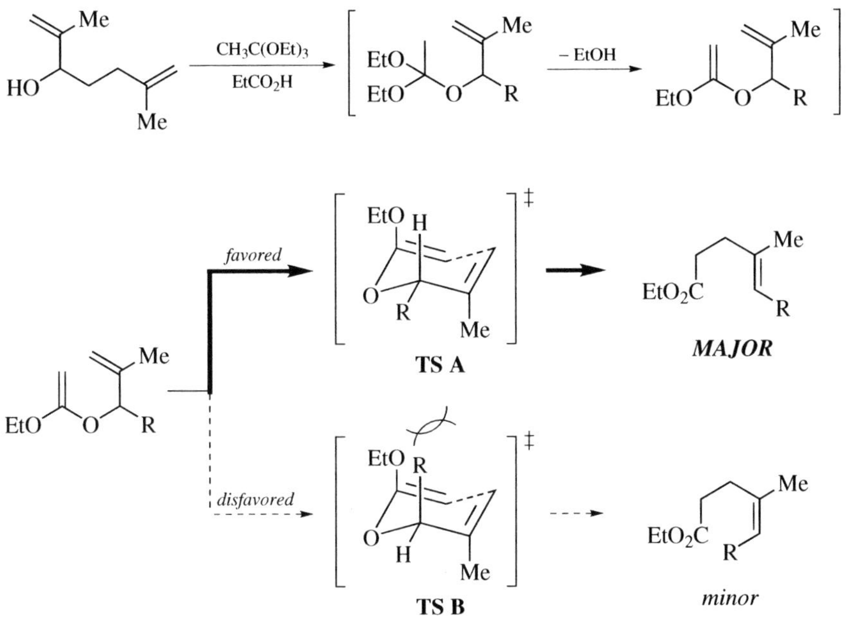

Scheme 1.2b

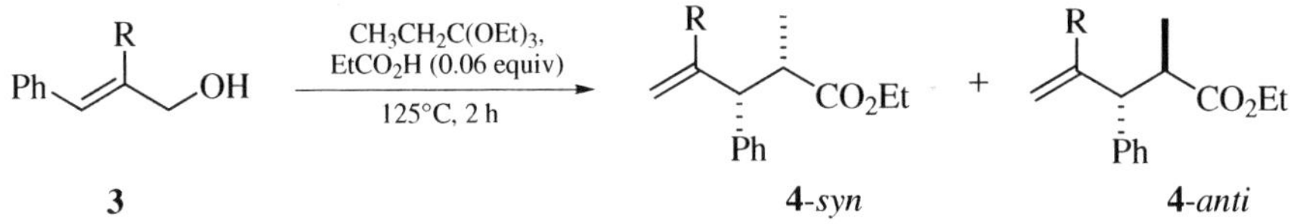

Scheme 1.2c

R	syn/anti	Yield, %
H	60:40	72
CH$_3$	85:15	65

in 60:40 selectivity, slightly favoring the *syn*-isomer. The *syn*-selectivity improved to 85:15 for the rearrangement reaction of the ketene acetal of (*E*)-2-methyl-3-phenyl-2-propen-1-ol (Scheme 1.2c).

The increased *syn/anti* selectivity observed in the Johnson–Claisen rearrangement of the ketene acetal from the trisubstituted alkene is an indication that the reaction goes through a six-membered chairlike transition state[4] (Scheme 1.2d). Transition state **B** is disfavored compared to state **A**, due to the developing 1,3-diaxial interactions between the two methyl groups in the transition state.

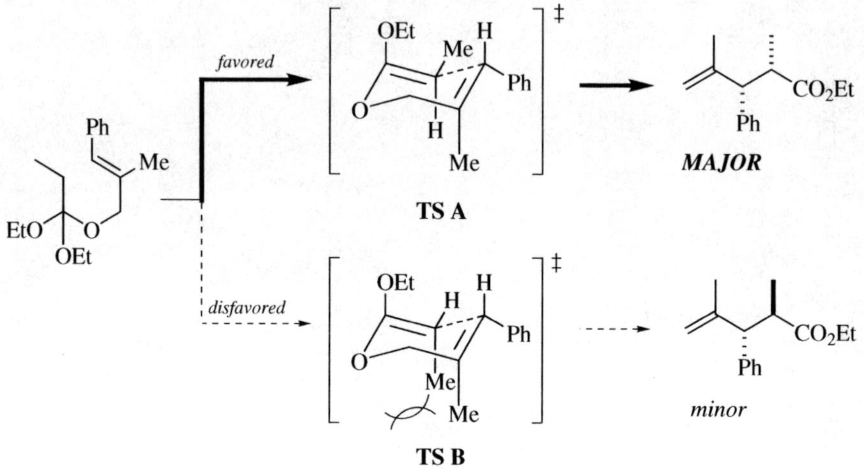

Scheme 1.2d

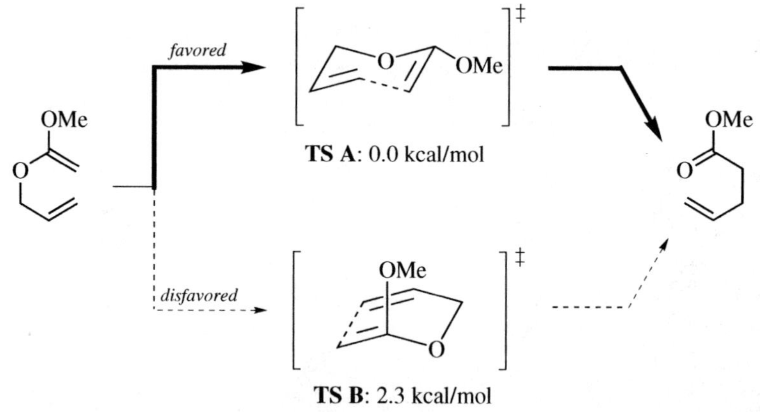

Scheme 1.2e

The preference for a chairlike transition state in the Johnson–Claisen rearrangement is supported by further Houk et al.'s computational studies (Scheme 1.2e).[5] For the rearrangement of the parent methyl ketene acetal, chairlike transition state **A** is favored over boatlike transition state **B** by 2.3 kcal/mol.

Johnson et al. used their newly developed orthoester Claisen reaction to achieve a highly stereoselective total synthesis of all-*trans* squalene (**5**)[1] (Scheme 1.2f). The diene diol **6** underwent Johnson–Claisen rearrangement when it was heated with ethyl orthoacetate in the presence of propionic acid for 3 h at 138 °C. The diene dialdehyde **7**, obtained by treatment of the resulting ester with lithium aluminum hydride followed by oxidation with Collins reagent, reacted with 2-propenyllithium to give the tetraene diol **8**. The tetraene dialdehyde **9**, which

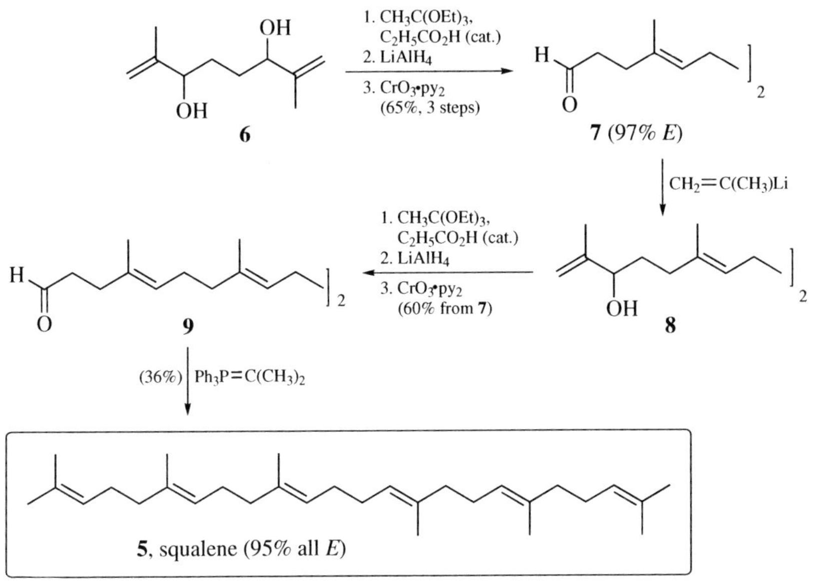

Scheme 1.2f

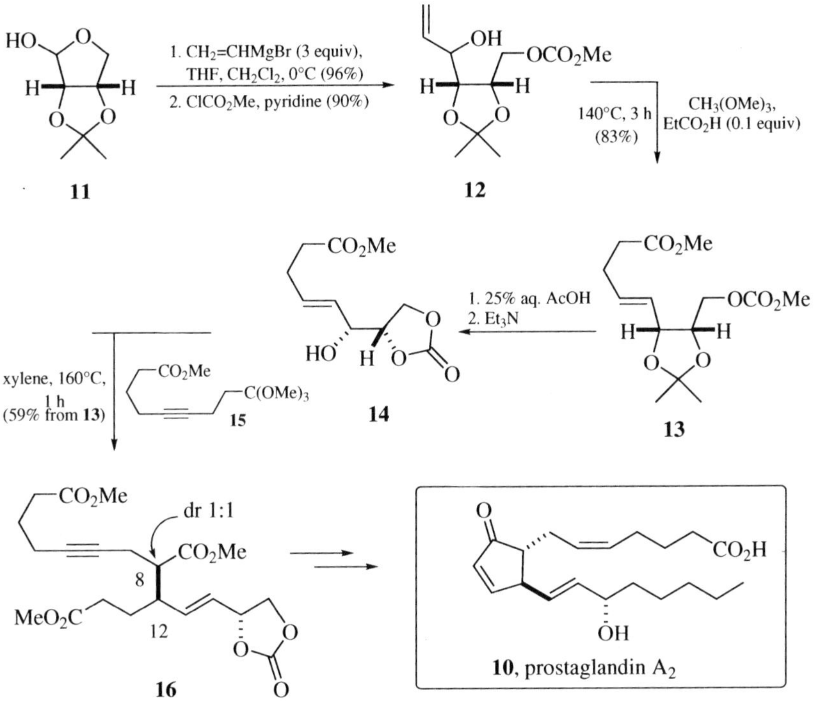

Scheme 1.2g

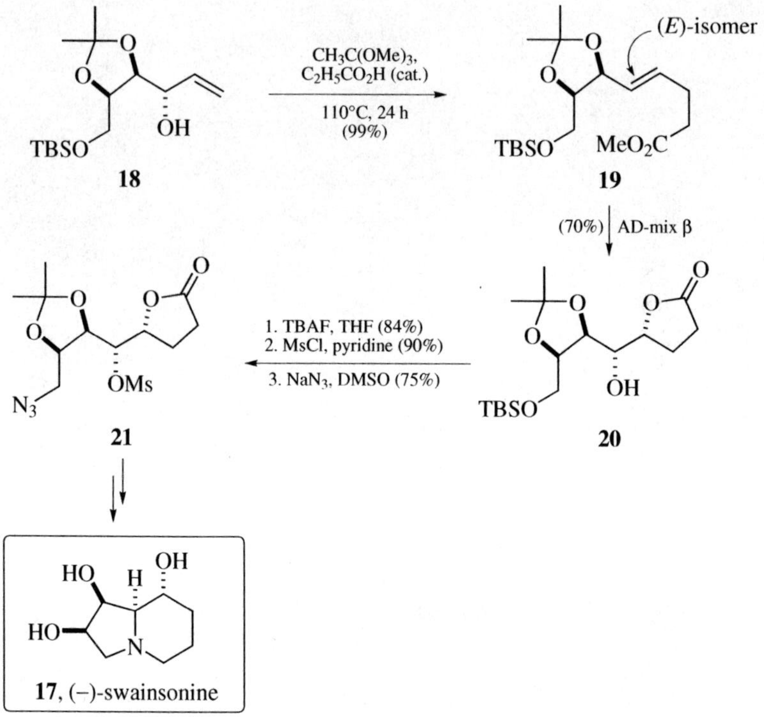

Scheme 1.2h

was accessed by the same reaction sequence as that for the conversion of **6** to **7**, afforded squalene upon treatment with isopropylidenetriphenylphosphorane in 36% yield.

In the total synthesis of prostaglandin A_2 (PGA$_2$; **10**), Stork and Raucher used two Johnson–Claisen rearrangements to obtain the key intermediates[6] (Scheme 1.2g). The first Johnson–Claisen rearrangement was carried out by heating the allylic alcohol **12**, derived from 2,3-isopropylidene-L-erythrose (**11**), with trimethyl orthoacetate to afford the unsaturated ester **13** with an (*E*)-geometry. Hydrolysis of the acetonide followed by treatment with triethylamine afforded the allylic alcohol **14**. The second Johnson–Claisen rearrangement of the allylic alcohol **14** with the orthoester **15** produced **16** in good yield. The chirality at the C_{12} center in **16** was secured by virtue of chirality transfer of a carbon–oxygen bond in **14** through a six-membered chairlike transition state in the rearrangement. The formation of a 1:1 mixture at the C_8 center is due to the nonstereoselective generation of the ketene acetal precursor of the rearrangement.

Pearson and Hembre synthesized a key intermediate (**19**) using the Johnson–Claisen rearrangement protocol in the total synthesis of the indolizidine

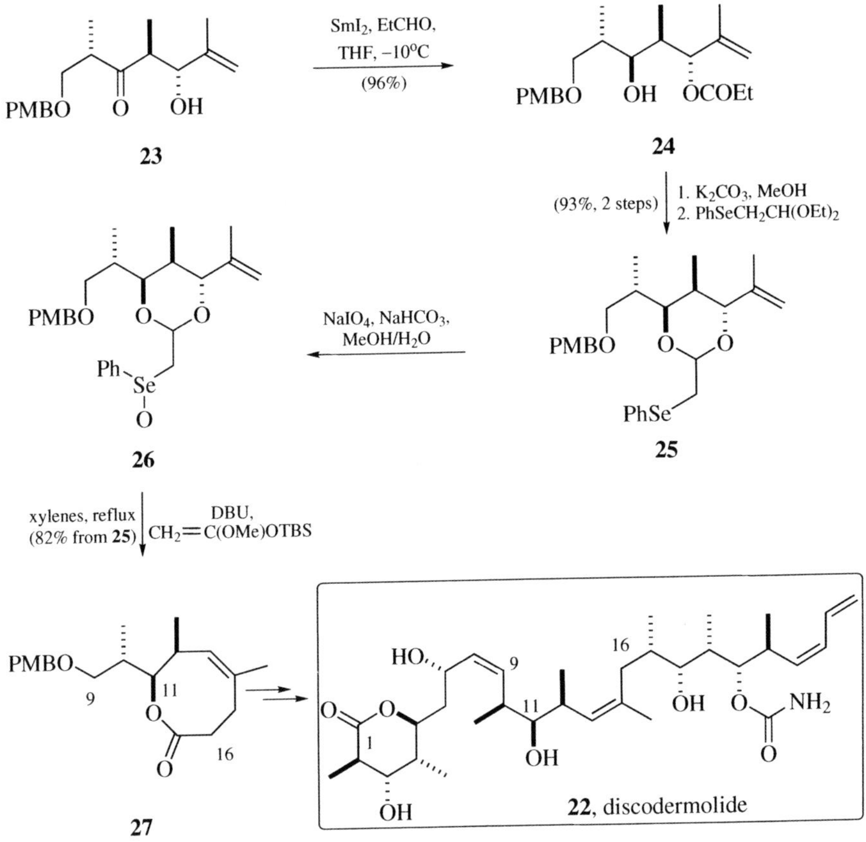

Scheme 1.2i

alkaloid (−)-swainsonine (**17**), a potent anticancer drug as an inhibitor of many mannosidases[7] (Scheme 1.2h). When the allylic alcohol **18** was heated in toluene with trimethyl orthoacetate in the presence of a catalytic amount of propionic acid with the constant removal of methanol, the γ,δ-unsaturated ester desired (**19**) was obtained in nearly quantitative yield with the (*E*)-alkene geometry. Oxidative lactonization via Sharpless dihydroxylation[8] of the alkene **19** provided the lactone desired (**20**) in 70% yield along with the other diastereomer in 9% isolated yield. Deprotection of the silyl ether, mesylation, and selective substitution of the less hindered primary mesylate of the diol produced the monoazide **21**. Further synthetic manipulations to the azide **21** provided 4.5 g of target molecule **17**, successfully demonstrating a practical synthesis of (−)-swainsonine.

In the total synthesis of a polyketide natural product, (+)-discodermolide (**22**), Paterson and co-workers synthesized the C_9–C_{16} fragment **27** using a Johnson–Claisen rearrangement[9] (Scheme 1.2i). Evans–Tishchenko reduction[10] of

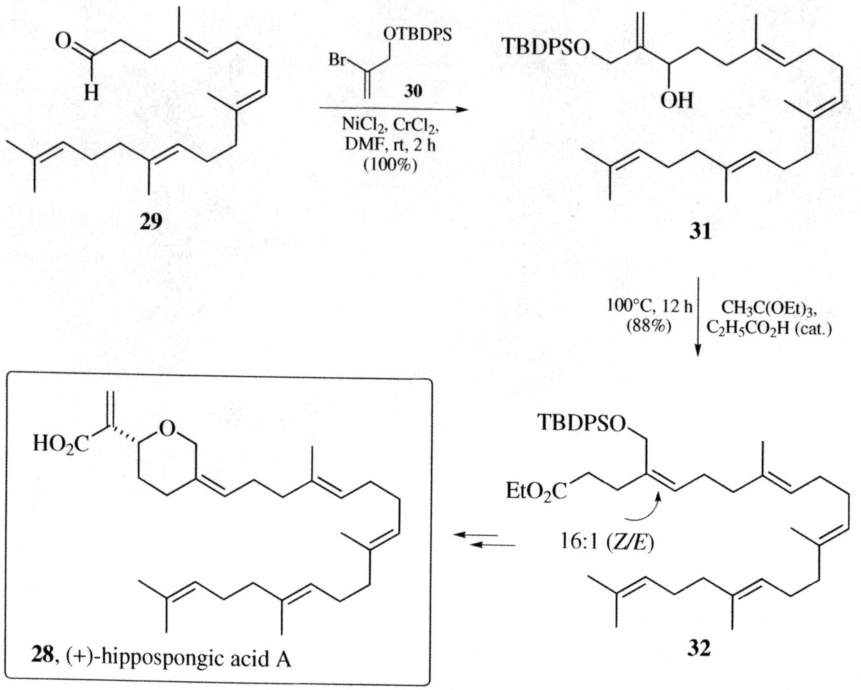

26   [3,3]-SIGMATROPIC REARRANGEMENTS

Scheme 1.2j

the β-hydroxy ketone **23** gave the *anti*-1,3-diol monoester **24** (see Chapter 4). Methanolysis followed by transacetalization afforded the selenide **25**. Oxidation of **25** with NaIO$_4$ resulted in the formation of the selenoxide **26**, which underwent β-elimination upon treatment with DBU. The ketene acetal thus generated underwent a highly stereoselective Johnson–Claisen rearrangement, via a six-membered chairlike transition state, to provide the eight-membered lactone **27** in excellent yield.

The natural product (+)-hippospongic acid A (**28**) shows a variety of biological activities, such as inhibition of gastrulation in starfish embryos and induction of apotosis in the human gastric cancer cell line. In the synthesis of **28**, Trost et al. prepared a key intermediate via a Johnson–Claisen rearrangement reaction[11] (Scheme 1.2j). The Nozaki–Hiyama–Kishi reaction[12] of the aldehyde **29** with the vinyl bromide **30** gave the allylic alcohol **31** in quantitative yield under mild reaction conditions. When the alcohol **31** was subjected to a Johnson–Claisen rearrangement at 100 °C, the product desired (**32**) was obtained in excellent yield with high stereoselectivity around the newly formed double bond. The stereochemical outcome of the rearrangement reaction is rationalized by a six-membered transition state.

REFERENCES

1. Johnson, W. S.; Werthemann, L.; Bartlett, W. R.; Brockson, T. J.; Li, T.-T.; Faulkner, D. J.; Petersen, M. R. *J. Am. Chem. Soc.* **1970**, *92*, 741.
2. Perrin, C. L.; Faulkner, D. J. *Tetrahedron Lett.* **1969**, 2783.
3. Daub, G. W.; Shanklin, P. L.; Tata, C. *J. Org. Chem.* **1986**, *51*, 3402.
4. Daub, G. W.; Edwards, J. P.; Okada, C. R.; Allen, J. W.; Maxey, C. T.; Wells, M. S.; Goldstein, A. S.; Dibley, M. J.; Wang, C. J.; Ostercamp, D. P.; Chung, S.; Cunningham, P. S.; Berliner, M. A. *J. Org. Chem.* **1997**, *62*, 1976.
5. Khaledy, M. M.; Kalani, M. Y. S.; Khong, K. S.; Houk, K. N.; Aviyente, V.; Neier, R.; Soldermann, N.; Velker, J. *J. Org. Chem.* **2003**, *68*, 572.
6. Stork, G.; Raucher, S. *J. Am. Chem. Soc.* **1976**, *98*, 1583.
7. Pearson, W. H.; Hembre, E. J. *J. Org. Chem.* **1996**, *61*, 7217.
8. Wang, Z.-M.; Zhang, X.-L.; Sharpless, K. B. *Tetrahedron Lett.* **1992**, *33*, 6407.
9. Paterson, I.; Florence, G. J.; Gerlach, K.; Scott, J. P.; Sereinig, N. *J. Am. Chem. Soc.* **2001**, *123*, 9535.
10. Evans, D. A.; Hoveyda, A. H. *J. Am. Chem. Soc.* **1990**, *112*, 6447.
11. Trost, B. M.; Machacek, M. R.; Tsui, H. C. *J. Am. Chem. Soc.* **2005**, *127*, 7014.
12. (a) Takai, K.; Kimura, K.; Kuroda, T.; Hiyama, T.; Nozaki, H. *Tetrahedron Lett.* **1983**, *24*, 5281; (b) Jin, H.; Uenishi, J.; Christ, W. J.; Kishi, Y. *J. Am. Chem. Soc.* **1986**, *108*, 5644.

1.3. Ireland–Claisen Rearrangement

In 1976, Ireland et al. reported that [3,3]-sigmatropic rearrangement of allylic esters as enolate anions or corresponding silylketene acetals produces the γ,δ- unsaturated acids in good yields and with excellent levels of diastereoselectivity. For example, rearrangement of the ester **1** afforded (*E*)-4-decenoic acid (**2**) with greater than 99% stereoselectivity and in high yield[1] (Scheme 1.3a). The *tert*-butyldimethylsilylketeneacetal **1OTBS**, generated by successive treatment of **1** with lithium diisopropylamide (LDA) and *tert*-butyldimethylsilyl chloride at −78 °C, undergoes rearrangement via a six-membered chairlike transition state[2] (Scheme 1.3b). An examination of nonbonded interactions readily indicates which of the two possible transition states will be favored. The equatorial disposition of the R group puts transition state **A** at lower energy, which results in predominant formation of the *E*-isomer.

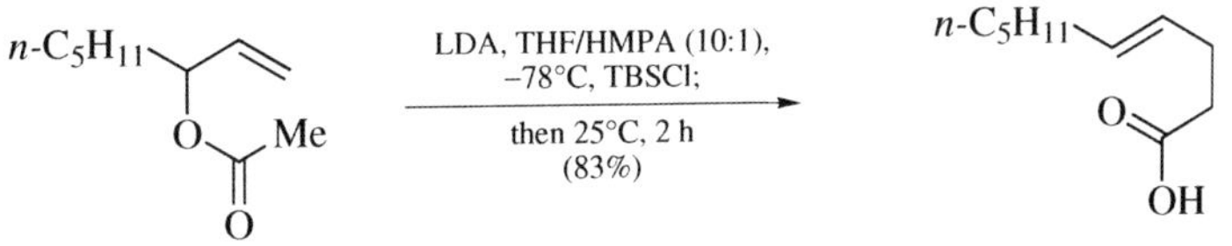

Scheme 1.3a

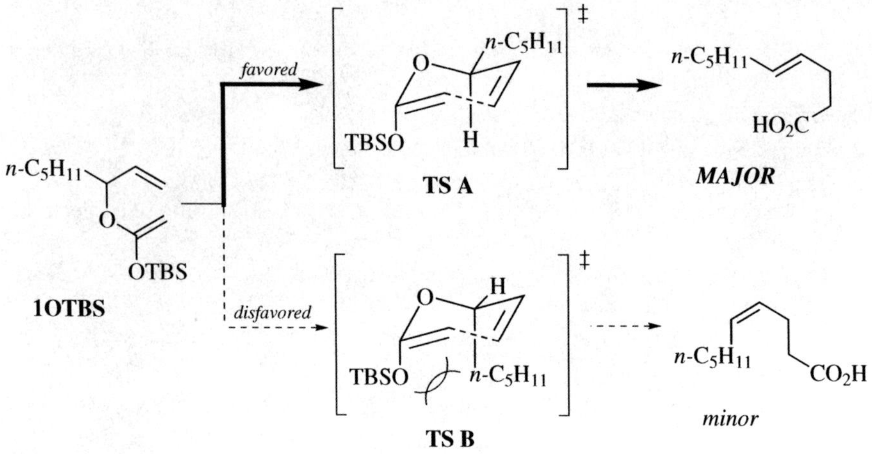

28 [3,3]-SIGMATROPIC REARRANGEMENTS

Scheme 1.3b

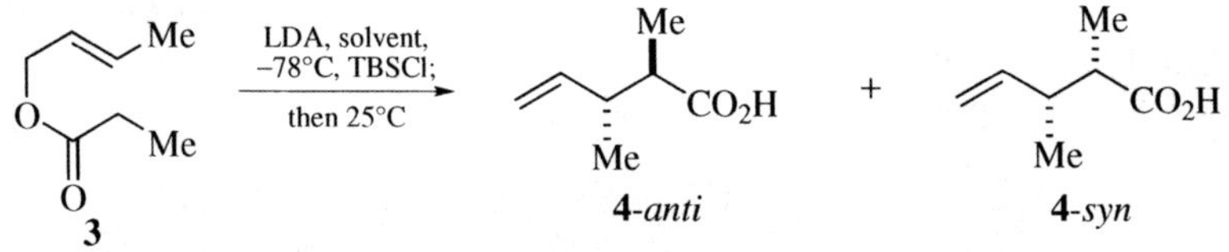

Scheme 1.3c

Solvent	anti/syn	Yield, %
THF	87:13	79
THF/HMPA (77:23)	19:81	73

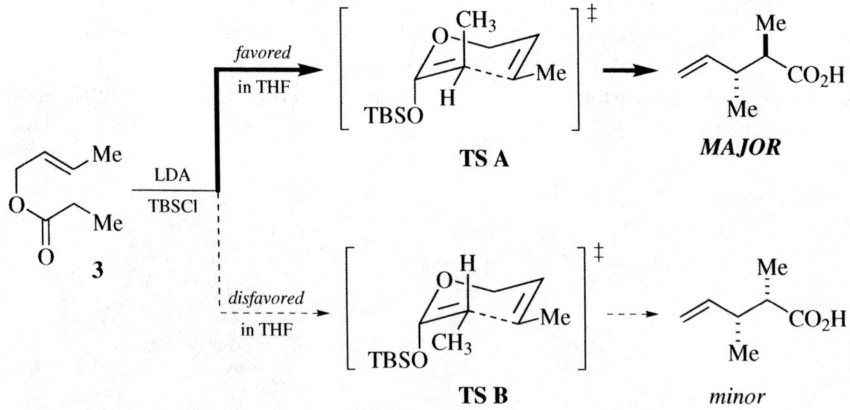

Scheme 1.3d

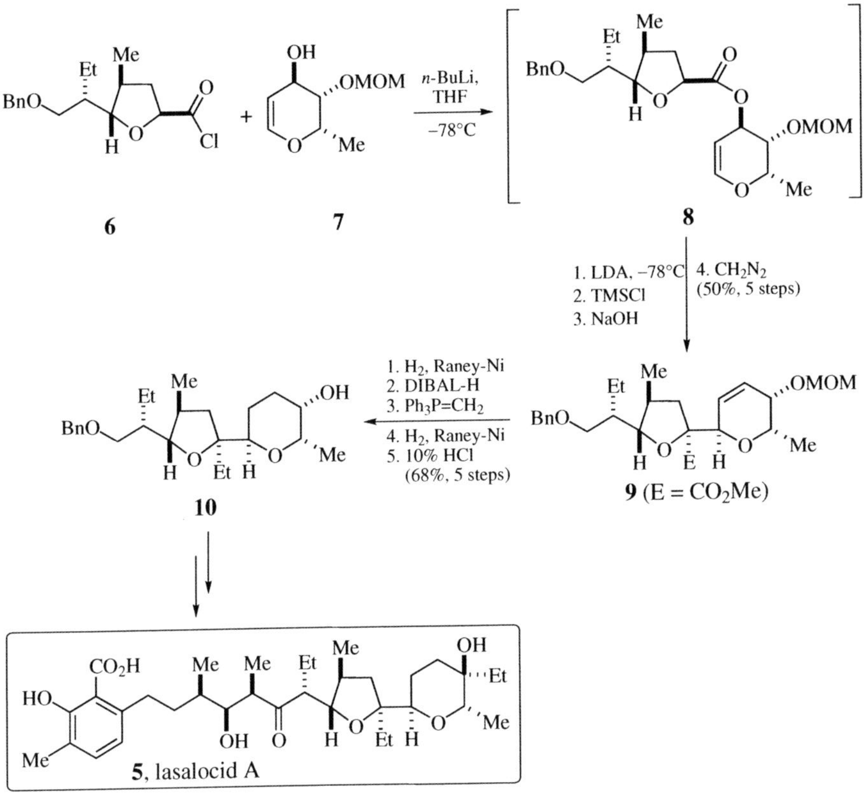

Scheme 1.3e

As the Ireland–Claisen rearrangement proceeds through a six-membered chair-like transition state, the stereochemistry about the newly formed carbon–carbon single bond can be predicted from the geometries of the double bonds in the starting 1,5-dienes.[1] For example, the stereochemical outcome of the rearrangement of (*E*)-crotyl propanoate (**3**) depends on the solvents used. The *anti*-isomer is obtained as a major isomer in THF, whereas the *syn*-isomer is the major product when the reaction was carried out with hexamethylphosphoramide (HMPA) as a cosolvent (Scheme 1.3c). Again, these results can be explained via a six-membered transition state (Scheme 1.3d). In THF, the (*E*)-enol ether is formed preferentially and subsequently undergoes a [3,3]-sigmatropic rearrangement via transition state **A**. When HMPA is used as a cosolvent along with THF, the (*Z*)-enol ether becomes a major isomer, resulting in the formation of **4**-*syn*.[1]

Claisen rearrangements of silylketene acetals have been used in numerous organic syntheses of natural products.[3] Ireland used his newly developed ester enolate Claisen rearrangement in the total synthesis of lasalocid A (X537A) (**5**), a polyether ionophore antibiotic natural product with a broad range of biological potency (Scheme 1.3e).[4] The ester **8**, which was prepared by the reaction

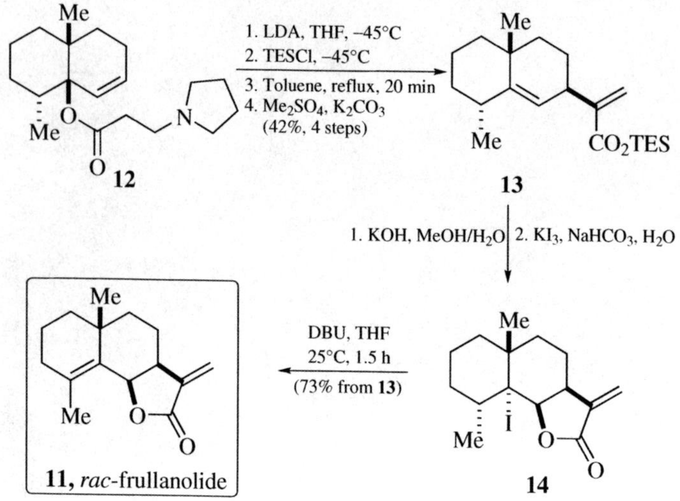

[3,3]-SIGMATROPIC REARRANGEMENTS

Scheme 1.3f

of the acyl chloride **6** from α-D-glucosaccharinic acid lactone and the glycal **7** from 6-deoxy-L-glucose, underwent Ireland–Claisen rearrangement to provide the tetrahydrofuran **9** in 50% yield after hydrolysis and esterification. The tetrahydrofuran **10** was eventually utilized as a key intermediate in Ireland's total synthesis of lasalocid A.[5]

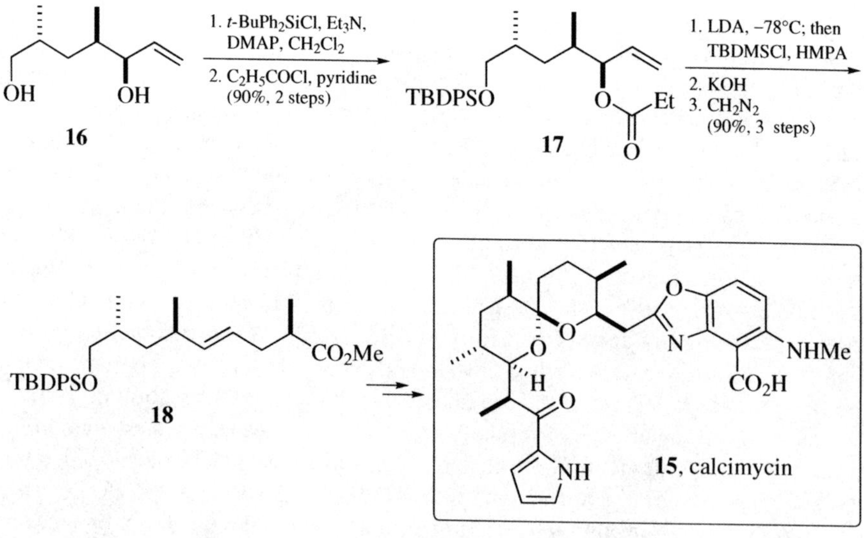

Scheme 1.3g

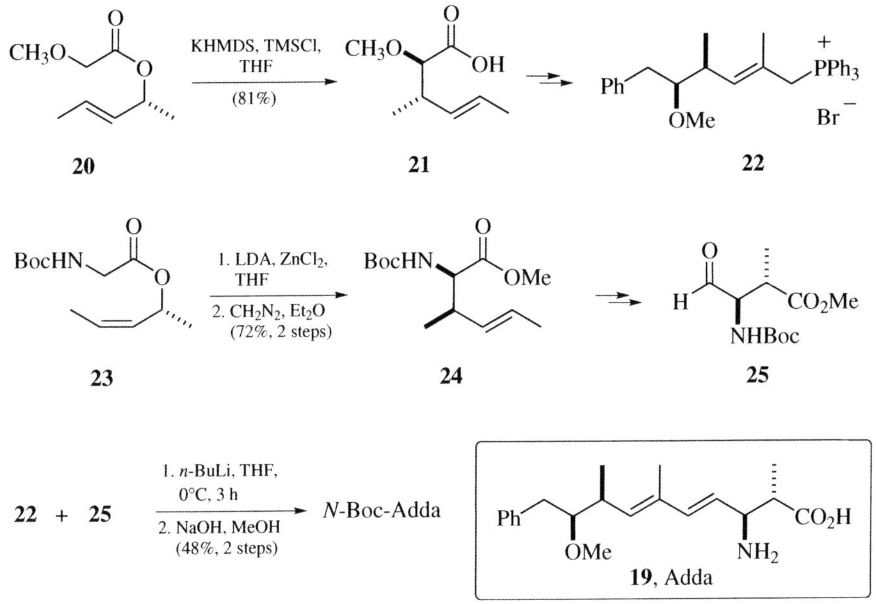

Scheme 1.3h

Still and Schneider employed Ireland–Claisen rearrangement in the total synthesis of (±)-frullanolide (**11**)[6] (Scheme 1.3f). The key step of the synthesis is efficient Ireland–Claisen rearrangement of the β-pyrrolidinopropionate ester **12**. The triethylsilylketene acetal rearranged in toluene at reflux and the pyrrolidine moiety was eliminated after stirring with a mixture of dimethyl sulfate and potassium carbonate in methanol to afford the α-substituted acrylic ester (**13**). Saponification followed by iodolactonization gave the iodolactone **14**, which upon treatment with DBU led to (±)-frullanolide.

In total synthesis of the structurally unique natural product calcimycin (**15**), Grieco and others used Ireland–Claisen rearrangement of the ester **17** to synthesize the key intermediate (**18**)[7] (Scheme 1.3g). Monosilylation of the diol **16** followed by treatment with propionyl chloride in pyridine gave rise to the ester **17** in 90% yield. Treatment of **17** with LDA in THF at −78 °C, subsequent addition of tert-butyldimethylsilyl chloride in HMPA, and brief heating of the resulting silylketene acetal provided the corresponding silyl ester. Subsequent hydrolysis of the silyl ester and esterification with diazomethane gave **18** in 90% yield from **17**.

The C_{20} amino acid (2*S*,3*S*,8*S*,9*S*,4*E*,6*E*)-3-amino-9-methoxy-2,6,8-trimethyl-10-phenyldeca-4,6-dienoic acid (Adda; **19**) is a molecule of interest to biologists and organic chemists as a component of the hepatotoxic cyclic peptides called *microcystins*. Kim and Toogood used Ireland–Claisen rearrangement in their successful synthesis of Adda[8] (Scheme 1.3h). The ester **20** underwent highly diastereoselective Ireland–Claisen rearrangement to provide the acid **21**. Conversion of this acid to the phosphonium bromide **22** was achieved in nine steps. Another Ireland–Claisen rearrangement of **23** in the presence of $ZnCl_2$[9] efficiently afforded the ester **24**. Wittig reaction of the two fragments **22** and **25**, followed by saponification, provided N-Boc-protected Adda.

REFERENCES

1. Ireland, R. E.; Mueller, R. H.; Willard, A. K. *J. Am. Chem. Soc.* **1976**, *98*, 2868.
2. (a) Wipf, P. in *Comprehensive Organic Synthesis*; Trost, B. M., Fleming, I., Eds.; Pergamon Press: Oxford, **1991**: Vol. 5, Chap. 7.2; (b) Ireland, R. E.; Wipf, P.; Xiang, J.-N. *J. Org. Chem.* **1991**, *56*, 3572.
3. For a review of the synthetic applications of Ireland–Claisen rearrangement, see Pereira, S.; Srebnick, M. *Aldrichimica Acta* **1993**, 26, 17.
4. Ireland, R. E.; Thaisrivongs, S.; Wilcox, C. S. *J. Am. Chem. Soc.* **1980**, *102*, 1155.
5. Ireland, R. E.; Anderson, R. C.; Badoud, R.; Fitzsimmons, B. J.; McGarvey, G. J.; Thaisrivongs, S.; Wilcox, C. S. *J. Am. Chem. Soc.* **1983**, *105*, 1988.
6. Still, W. C.; Schneider, M. J. *J. Am. Chem. Soc.* **1977**, *99*, 948.
7. (a) Grieco, P. A.; Williams, E.; Tanaka, H.; Gilman, S. *J. Org. Chem.* **1980**, *45*, 3537; (b) Martinez, G. R.; Grieco, P. A.; Williams, E.; Kanai, K.-I.; Srinivasan, C. V. *J. Am. Chem. Soc.* **1982**, *104*, 1436.
8. Kim, H. Y.; Toogood, P. L. *Tetrahedron Lett.* **1996**, *37*, 2349.
9. Kazmaier, U. *Angew. Chem. Int. Ed.* **1994**, *33*, 998.

1.4. Cope Rearrangement

Because of the concerted nature of the mechanism of Cope rearrangement, chirality at C_3 in the starting material leads to enantiospecific formation of the new chiral center in the product. For example, Cope rearrangement of (3*R*,5*E*)-3-methyl-3-phenyl-1,5-heptadiene (**1**) at 250 °C resulted in an 87:13 mixture of *trans*- and *cis*-3-methyl-6-phenyl-1,5-heptadiene in quantitative yield[1] (Scheme 1.4a). The optical purity of the starting material is 95% ee, and those of the products are 91% ee for **2E** and 89% ee for **2Z**. Thus, the optical integrity of the starting material is preserved during thermal rearrangement, suggesting that the rearrangement is concerted.

The stereochemical outcome of the above reaction is explained in terms of a six-membered chairlike transition state[1] (Scheme 1.4b). The 87:13 preference for

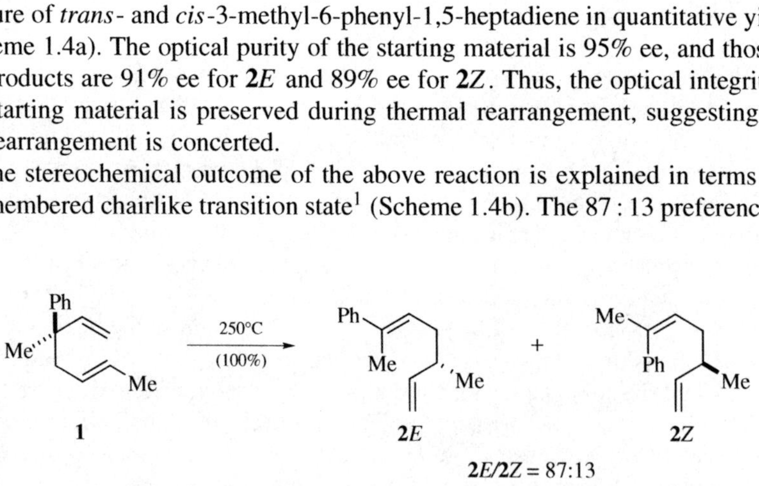

Scheme 1.4a

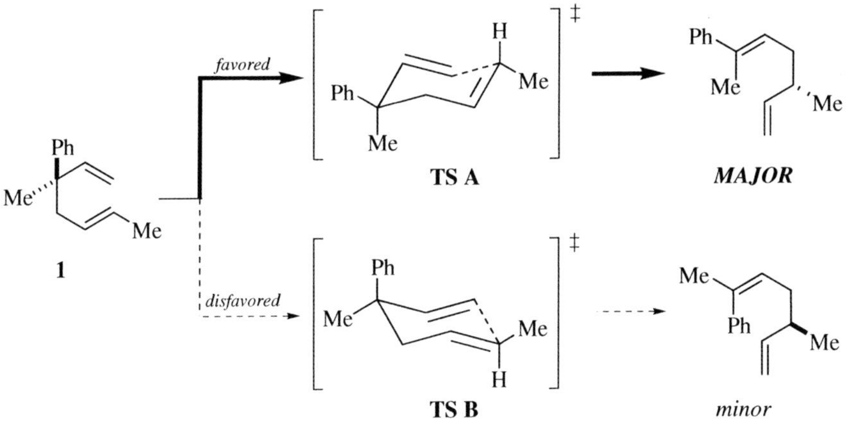

Scheme 1.4b

2*E* corresponds to a free-energy difference of about 2 kcal/mol between transition states **A** and **B**. Based on the A-values of the monosubstituted cyclohexanes, it was understood that transition state **A** in which the phenyl subsitituent group occupies an equatorial position would be favored over **B**.

Raucher et al. used a tandem Cope–Claisen rearrangement during total synthesis of the germacrane sesquiterpene (+)-dihydrocostunolide (**3**)[2] (Scheme 1.4c).

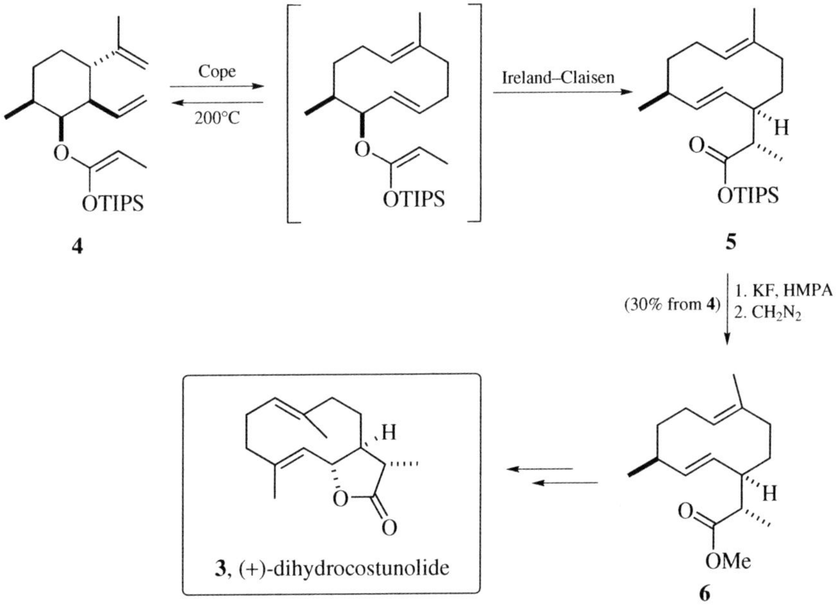

Scheme 1.4c

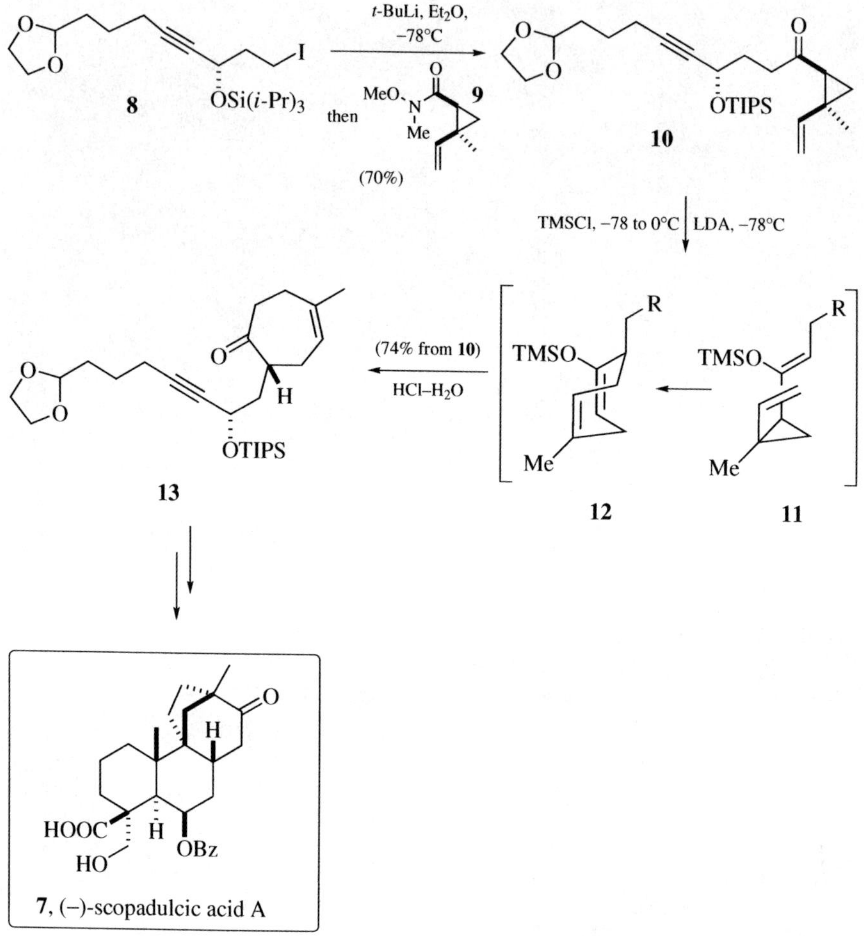

[3,3]-SIGMATROPIC REARRANGEMENTS

Scheme 1.4d

A solution of the silylketene acetal **4** in dodecane was subjected to thermolysis at 200 °C for 140 minutes. The Cope–Claisen rearrangement product **5** was then treated with KF in HMPA followed by esterification to afford the methyl ester **6**.

Fox et al. used the Cope rearrangement in total synthesis of the structurally unique tetracyclic diterpene acid (−)-scopadulcic acid A (**7**), which exhibits a broad range of pharmacological activities[3] (Scheme 1.4d). Lithiation of the optically active iodide **8** with *t*-BuLi followed by condensation of the resulting organolithium species with the amide **9** afforded the cyclopropyl ketone **10**. Compound **10** was then treated sequentially with LDA and TMSCl to provide a silyl enol ether intermediate (**11**), which underwent Cope rearrangement to furnish the silyloxy cycloheptadiene **12**. Hydrolysis of **12** then resulted in the cycloheptenone **13** as a single stereoisomer in 74% overall yield.

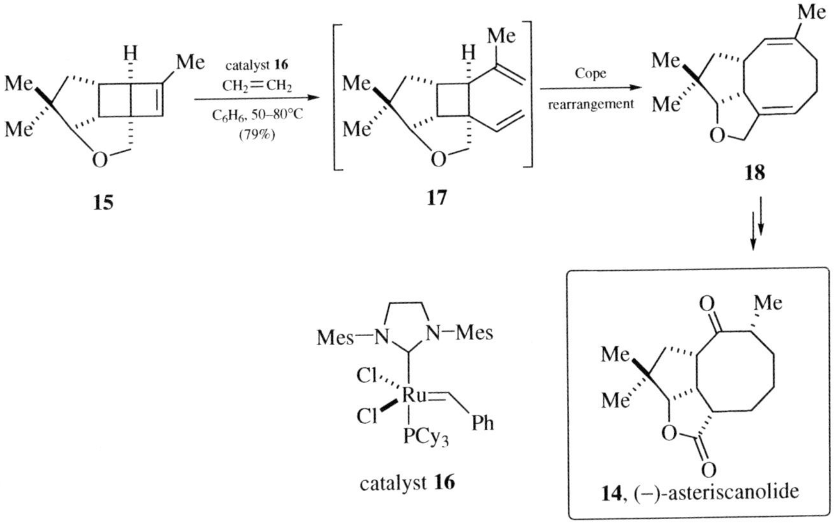

Scheme 1.4e

The Cope rearrangement was used in the total synthesis of (−)-asteriscanolide (**14**), a novel sesquiterpene natural product[4] (Scheme 1.4e). Ring-opening metathesis of the cyclobutene **15** with ethylene in the presence of the ruthenium catalyst **16**[5] proceeded smoothly to provide the cyclooctadiene **18** via Cope rearrangement of the intermediate dialkenyl cyclobutane (**17**).

When the vinyldiazoacetate **19**, which can be prepared from benzaldehyde in a one-pot process,[6] was treated in 2,2-dimethylbutane (DMB) with dirhodium tetrakis[(S)-N-(dodecylbenzenesulfonyl)prolinate] [Rh$_2$(S-DOSP)$_4$] in the presence of 4-methyl or 4-trimethylsilyloxy-1,2-dihydronaphthalene (**20**), the product **21** was obtained with exceptionally high levels of enantio- and diastereo selectivity[7] (Scheme 1.4f).

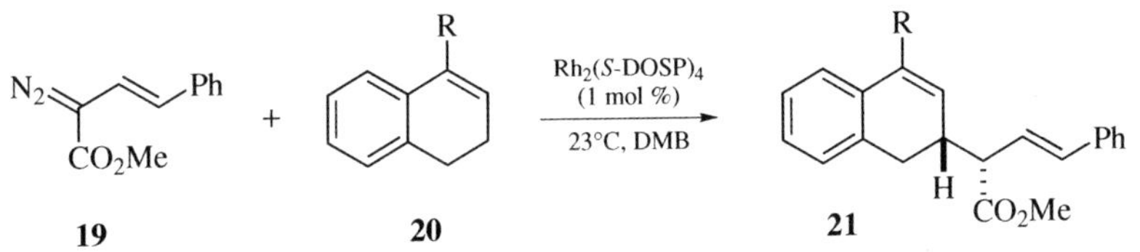

R	ee, %	de, %	Yield, %
Me	99	>98	92
OTMS	98	>98	55

Scheme 1.4f

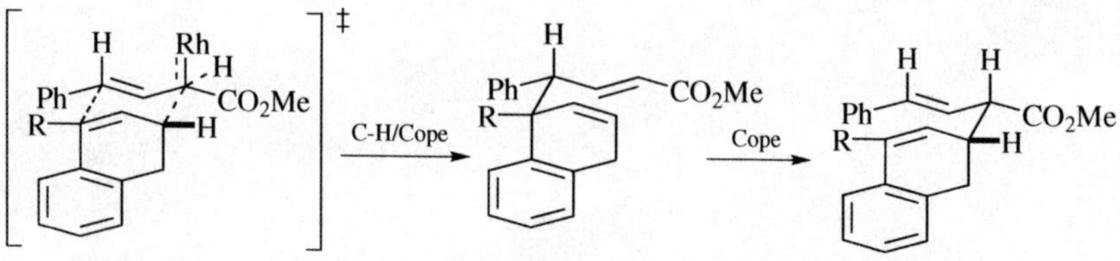

Scheme 1.4g

The highly enantioselective reaction is explained in terms of a double Cope rearrangement event[7] (Scheme 1.4g). The substrate **20** is approaching from the front side, due to the chiral environment posed by the D_2-symmetric rhodium catalyst.[8] The Cope rearrangement then presumably occurs to form **20a** through chairlike transition state **A**. Another Cope rearrangement of the 1,5-diene **20a** affords the product **21** in a highly stereoselective manner.

REFERENCES

1. Hill, R. K.; Gilman, N. W. *J. Chem. Soc. Chem. Commun.* **1967**, 619.
2. Raucher, S.; Chi, K.-W.; Hwang, K.-J.; Burks, J. E., Jr. *J. Org. Chem.* **1986**, *51*, 5503.
3. Fox, M. E.; Li, C.; Marino, J. P., Jr.; Overman, L. E. *J. Am. Chem. Soc.* **1999**, *121*, 5467.
4. Limanto, J.; Snapper, M. L. *J. Am. Chem. Soc.* **2000**, *122*, 8071.
5. (a) Scholl, M.; Ding, S.; Lee, C. W.; Grubbs, R. H. *Org. Lett.* **1999**, *1*, 953; (b) Trnka, T. M.; Grubbs, R. H. *Acc. Chem. Res.* **2001**, *34*, 18; (c) Schrock, R. R. *Chem. Rev.* **2002**, *102*, 145.
6. Davies, H. M. L.; Yang, J.; Manning, J. R. *Tetrahedron: Asymmetry* **2006**, *17*, 665.
7. Davies, H. M. L.; Jin, Q. *J. Am. Chem. Soc.* **2004**, *126*, 10862.
8. Nowlan, D. T., III; Gregg, T. M.; Davies, H. M. L.; Singleton, D. A. *J. Am. Chem. Soc.* **2003**, *125*, 15902.

1.5. Anionic Oxy-Cope Rearrangement

In 1964, Berson and Jones reported that heating **1** in the gas phase at 320 °C gave *cis*-2-octalone (**2**) in 50% yield[1] (Scheme 1.5a).

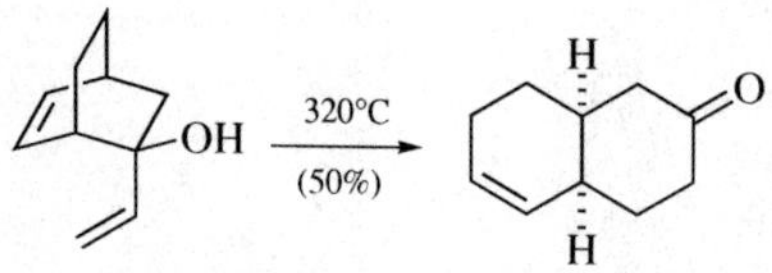

Scheme 1.5a

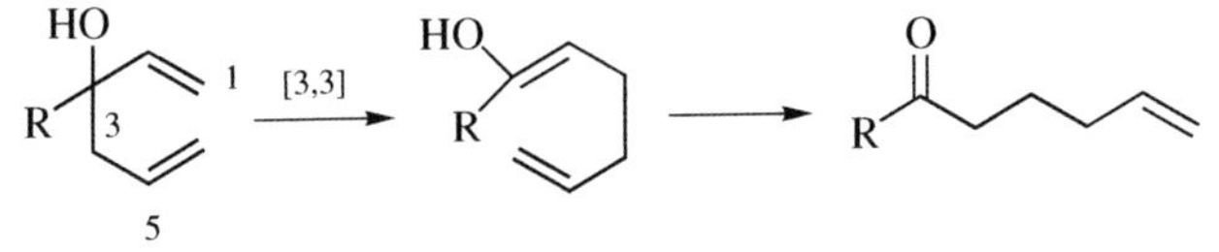

Scheme 1.5b

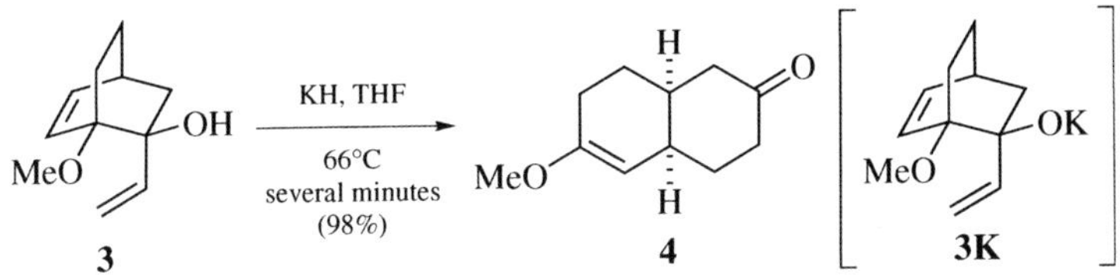

Scheme 1.5c

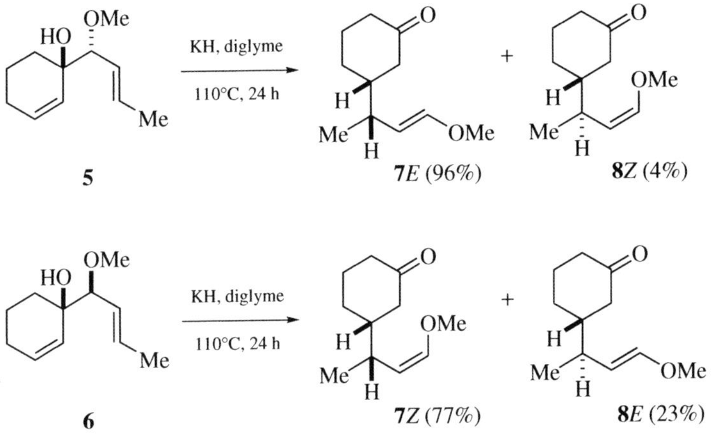

Scheme 1.5d

Berson proposed the term *oxy-Cope rearrangement* for the reaction, as the reaction is a [3,3]-sigmatropic Cope rearrangement of 3-hydroxy-1,5-hexadiene[1] (Scheme 1.5b). The oxy-Cope rearrangement would be a synthetically useful route to access δ,ε-unsaturated carbonyl compounds from the corresponding secondary or tertiary alcohols if the reaction conditions were mild. In 1975, Evans and Golob discovered that the Cope rearrangement of 3-hydroxy-1,5-hexadienes proceeds extremely fast in the presence of potassium hydride. For example, heating the potassium alkoxide **3K** at 66 °C for several minutes in anhydrous THF completed the Cope rearrangement to afford the methoxy ketone in superb yield[2] (Scheme 1.5c). From kinetic experiments it was determined that at 25 °C rearrangement of **3K** occurred 10^{12} times faster than **3** in the presence of 1.1 equivalents of 18-crown-6.

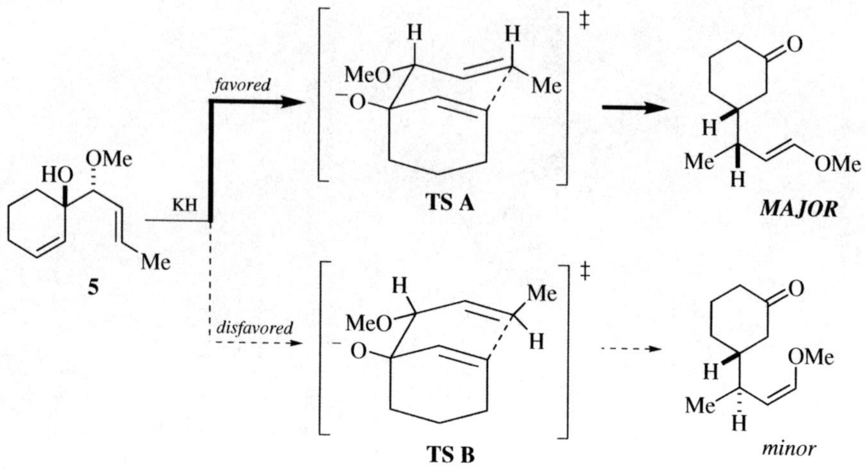

Scheme 1.5e

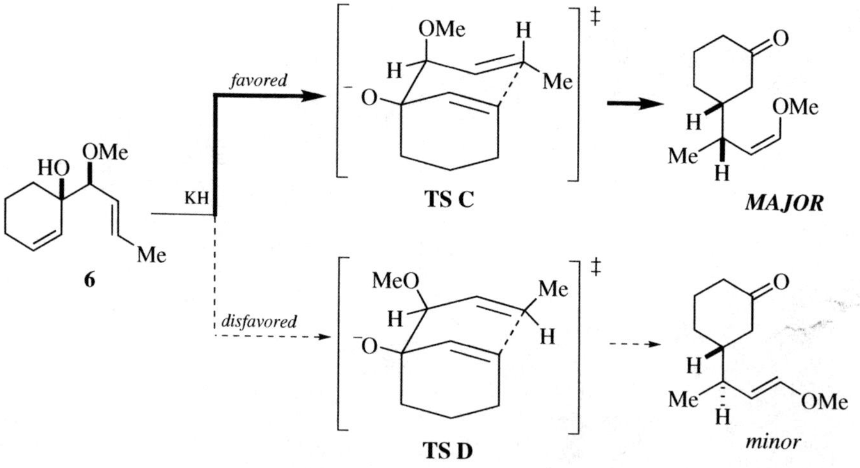

Scheme 1.5f

To gain insights into the possible transition-state geometry of the sigmatropic process, Evans et al. rearranged the diastereomeric dienols **5** and **6**[3] (Scheme 1.5d). When a mixture of **5** and KH was heated in diglyme at 110 °C for a day, the rearranged products were obtained in 78% yield and with high diastereoselectivity. Under similar reaction conditions, the dienol **6** also underwent anionic oxy-Cope rearrangement, but with poor diastereoselectivity. The striking difference in the stereoselectivity observed in the rearrangement of **5** and **6** suggests that a six-membered transition state is operating in these reactions. In the rearrangement of **5**, the major product **7E** is formed via chairlike transition state **A**,

Molecules	Bond Energy, kcal/mol
H—CH$_2$OH	90.7
H—CH$_2$OK	79.0

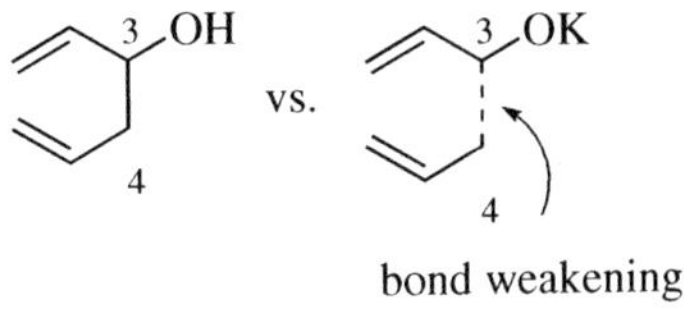

Scheme 1.5g

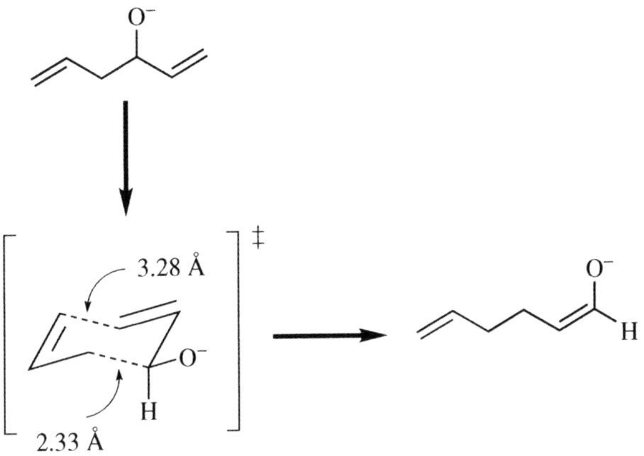

TS A

Scheme 1.5h

and the minor product **8Z** is produced via the boatlike transition state **B** (Scheme 1.5e). The 96 : 4 selectivity indicates that transition state **A** is more stable than **B** by 2.2 kcal/mol.

For the rearrangement of **6**, the major isomer is once again formed through six-membered chairlike transition state **C**[3] (Scheme 1.5f). Because the methoxy substituent in transition state **C** is now in an axial position, the free-energy gap between transition states **C** and **D** becomes narrow, resulting in diminished 3 : 1 diastereoselectivity.

To explain the marked rate enhancement in the anionic oxy-Cope rearrangement, Evans and others conducted theoretical calculations of the carbon–hydrogen bond strengths for methanol and potassium methoxide[4] (Scheme 1.5g). The computations indicate that the carbon–hydrogen bond in KOMe is significantly weaker than that in MeOH. From these studies, it is concluded that weakening of the C_3–C_4 bond is responsible for the rate acceleration in the anionic oxy-Cope rearrangement.

Houk and others performed a computational study with density functional and ab initio calculations to better understand the reaction mechanism of the

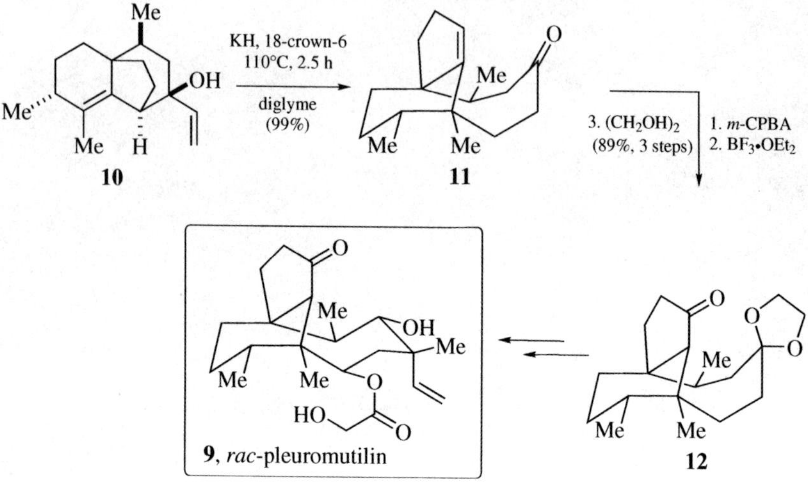

Scheme 1.5i

anionic oxy-Cope rearrangement[5] (Scheme 1.5h). The reaction proceeds via a concerted reaction pathway with an activation energy of 9.9 kcal/mol. Although the six-membered transition-state structure is dissociative, no intermediate is found over the entire course of the rearrangement.

Anionic oxy-Cope rearrangement has been used extensively in the total synthesis of natural products.[6] For example, Boeckman et al. employed a remarkably facile anionic oxy-Cope rearrangement in the total synthesis of (±)-pleuromutilin (**9**), which is utilized as an animal food additive to control dysentery in swine and poultry[7] (Scheme 1.5i). The crucial anionic oxy-Cope rearrangement of the β,β-disubstituted alcohol **10** proceeded cleanly at 110 °C on exposure to potassium hydride and 18-crown-6 ether to afford the ketone **11** in 99% yield. Epoxidation of cyclopentene ring and rearrangement of the resulting epoxide followed by selective ketalization gave the ketal **12**.

The anionic oxy-Cope rearrangement was the key step in Lee et al.'s synthesis of (+)-dihydromayurone (**13**)[8] (Scheme 1.5j). When the allylic alcohol **14** was treated with potassium hydride and 18-crown-6 ether to effect the crucial anionic oxy-Cope rearrangement, the aldehyde **15** was obtained in high enantioselectivity. The highly efficient transfer of chirality from the secondary allylic alcohol center to the quaternary carbon center in **15** is indicative of transition state **14TS**, in which the carbon–oxygen bond adopts an equatorial position in the chairlike transition state. The aldehyde **15** was oxidized to the corresponding carboxylic acid, which was in turn converted to the diazo ketone **16** via the corresponding acyl chloride. The target molecule (**13**) was then obtained readily in high yield when the diazo ketone **16** was treated with a catalytic amount of rhodium acetate in benzene.

In the total synthesis of (−)-salsolene oxide (**17**), an architecturally unusual sesquiterpene with an unsaturated bicyclo[5.3.1]undecane core and trisubstituted

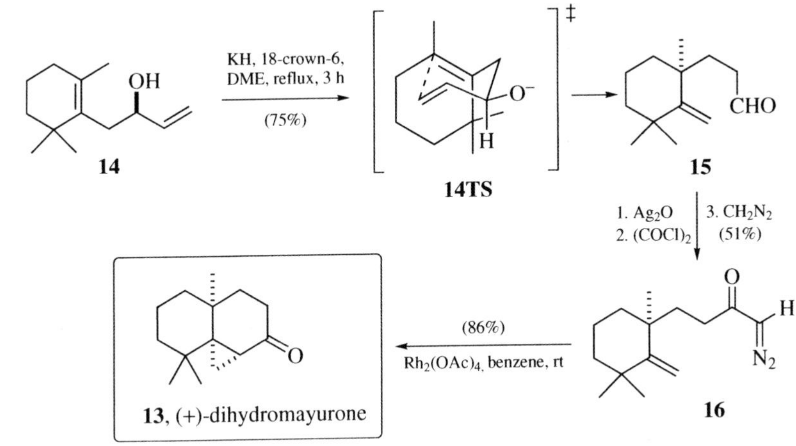

Scheme 1.5j

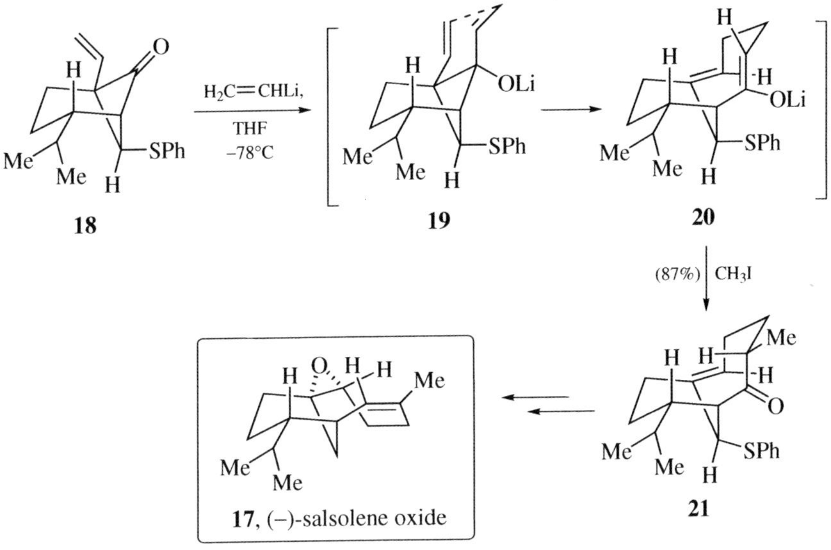

Scheme 1.5k

oxirane, Paquette and co-workers utilized the anionic oxy-Cope rearrangement to synthesize a key intermediate (**21**)[9] (Scheme 1.5k). The thiophenyl substituent in the bicyclo ketone **18** directed the 1,2-addition of vinyllithium to the *exo*-face of **18**. The ring strain in 1,2-divinylcyclobutanoxide (**19**) was sufficient to promote a facile [3,3]-sigmatropic rearrangement under the reaction conditions. The enolate anion **20** was therefore generated stereoselectively via a six-membered chairlike transition state. Direct methylation of **20** with excess methyl iodide furnished **21**.

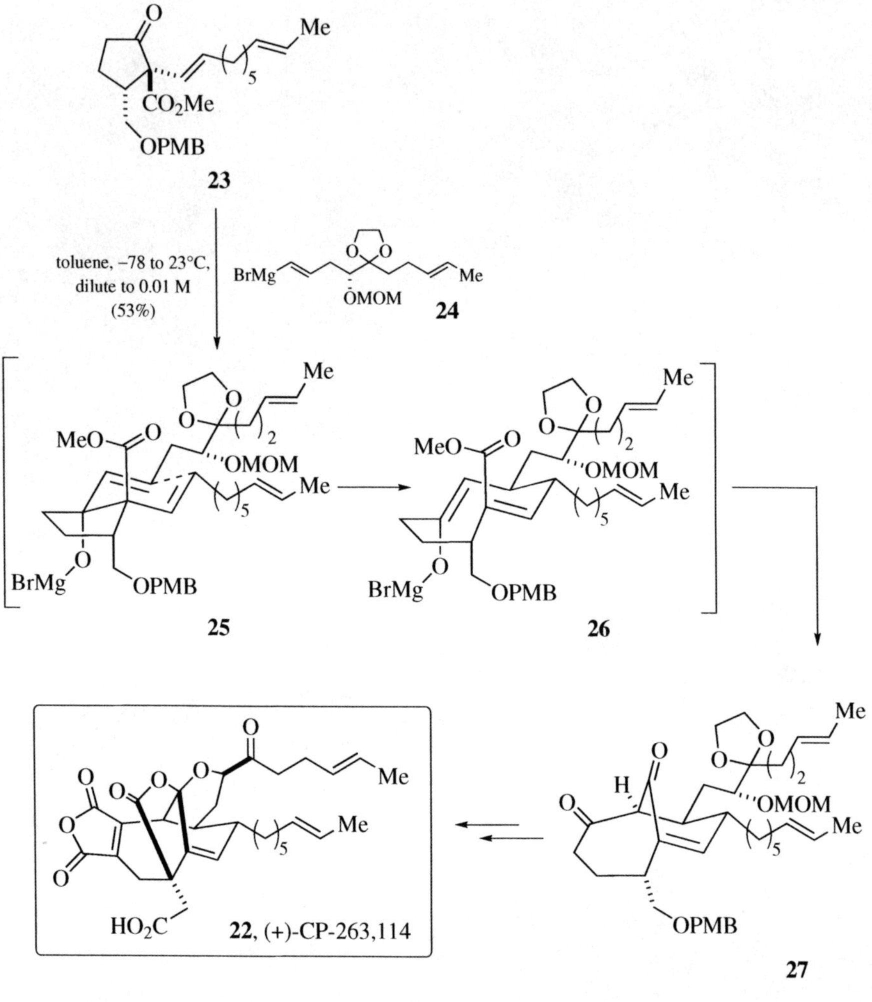

[3,3]-SIGMATROPIC REARRANGEMENTS

Scheme 1.5l

Shair et al. employed anionic oxy-Cope rearrangement in their synthesis of (+)-CP-263,114 (**22**), a fungal metabolite with the ability to inhibit squalene synthase and Ras farnesyltransferase[10] (Scheme 1.5l). The addition of the Grignard reagent **24** into the cyclopentanone (+)-**23** provided a bromomagnesium alkoxide **25**, that underwent anionic oxy-Cope rearrangement to furnish the cyclononadiene **26**. An in situ transannular cyclization of **26** delivered **27**.

REFERENCES

1. Berson, J. A.; Jones, M., Jr. *J. Am. Chem. Soc.* **1964**, *86*, 5019.
2. Evans, D. A.; Golob, A. M. *J. Am. Chem. Soc.* **1975**, *97*, 4765.

3. Evans, D. A.; Balliargeon, D. J.; Nelson, J. V. *J. Am. Chem. Soc.* **1978**, *100*, 2242.
4. (a) Evans, D. A.; Baillargeon, D. J. *Tetrahedron Lett.* **1978**, *19*, 3315; (b) Evans, D. A.; Baillargeon, D. J. *Tetrahedron Lett.* **1978**, *19*, 3319; (c) Steigerwald, M. L.; Goddard, W. A., III; Evans, D. A. *J. Am. Chem. Soc.* **1979**, *101*, 1994.
5. (a) Yoo, H.-Y.; Houk, K. N.; Lee, J. K.; Scialdone, M. A.; Meyers, A. I. *J. Am. Chem. Soc.* **1998**, *120*, 205; (b) Haeffner, F.; Houk, K. N.; Reddy, R.; Paquette, L. A. *J. Am. Chem. Soc.* **1999**, *121*, 11880; (c) Haeffner, F.; Houk, K. N.; Schulze, S. M.; Lee, J. K. *J. Org. Chem.* **2003**, *68*, 2310.
6. For reviews, see (a) Paquette, L. A. *Tetrahedron* **1997**, *53*, 13971. (b) Paquette, L. A. *Angew. Chem. Int. Ed.* **1990**, *29*, 609.
7. Boeckman, R. K., Jr.; Springer, D. M.; Alessi, T. R. *J. Am. Chem. Soc.* **1989**, *111*, 8284.
8. Lee, E.; Shin, I.-J.; Kim, T.-S. *J. Am. Chem. Soc.* **1990**, *112*, 260.
9. Paquette, L. A.; Sun, L.-Q.; Watson, T. J. N.; Friedrich, D.; Freeman, B. T. *J. Am. Chem. Soc.* **1997**, *119*, 2767.
10. Chen, C.; Layton, M. E.; Sheehan, S. M.; Shair, M. D. *J. Am. Chem. Soc.* **2000**, *122*, 7424.

1.6. Aza-Cope–Mannich Reaction

The cationic aza-Cope rearrangement was discovered in 1950 when the α-allylbenzylamine **1** was treated with formaldehyde and formic acid to give two unexpected products, 1-dimethylamino-3-butene (**2**) and benzaldehyde[1] (Scheme 1.6a). It was postulated that the cleavage reaction presumably occurred via a pathway similar to that in the Cope rearrangement. The iminium ion **3** would undergo [3,3]-sigmatropic rearrangement to another iminium ion (**4**), which upon hydrolysis produces homoallylamine and benzaldehyde. The aza-Cope rearrangement is synthetically useful because the rearrangement occurs under mild reaction conditions, and [3,3]-sigmatropic rearrangement typically proceeds with a high level of stereocontrol.

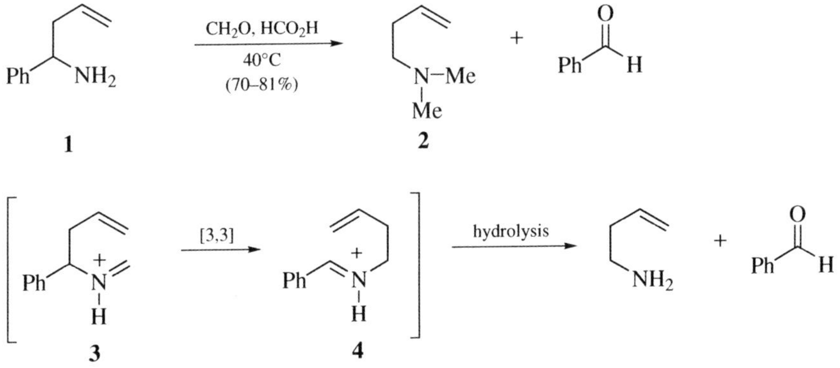

Scheme 1.6a

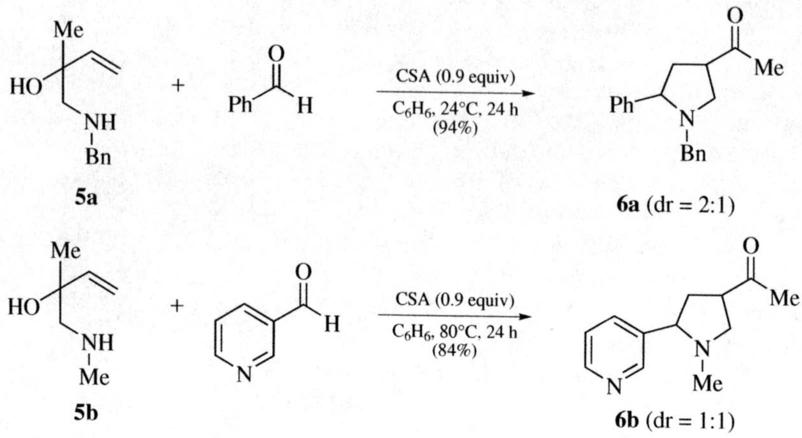

44   [3,3]-SIGMATROPIC REARRANGEMENTS

Scheme 1.6b

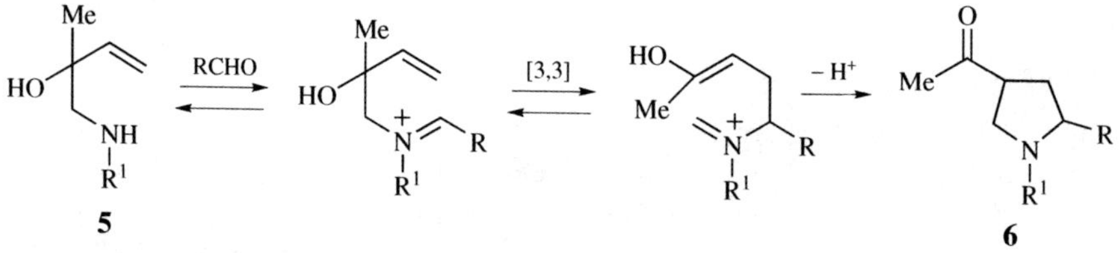

Scheme 1.6c

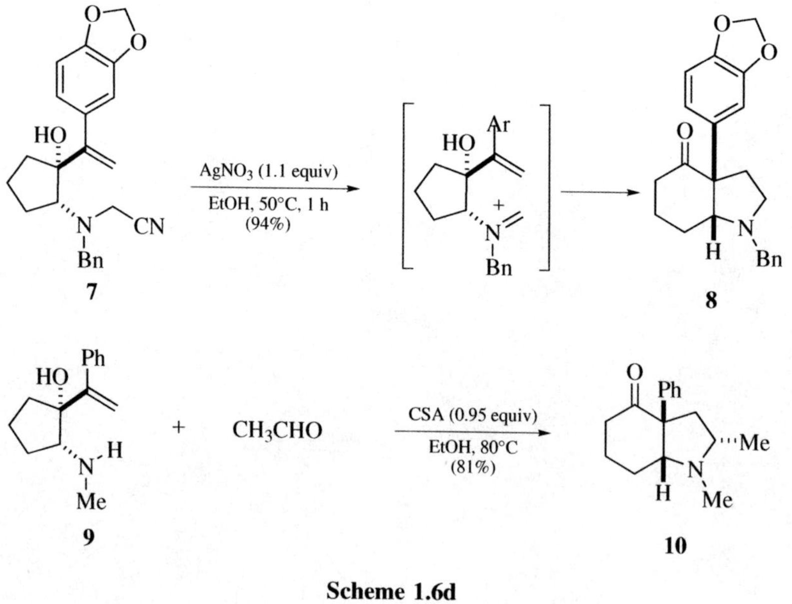

Scheme 1.6d

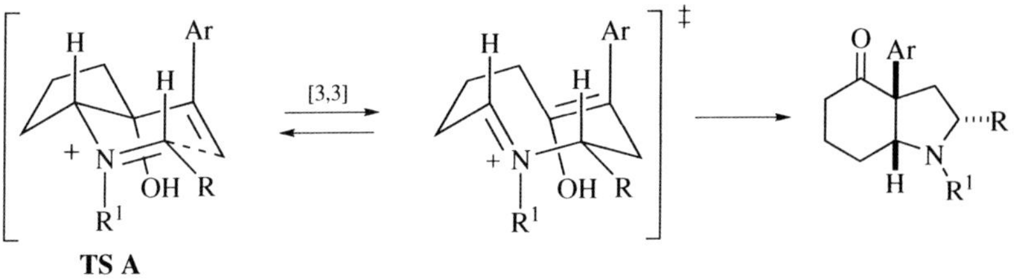

Scheme 1.6e

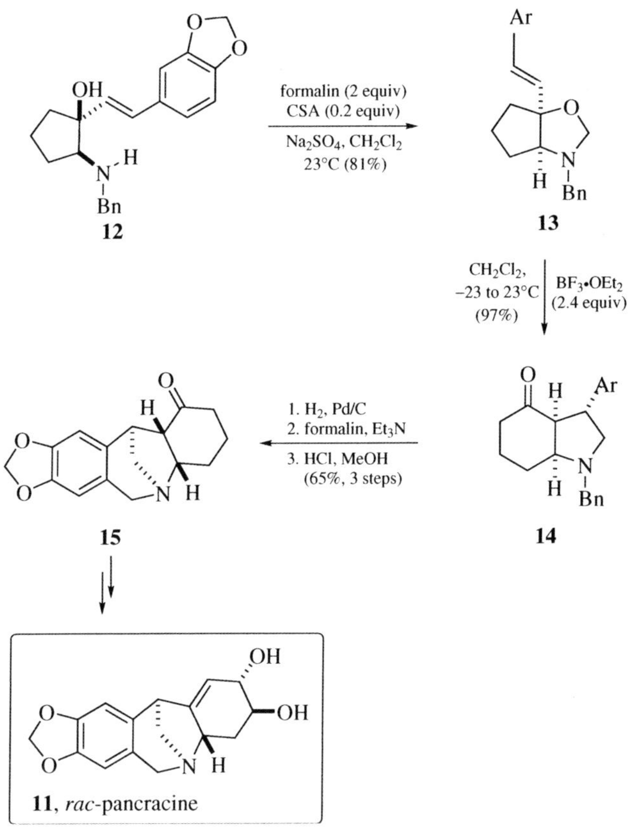

Scheme 1.6f

To control the equilibrium position of the rearrangement, Overman and others introduced a nucleophilic hydroxyl group at the C_2 position to capture the rearranged iminium ion[2] (Scheme 1.6b). Although the levels of diastereoselectivity for the formation of pyrrolidines **6a** and **6b** are low, the tandem cationic aza-Cope–Mannich cyclization provides a variety of substituted 3-acylpyrrolidines in high yields under mild reaction conditions. The first step in the reaction is the formation of the iminium ion, which undergoes a facile [3,3]-sigmatropic rearrangement to provide an enol iminium intermediate[2] (Scheme 1.6c). Intramolecular attack of the enol on the rearranged iminium ion then produces pyrrolidine.

The chemistry can be extended to the construction of more complex ring systems.[3] When cyclic amino alcohols are subjected to the tandem cationic aza-Cope–Mannich reaction, pyrrolidine-annulated bicyclic products are formed in which the starting ring is now expanded by one number[4] (Scheme 1.6d). For example, when the tandem aza-Cope–Mannich cyclization reactions were performed on the cyclopentanols **7** and **9**, the *cis*-octahydroindoles **8** and **10** were formed, respectively, in high yields as a single diastereomer (Scheme 1.6d). The exclusive formation of a single diastereomer is rationalized in terms of chair-like transition state **A**, in which the *E*-iminium ion isomer rapidly undergoes [3,3]-sigmatropic rearrangement (Scheme 1.6e).

The highly diastereoselective intramolecular aza-Cope–Mannich reaction was used in the total synthesis of (±)-pancracine (**11**), an alkaloid natural product[5]

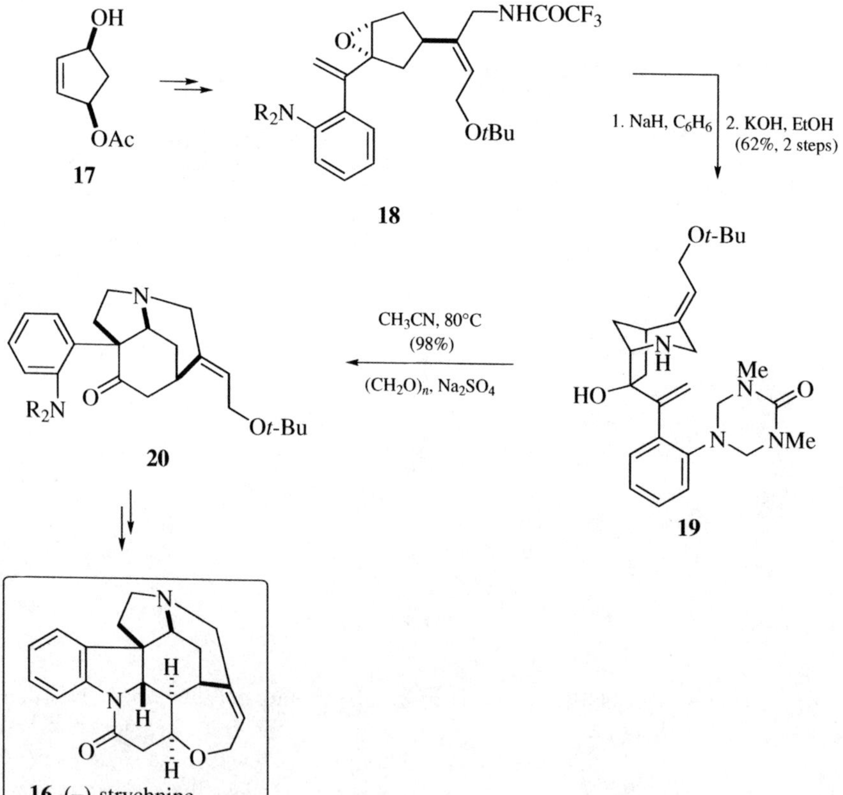

Scheme 1.6g

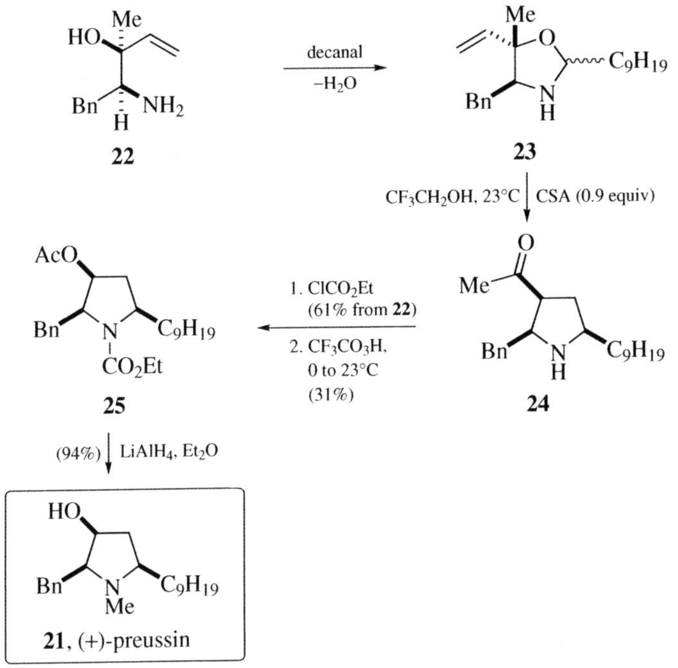

Scheme 1.6h

(Scheme 1.6f). Reaction of the E-allylic alcohol **12** with formaldehyde in the presence of acid catalyst and sodium sulfate gave the oxazolidine **13**. Exposure of **13** to 2.4 equivalents of $BF_3 \cdot OEt_2$ provided a key intermediate hydroindolone (**14**) in 97% yield as a single diastereomer. Hydrogenolysis of **14** followed by Pictet–Spengler cyclization afforded the methanomorphanthridine ketone **15** in 65% yield.

(−)-Strychnine (**16**), an alkaloid natural product isolated in 1818 from *Strychnos ignatii*, represents one of the most challenging target molecules in organic synthesis. In 1993, Overman used his highly efficient cationic aza-Cope–Mannich cyclization reaction to accomplish the total synthesis of (−)-strychnine[6] (Scheme 1.6g). Treatment of **18**, prepared from $(1R,4S)$-(+)-4-hydroxy-2-cyclopentenyl acetate (**17**), with NaH, followed by removal of the trifluoroacetyl group, provided the azabicyclooctane **19**. The crucial aza-Cope–Mannich cyclization was accomplished in essentially quantitative yield in an 800-mg scale reaction to afford the diamine **20**. Further elaboration of the intermediate **20** led to the first asymmetric synthesis of (−)-strychnine (**16**).[7]

Deng and Overman employed the aza-Cope–Mannich reaction in the enantioselective total synthesis of (+)-preussin (**21**), a potent antifungal agent possessing a pyrrolidine skeleton[8] (Scheme 1.6h). Conversion of the amino alcohol **22** to the oxazolidine derivative **23** was readily accomplished by reacting with decanal in hot benzene with removal of water using a Dean–Stark trap. Treatment of

23 with 0.9 equivalent of camphorsulfonic acid (CSA) in CF_3CH_2OH at 23°C yielded the desired all-*cis* pyrrolidine **24** as the major product. Compound **24** was treated directly with ethyl chloroformate followed by Baeyer–Villiger oxidation with trifluoroperoxyacetic acid to afford the product **25**. Finally, reduction of **25** with $LiAlH_4$ in Et_2O provided (+)-preussin (**21**) in 94% yield.

REFERENCES

1. Morowitz, R. M.; Geissman, T. A. *J. Am. Chem. Soc.* **1950**, *72*, 1518.
2. Overman, L. E.; Kakimoto, M.-A.; Okazaki, M. E.; Meier, G. P. *J. Am. Chem. Soc.* **1983**, *105*, 6622.
3. (a) Overman, L. E. *Acc. Chem. Res.* **1992**, *25*, 352; (b) Overman, L. E. *Aldrichimica Acta* **1995**, *28*, 107.
4. Overman, L. E.; Mendelson, L. T.; Jacobsen, E. J. *J. Am. Chem. Soc.* **1983**, *105*, 6629.
5. Overman, L. E.; Shim, J. *J. Org. Chem.* **1991**, *56*, 5005.
6. Knight, S. D.; Overman, L. E.; Pairaudeau, G. *J. Am. Chem. Soc.* **1993**, *115*, 9293.
7. Knight, S. D.; Overman, L. E.; Pairaudeau, G. *J. Am. Chem. Soc.* **1995**, *117*, 5776.
8. Deng, W.; Overman, L. E. *J. Am. Chem. Soc.* **1994**, *116*, 11241.

2 Aldol Reactions

GENERAL CONSIDERATIONS

The *aldol reaction* is an addition of metal enolates to aldehydes or ketones to form β-hydroxy carbonyl compounds.[1] The simplest aldol reaction would be the reaction of acetaldehyde lithium enolate with formaldehyde (Scheme 2.I). As the transition state of this reaction involves six atoms, the aldol reaction is another example where a six-membered transition state is presumed to be operating. The transition state of the aldol reaction is very similar to those of Claisen and Cope rearrangements, and therefore the remarkable facility of the lithium enolate reaction is attributed to the stability of an aromatic transition state.[2]

The aldol reaction is one of the most synthetically useful methods in organic synthesis to form a carbon–carbon bond in a stereoselective and predictable manner.[3] In general, the *Z*-enolates react with aldehydes to give *syn*-products preferentially, whereas the *E*-enolates produce *anti*-adducts as a major diastereomer (Scheme 2.II). One of the simplest explanations for this diastereoselectivity would be the Zimmerman–Traxler six-membered chairlike transition-state model, in which the metal cation is chelated by the two oxygens of the reacting molecules, and the alkyl (R) group of the aldehyde prefers to be equatorial[4] (Scheme 2.III). In transition-state *Z*-chair **B**, the R group occupies an axial position, resulting in an unfavorable 1,3-diaxial interaction[5] with the R^1 group of the enolate. This model clearly explains the fact that low stereoselectivity is observed when R^1 is a small group.

E-Enolates often react with lower stereoselectivity than those of the corresponding *Z*-enolates. A classic example to illustrate this point is a study carried out by Heathcock et al.[6] (Scheme 2.IV). When the carbonyl compounds **1** were deprotonated with lithium diisopropylamide (LDA) and the resulting enolates were subsequently treated with benzaldehyde at $-72°$ C, the aldol products desired (**2**) were obtained in 83 to 99% yield. The *Z*-enolates derived from *t*-butyl and 1-adamantyl ethyl ketones afforded *syn*-products in excellent levels of diastereoselectivity. The fact that the *syn*/*anti* ratios directly reflect the isomeric purity of the reacting enolates hints that the *Z*-enolates in these cases undergo aldol reaction through a chairlike six-membered transition state (Scheme 2.III,

---

Six-Membered Transition States in Organic Synthesis, By Jaemoon Yang
Copyright © 2008 John Wiley & Sons, Inc.

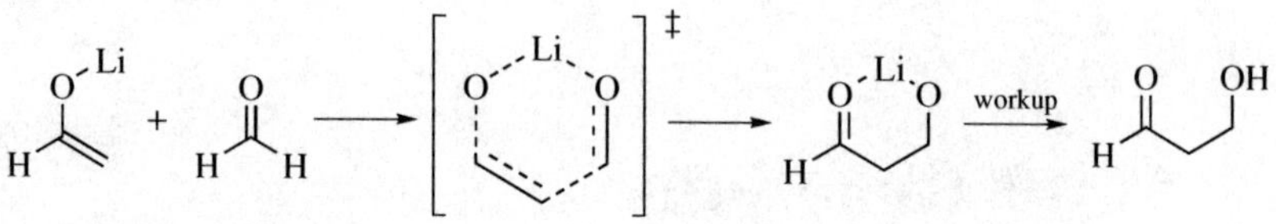

Scheme 2.I

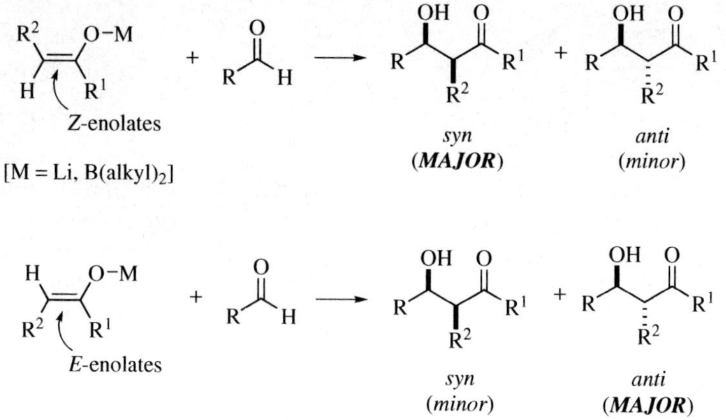

Scheme 2.II

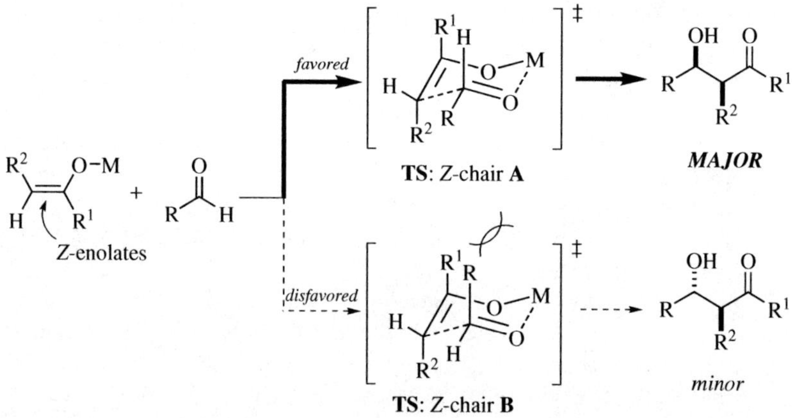

Z-enolates [M = Li, B(alkyl)$_2$]: syn (MAJOR) + anti (minor)

E-enolates: syn (minor) + anti (MAJOR)

Scheme 2.III

Z-enolates: favored → TS: Z-chair A → MAJOR; disfavored → TS: Z-chair B → minor

TS: Z-chair **A**). On the other hand, the *E*-enolates of 3-pentanone and methyl propanoate gave products slightly favoring a *syn*-diastereomer.

If chairlike transition state **TS**: *E*-chair **A** were the highly favored state for the aldol reaction of the lithium *E*-enolate of methyl propanoate, we would expect an *anti*-isomer as a major product (Scheme 2.V). To accommodate situations

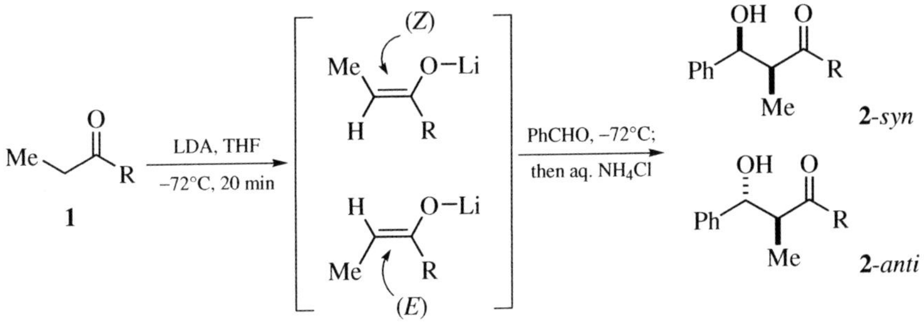

GENERAL CONSIDERATIONS 51

R	Z/E	syn/anti
t-C$_4$H$_9$	>98:2	>98:2
1-adamantyl	>98:2	>98:2
2,4,6-(CH$_3$)$_3$C$_6$H$_2$	5:95	8:92
C$_2$H$_5$	30:70	64:36
OCH$_3$	5:95	62:38

Scheme 2.IV

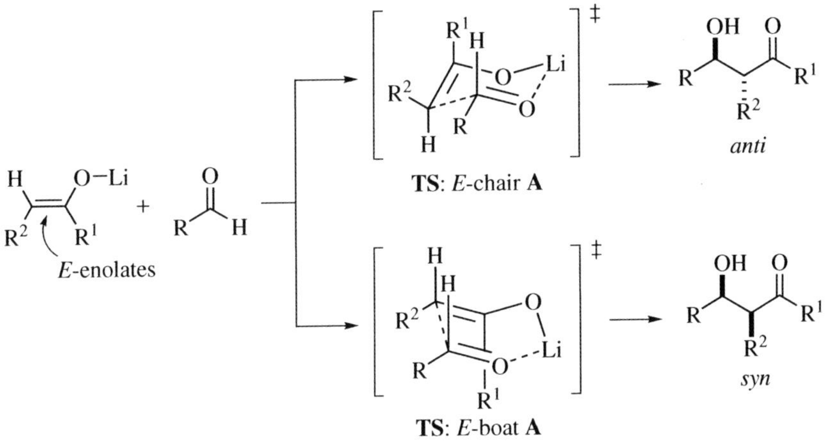

Scheme 2.V

where a significant amount of *syn*-isomer is formed from the *E*-enolate, many alternative stereochemical models, including a boatlike transition state, have been proposed.[7]

To achieve a stereoselective aldol reaction that does not depend on the structural type of the reacting carbonyl compounds, many efforts have been made to use boron enolates. Based on early studies by Mukaiyama et al.[8a] and Fenzl and Köster,[8b] in 1979, Masamune and others reported a highly diastereoselective aldol reaction involving dialkylboron enolates (enol borinates)[9]

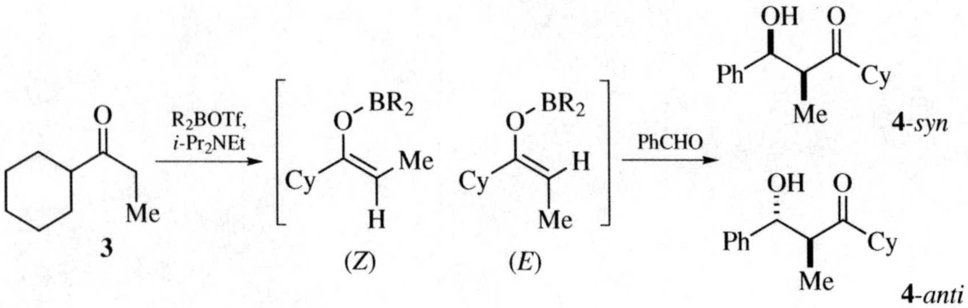

R$_2$BOTf	Z/E	syn/anti	Yield, %
n-Bu$_2$BOTf	—	43:57	—
(9-BBN)OTf	>95:<5	>97:3	79
(c-C$_5$H$_9$)$_2$BOTf	12:88	14:86	88

Scheme 2.VI

(Scheme 2.VI). When the cyclohexyl ethyl ketone **3** was converted into the corresponding enol borinate by using dibutylboron trifluoromethanesulfonate (triflate) and N,N-diisopropylethylamine (DIPEA, Hünig's base) according to Mukaiyama and Inoue's protocol,[10] and the resulting enolate was allowed to react with benzaldehyde, the aldol products were obtained with little selectivity. The situation changed dramatically when different dialkylboron triflates were employed. With the use of 9-borabicyclo[3.3.1]nonyl (9-BBN) triflate, the *syn*-isomer was obtained with superior (*syn/anti* = 97:3) selectivity. The *anti*-isomer, on the other hand, was readily accessible by the use of dicyclopentylboron triflate, although the stereoselectivity in this case was not as high as that observed in the *syn*-isomer. Thus, Masamune demonstrated elegantly that by a proper selection of reagents, a flexible synthesis of either *syn*- or *anti*-aldol products is possible in a highly stereoselective manner starting from the same ketone (**3**).

In 1979, Evans and others independently disclosed a highly diastereoselective aldol reaction employing a boron enolate to improve the kinetic diastereoselectivity in the aldol reaction.[11] A variety of boron enolates, generated from the ethyl ketone **5** by using an equimolar quantity of di-*n*-butylboron triflate (*n*-Bu$_2$BOTf) and Hünig's base, undergo aldol condensation to give product **6** in excellent levels of diastereoselectivity (Scheme 2.VII). The high level of diastereoselection observed with Z-boron enolates can be explained in terms of six-membered chairlike transition states[11] (Scheme 2.VIII). Transition state **B** is destabilized relative to **A** due to two distinct energetically unfavorable 1,3-diaxial interactions: one between the phenyl and R groups and the other between the phenyl and boron butyl groups. One major reason for the huge improvement with boron enolates over lithium enolates is the creative exploitation of *metal-centered steric effects*. Thus, the inherent 1,3-diaxial interactions in transition state **B** are maximized by

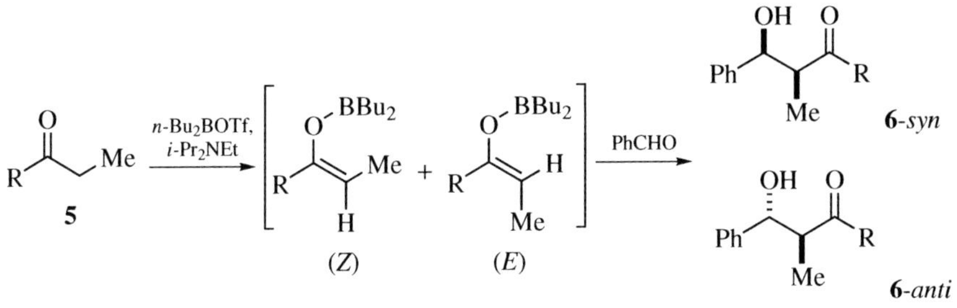

GENERAL CONSIDERATIONS   53

Scheme 2.VII

RCOCH$_2$CH$_3$	Enolate Formation	Z/E	syn/anti	Yield, %
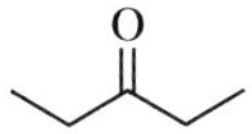 (Et-C(O)-Et)	−78°C, 30 min	>99:1	>97:3	77
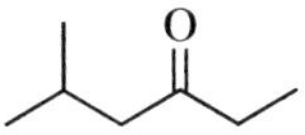 (i-Pr-C(O)-Et)	−78°C, 30 min; 0°C, 30 min	>99:1	>97:3	82
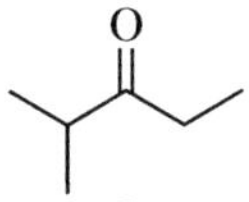 (i-Pr-C(O)-Et)	−78°C, 30 min; 0°C, 60 min	45:55	44:56	92
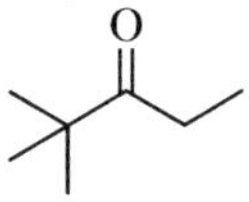 (t-Bu-C(O)-Et)	35°C, 120 min	>99:1	>97:1	65
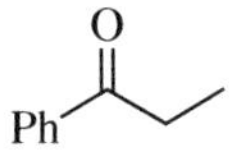 (Ph-C(O)-Et)	25°C, 60 min	>99:1	>97:3	82

shortening the metal–oxygen bond lengths, as the boron–oxygen bond (1.4 Å) is much shorter than the lithium–oxygen bond[11] (1.9 to 2.2 Å).

To quantify the energy difference between various transition-state structures of the aldol reactions, theoretical studies have been performed. In an attempt to discover the working models of stereoselectivity, in 1988 Houk and others performed a computational study of the transition structures for aldol reactions[12] (Scheme 2.IX). As a model system, calculations were carried out on the hypothetical reaction of propanal dihydroborinate and formaldehyde. The Z-enol dihydroborinate **7Z** prefers chairlike transition structure **A** over twist-boat transition structure **B** by 4.0 kcal/mol because of the destabilizing interactions between the methyl and the hydrogen on boron in twist-boat transition structure **B**. Introduction of alkyl groups larger than the methyl in the Z-enol borinate **7Z** or replacement of the hydrogens on boron by alkyl groups should favor chairlike transition structure **A** even more.

On the other hand, interaction between the pseudoaxial hydrogen of the enolate and the hydrogen of the boron in transition state **B** is relatively small for E-enolate

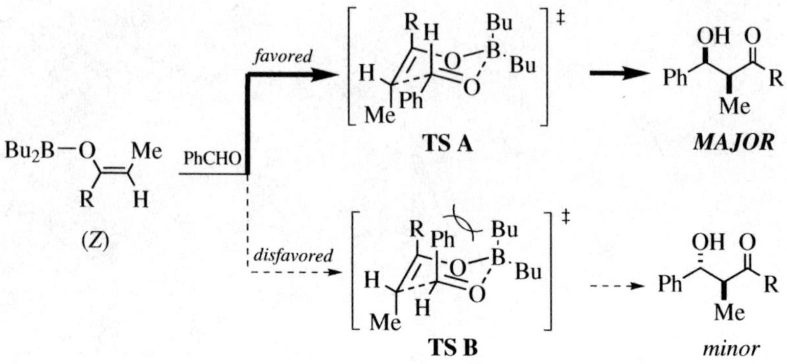

Scheme 2.VIII

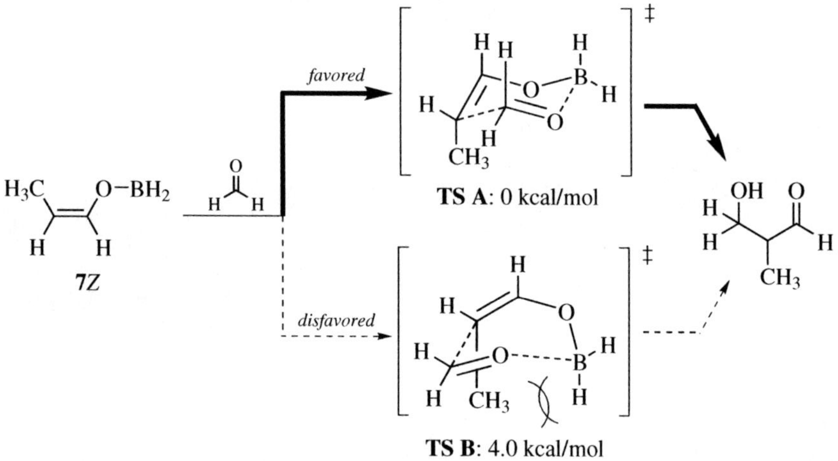

Scheme 2.IX

7E, so that *E*-enolate prefers the chair conformation less and there should be strong competition between transition structures **A** and **B**. Consequently, the *E*-enolates give lower stereoselectivity because the minor product can be formed through a twist-boat structure of comparable energy (Scheme 2.X).

Gennari et al. developed a computational model to reproduce the experimental *syn/anti* setereoselectivity for the aldol reactions of *Z* and *E* enol borinates of butanone with acetaldehyde.[13] For the reaction of *Z*-enol borinate **8Z**, the chair transition state **TS: Z-chair A** dominates over other three-transition states (Scheme 2.XI). When a Boltzmann distribution was calculated for the competing transition structures, a complete *syn/anti* selectivity of 99:1 was predicted. The aldol reaction of *E*-enol borinate **8E** with acetaldehyde is, however, calculated to have four transition structures of similar energy (Scheme 2.XII). Although

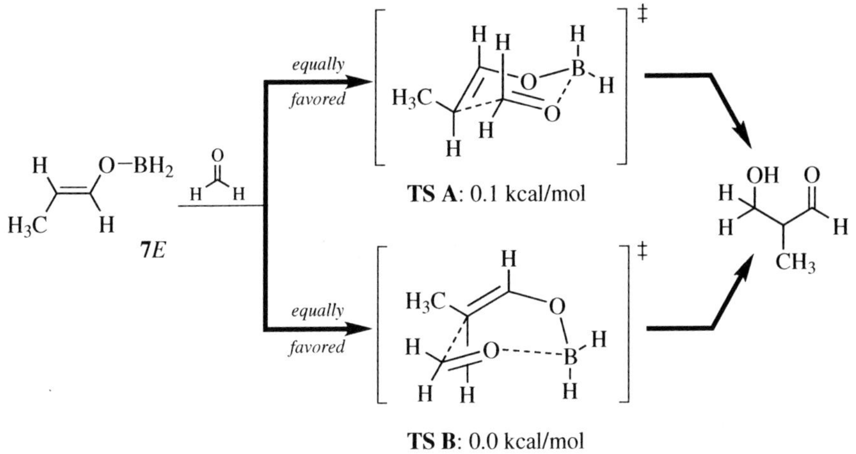

GENERAL CONSIDERATIONS 55

Scheme 2.X

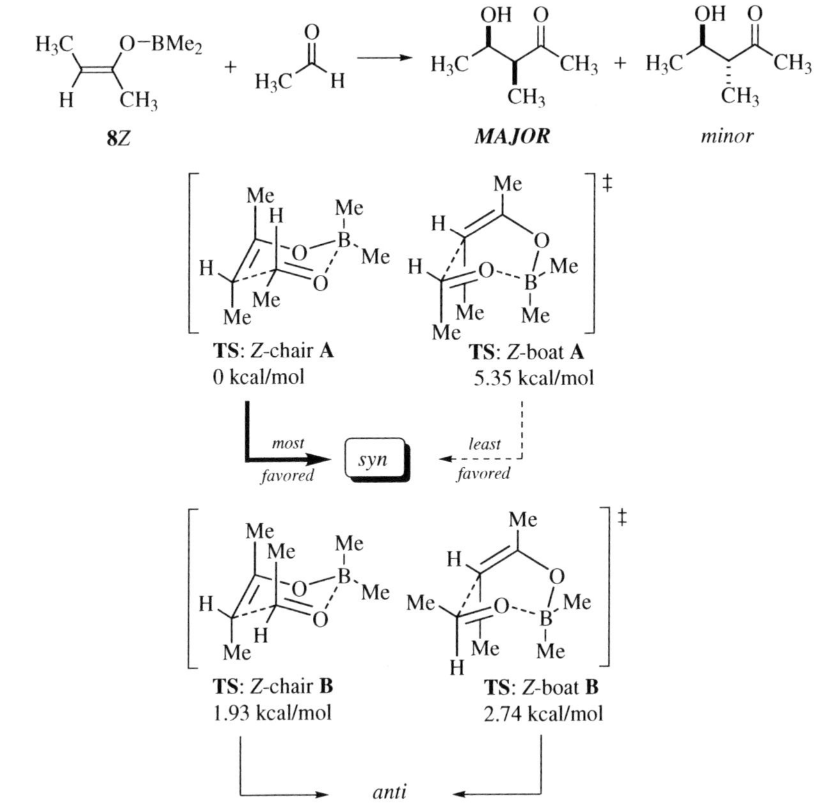

TS A: 0.1 kcal/mol

TS B: 0.0 kcal/mol

Scheme 2.XI

TS: Z-chair **A**
0 kcal/mol

TS: Z-boat **A**
5.35 kcal/mol

most favored → syn ← least favored

TS: Z-chair **B**
1.93 kcal/mol

TS: Z-boat **B**
2.74 kcal/mol

→ anti ←

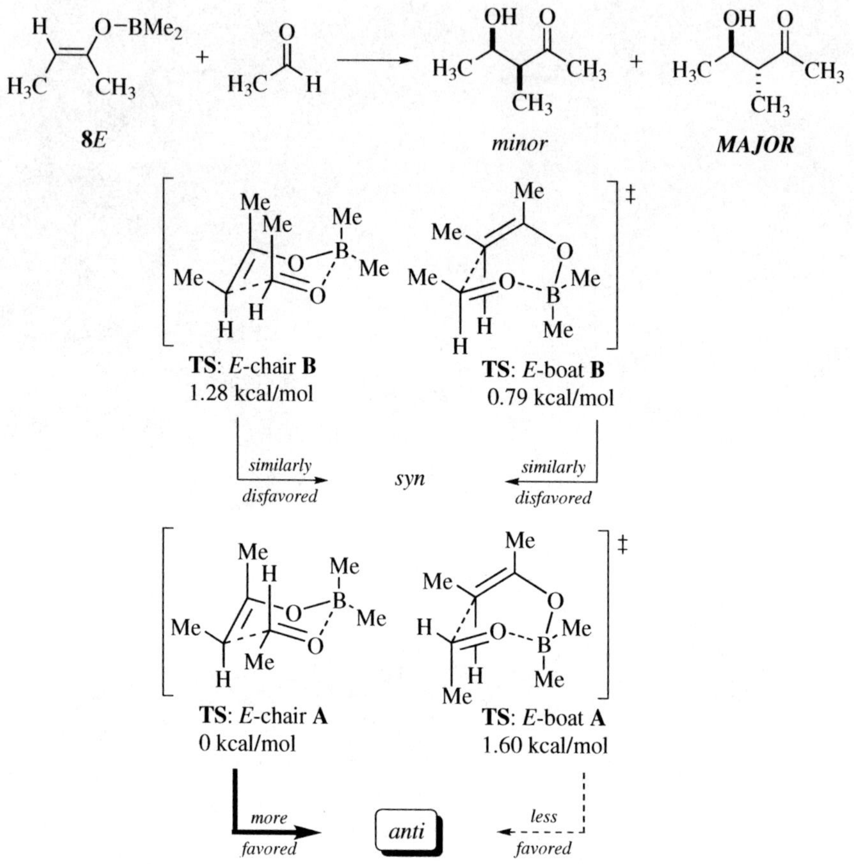

Scheme 2.XII

[Reaction: 8E (H, H₃C, CH₃, O–BMe₂) + H₃C-CHO → minor (syn: H₃C-CH(OH)-CH(CH₃)-C(O)-CH₃) + MAJOR (anti: H₃C-CH(OH)-CH(CH₃)-C(O)-CH₃)]

TS: *E*-chair B
1.28 kcal/mol

TS: *E*-boat B
0.79 kcal/mol

↓ similarly disfavored — *syn* — similarly disfavored ↓

TS: *E*-chair A
0 kcal/mol

TS: *E*-boat A
1.60 kcal/mol

↓ more favored — *anti* — less favored ↓

Scheme 2.XII

chairlike transition state **TS:** *E*-chair **A** has the lowest energy, twist-boat transition state **TS:** *E*-boat **B** is only 0.79 kcal/mol higher than **TS:** *E*-chair **A** and can be a competitive transition structure. The overall outcome is a reduced diastereoselectivity (*syn/anti* = 14 : 86) for the *E*-enol borinate.

REFERENCES

1. For a general introduction to the aldol reaction, see March, J. *Advanced Organic Chemistry*, 4th ed.; Wiley: New York, **1992**; pp. 937–944.
2. Kleschick, W. A.; Buse, C. T.; Heathcock, C. H. *J. Am. Chem. Soc.* **1977**, *99*, 247.
3. For reviews of the aldol reaction, see (a) Palomo, C.; Oiarbide, M.; Garcia, J. M. *Chem. Soc. Rev.* **2004**, *33*, 65; (b) Alcaide, B.; Almendros, P. *Eur. J. Org. Chem.* **2002**, *10*, 1595; (c) Mahrwald, R. *Chem. Rev.* **1999**, *99*, 1095; (d) Nelson, S. G. *Tetrahedron: Asymmetry* **1998**, *9*, 357; (e) Cowden, C. J.; Paterson, I. *Org.*

React. **1997**, *51*, 1; (f) Heathcock, C. H. in *Comprehensive Organic Synthesis*; Heathcock, C. H., Ed.; Pergamon Press: New York, **1991**; Vol. 2, p. 181; (g) Evans, D. A.; Nelson, J. V.; Taber, T. in *Topics in Stereochemistry*, Wiley: New York, **1982**; Vol. *13*, p. 1.

4. Juaristi, E. *Introduction to Stereochemistry and Conformational Analysis*, Wiley: New York, **1991**; Chap. 13.

5. Eliel, E. L.; Willen, S. H. *Stereochemistry of Organic Compounds*, Wiley: New York, **1994**; Chap. 11.

6. Heathcock, C. H.; Buse, C. T.; Kleschick, W. A.; Pirrung, M. C.; Sohn, J. E.; Lampe, J. *J. Org. Chem.* **1980**, *45*, 1066.

7. (a) For an insightful description of the differences among many different types of transition states, see Denmark, S. E.; Henke, B. R. *J. Am. Chem. Soc.* **1991**, *113*, 2177; (b) Evans, D. A.; McGee, C. R. *Tetrahedron Lett.* **1980**, *21*, 3975.

8. (a) Mukaiyama, T.; Inomata, K.; Muraki, M. *J. Am. Chem. Soc.* **1973**, *95*, 967; (b) Fenzl, W.; Köster, R. *Liebigs Ann. Chem.* **1975**, 1322.

9. Masamune, S.; Mori, S.; Horn, D. V.; Brooks, D. W. *Tetrahedron Lett.* **1979**, *20*, 2229.

10. Mukaiyama, T.; Inoue, T. *Chem. Lett.* **1976**, 559.

11. (a) Evans, D. A.; Vogel, E.; Nelson, J. V. *J. Am. Chem. Soc.* **1979**, *101*, 6120; (b) Evans, D. A.; Nelson, J. V.; Vogel, E.; Taber, T. R. *J. Am. Chem. Soc.* **1981**, *103*, 3099.

12. (a) Li, Y.; Paddon-Row, M. N.; Houk, K. N. *J. Am. Chem. Soc.* **1988**, *110*, 3684; (b) Li, Y.; Paddon-Row, M. N.; Houk, K. N. *J. Org. Chem.* **1990**, *55*, 481.

13. Bernardi, A.; Capelli, A. M.; Gennari, C.; Goodman, J. M.; Paterson, I. *J. Org. Chem.* **1990**, *55*, 3576.

REACTIONS

2.1. Asymmetric *Syn*-Aldol Reaction

When Evans and others in 1979 carried out the reaction of the boron enolate of the chiral ethyl ketone **1**, synthesized from (*S*)-proline, to study the reaction of chiral enolate in the aldol reaction, the *syn*-diastereomers were formed in high stereoselectivity[1] (Scheme 2.1a). Another significant observation was that the diastereoselectivity (i.e., the ratio of **2**-*syn* to **2'**-*syn*), of the reaction is high. This experiment is significant, as it was one of the first enantioselective aldol reactions to afford optically active products starting from chiral enolates. To explain the results, two diastereomeric six-membered chairlike transition states, **A** and **B**, are compared (Scheme 2.1b). One expects transition state **A** to be preferred over **B** as a consequence of the influence of metal-center steric parameters. In other words, the steric repulsion between R_S and the butyl group is less than that between R_L and the butyl group.

Asymmetric *syn*-aldol condensation reactions employing chiral auxiliaries were reported in 1981 by both Masamune et al.[2] and Evans et al.[3] Masamune et al. introduced boron enolates obtained from (*S*)-mandelic acid, which underwent a highly diastereoselective *syn*-aldol reaction[2] (Scheme 2.1c). There are several salient features of Masamune et al.'s auxiliary. First, no trace of *anti*-aldol products was found in the reaction mixture, meaning that an exclusive formation of Z-enolates occurred. Second, the ratios of **5a** to **5b** are impressively high and depend largely on the size of the ligands attached to the boron. Third, in the case of α-branched aldehydes such as isobutyraldehyde, the use of either **4a** or **4b** is

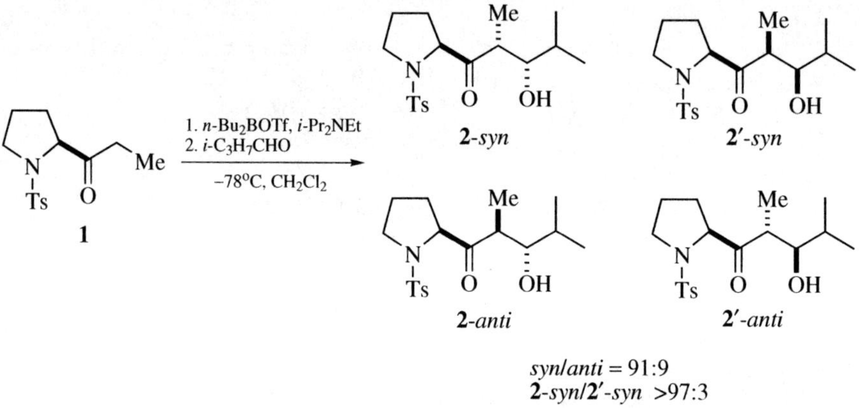

Scheme 2.1a

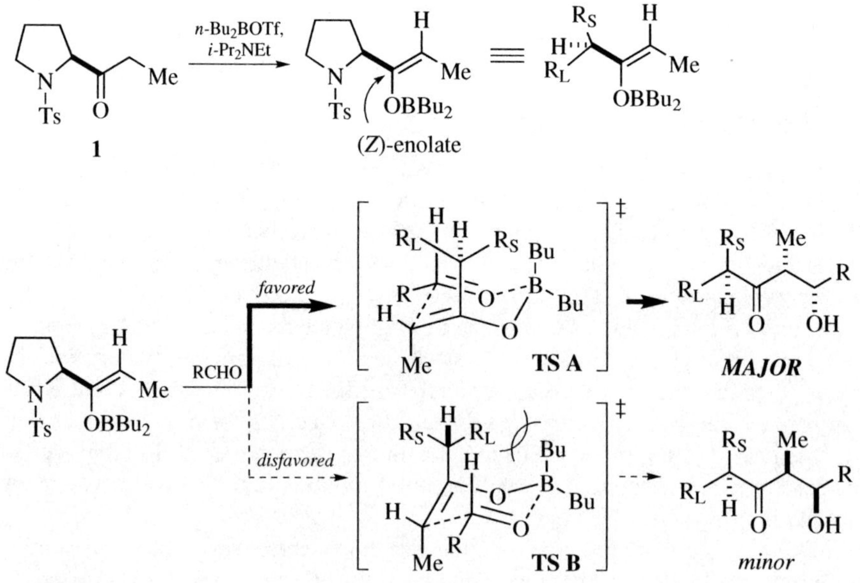

Scheme 2.1b

(*S*)-mandelic acid

R′	5a:5b (4a)	5a:5b (4b)	5a:5b (4c)
C_2H_5	17:1	50:1	>100:1
i-C_3H_7	>100:1	>100:1	no reaction
C_6H_5	14:1	40:1	75:1
$BnO(CH_2)_2$	16:1	28:1	100:1

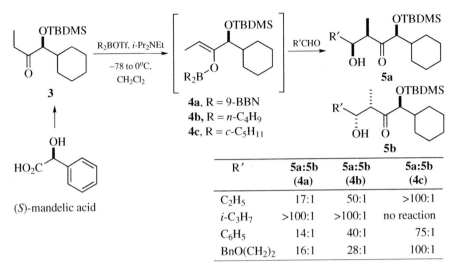

Scheme 2.1c

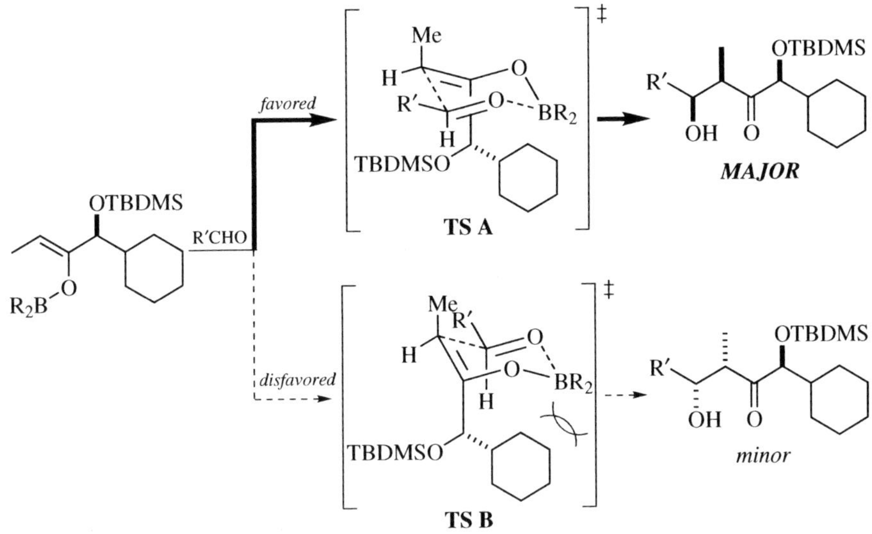

Scheme 2.1d

recommended to achieve excellent levels of diastereoselection, as the aldol reaction does not proceed with **4c**, due possibly to steric congestion in the transition state (Scheme 2.1d).

Masamune et al. used this aldol strategy to achieve the total synthesis of 6-deoxyerythronolide B (**6**), a common biosynthetic precursor leading to all the erythromycins presently known[4] (Scheme 2.1e). The highlight of the synthesis is

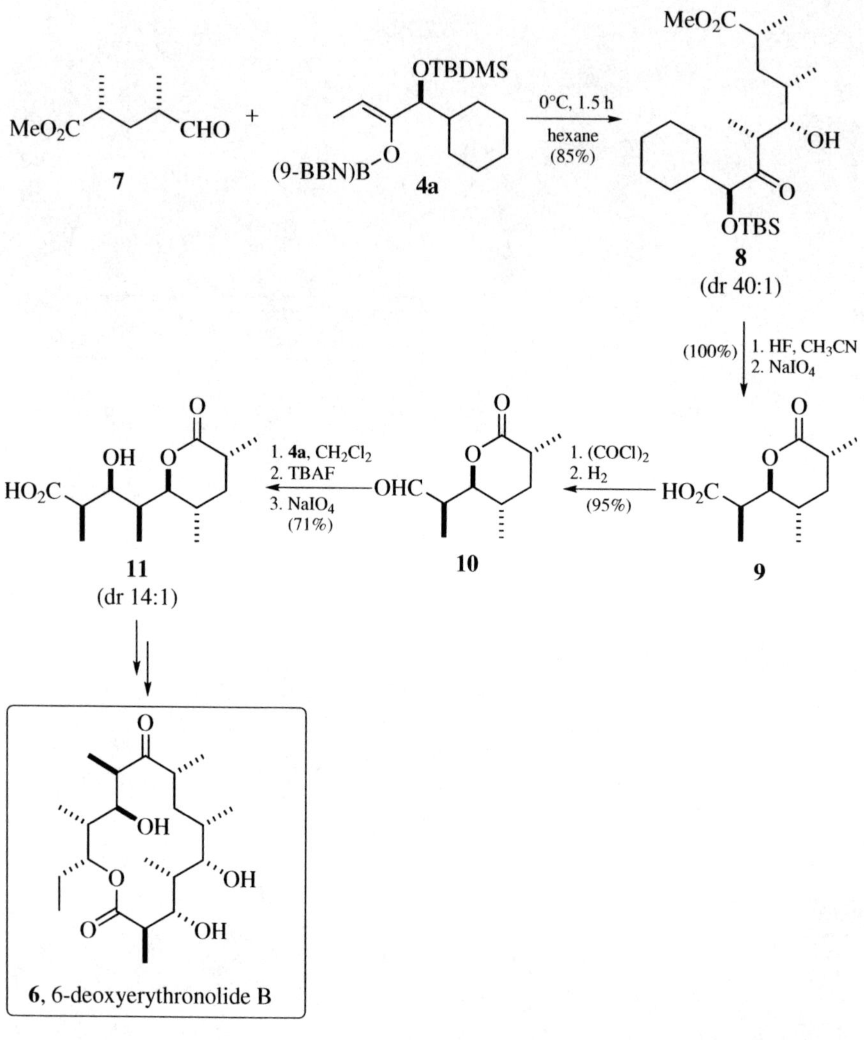

Scheme 2.1e

the iterative use of the highly efficient diastereoselective aldol condensation with the chiral boron reagent (*S*)-**4a**. Thus, the aldol condensation of the aldehyde **7** with the chiral reagent **4a** provided the *syn*-aldol product **8** in 85% yield and 40 : 1 stereoselectivity. Successive treatment of **8** with hydrogen fluoride and sodium metaperiodate gave the Prelog–Djerassi lactonic acid **9**. Another aldol reaction of the aldehyde **10** with **4a** once again proceeded smoothly to afford **11** after treatment of the aldol product with tetra-*n*-butylammonium fluoride followed by sodium metaperiodate.

In the same year that Masamune et al. published their chiral auxiliary-based stereoselective aldol reaction, Evans et al. synthesized two recyclable chiral

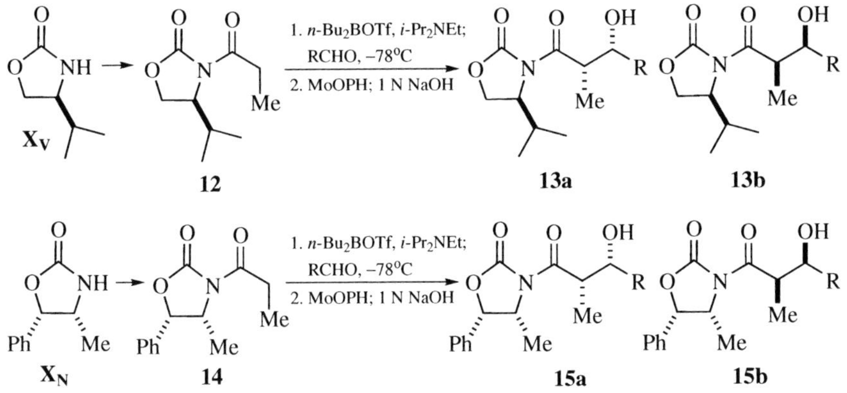

R	13a/13b	Yield, %	15a/15b	Yield, %
$i\text{-}C_3H_7$	497:1	69	<1:500	78
$n\text{-}C_4H_9$	141:1	68	<1:500	71
C_6H_5	>500:1	81	<1:500	60

Scheme 2.1f

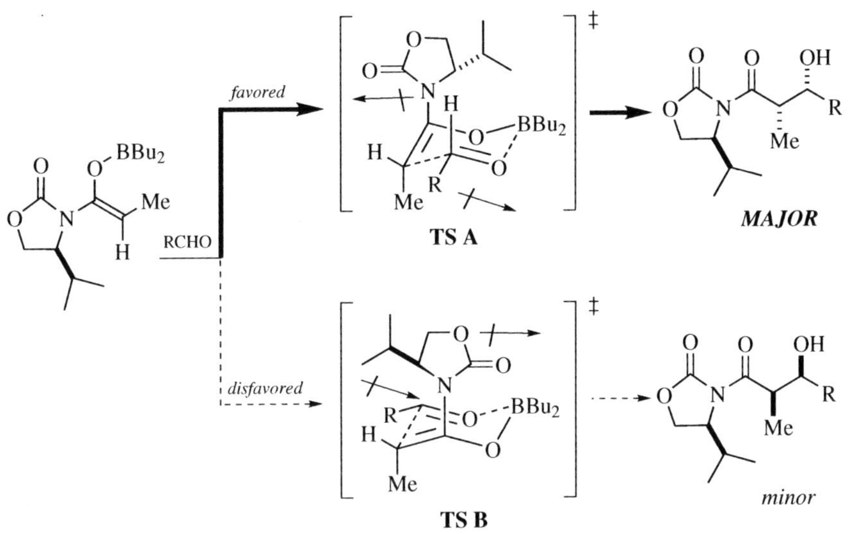

Scheme 2.1g

oxazolidinones, $\mathbf{X_V}$ and $\mathbf{X_N}$, prepared from (S)-valinol or (1S,2R)-norephedrine, which served as efficient chiral auxiliaries for highly diastereoselective *syn*-aldol condensations via the boron enolates[3] (Scheme 2.1f). The diastereoselectivity is exceptionally high and the substrate scope is quite general; both alkyl and aryl aldehydes give satisfactory results.[5] To explain the enantiofacial stereoselectivity,

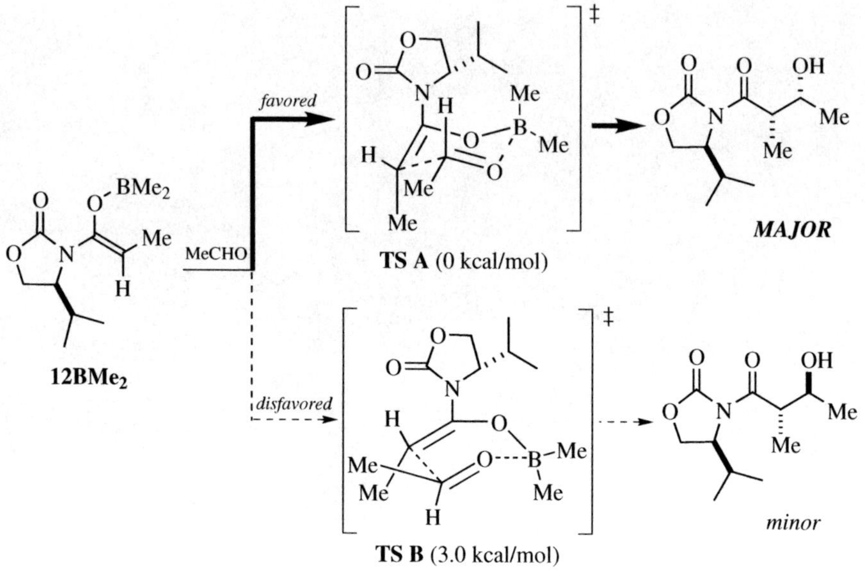

Scheme 2.1h

the dipole alignment in two diastereoisomeric chairlike six-membered transition states were compared[6] (Scheme 2.1g). Transition state **A**, in which the two carbonyl dipoles of aldehyde and oxazolidinone are opposed to each other, would be a conformer of lower energy than transition state **B**, where two dipoles are in the same direction.

Makino and others carried out a computational study on the Evans aldol reaction of dimethylborinate **12BMe$_2$** with acetaldehyde[7] (Scheme 2.1h). The AM1 semiempirical calculations indicate that six-membered chairlike transition state **A**, which would lead to formation of the major *syn*-isomer, is more stable than **B** by 3.0 kcal/mol, providing a theoretical confirmation of the experimental observations.

In the total synthesis of (+)-trienomycins A and F, Smith et al. used an Evans aldol reaction technology to construct a 1,3-diol functional group[8] (Scheme 2.1i). Asymmetric aldol reaction of the boron enolate of **14** with methacrolein afforded exclusively the desired *syn*-diastereomer (**17**) in high yield. Silylation, hydrolysis using the lithium hydroperoxide protocol, preparation of Weinreb amide mediated by carbonyldiimidazole (CDI), and DIBAL-H reduction cleanly gave the aldehyde **18**. Allylboration via the Brown protocol[9] (see Chapter 3) then yielded a 12.5:1 mixture of diastereomers, which was purified to provide the alcohol desired (**19**) in 88% yield. Desilylation and acetonide formation furnished the diene **20**, which contained a C_9-C_{14} subunit of the TBS ether of (+)-trienomycinol.

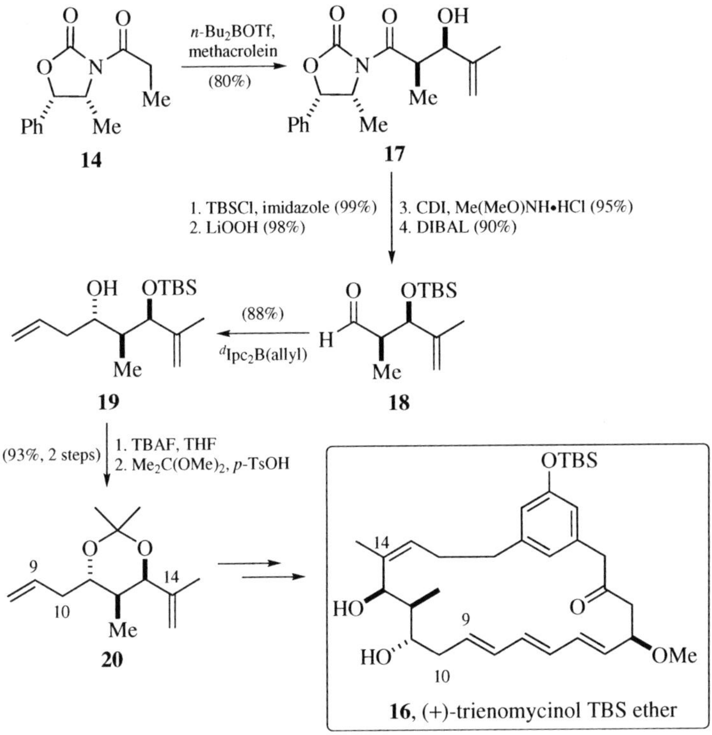

Scheme 2.1i

Glucolipsin A (**21**) is a macrocyclic dilactone natural product that exhibits glucokinase-activating properties. Fürstner et al. employed an Evans aldol strategy to synthesize the *syn*-aldol intermediate **23**[10] (Scheme 2.1j). The aldol reaction of the boron enolate of **14** with 14-methylpentadecanal (**22**) delivered the *syn*-aldol product **23** in essentially diastereomerically pure form (99% de) after purification. Subsequent glycosidation of the alcohol **23** with trichloroacetimidate (**24**) was facilitated by catalytic amounts of TMSOTf (20 mol %) to afford the key intermediate (**25**) in moderate yield.

Gage and Evans later introduced another oxazolidinone, X_P, derived from (*S*)-phenylalanine[11] (Scheme 2.1k). The new chiral auxiliary, readily prepared in either enantiomeric form from the corresponding phenylalanine, is more convenient than the valine-derived oxazolidinone X_V because the oxazolidinone X_P can be purified easily by direct crystallization.[11] Another practical advantage of X_P over the valine-derived oxazolidinone X_V is that the oxazolidinone X_P contains an ultraviolet chromophore, which facilitates thin-layer chromatographic or high-performance liquid chromatographic analysis when it is employed as a chiral auxiliary. The aldol reaction of *N*-propionyloxazolidinone (**26S**) with benzaldehyde proceeds in an extremely stereoselective manner[12] to provide the β-hydroxy acid **28** as a single diastereomer in high yield after removal of chiral auxiliary using the lithium hydroperoxide protocol.

Evans and others used this methodology to complete numerous total syntheses of natural products. For example, they built the spiroketal subunit **29** in the asymmetric synthesis of the macrolide antibiotic rutamycin B[13] (Scheme 2.1l). The

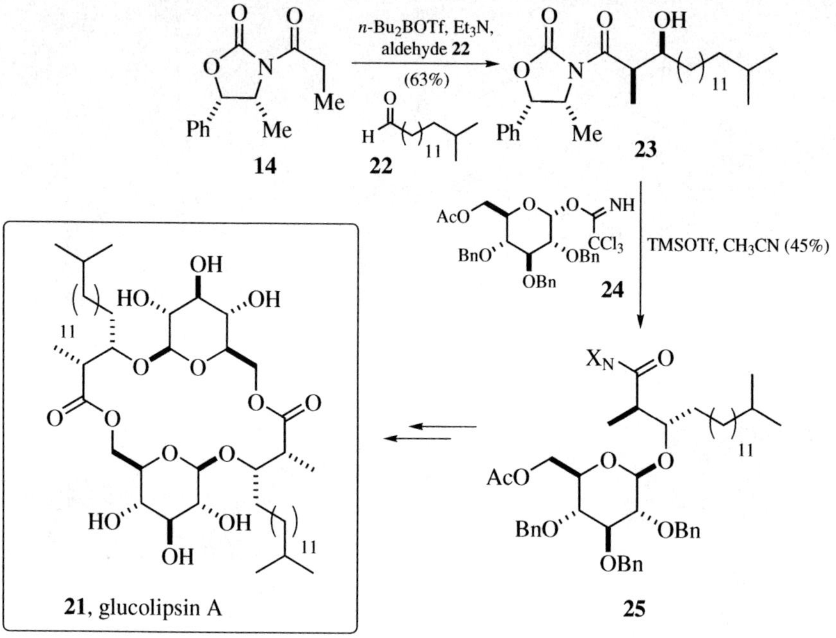

Scheme 2.1j

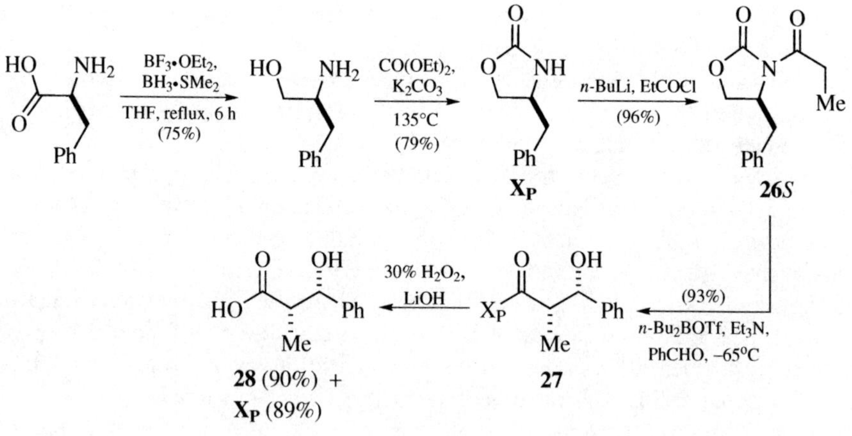

Scheme 2.1k

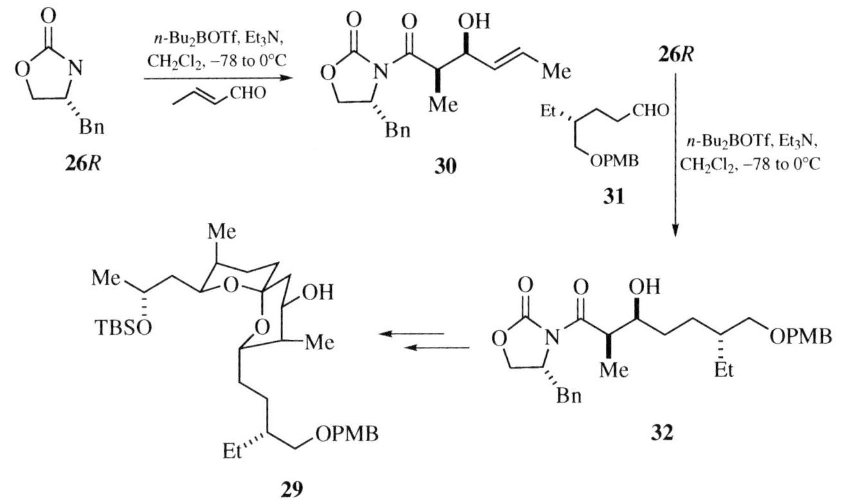

Scheme 2.1l

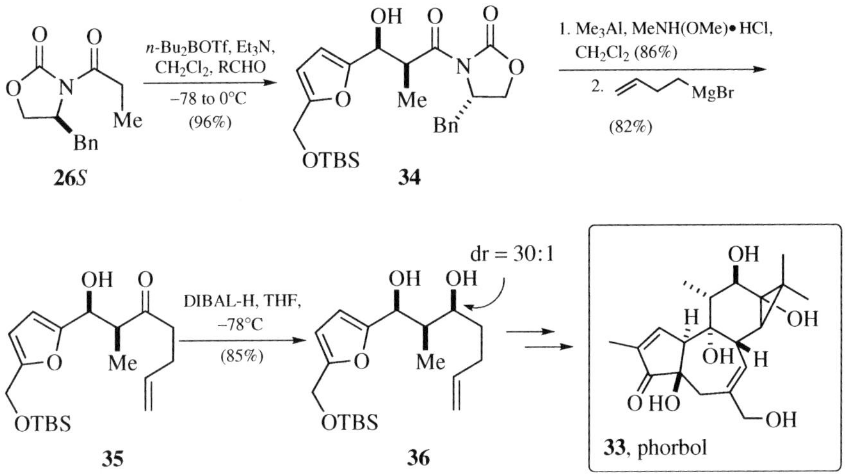

Scheme 2.1m

propionamide **26R**, prepared from the (R)-phenylalanine-derived imide chiral auxiliary, underwent highly diastereoselective aldol reactions with crotonaldehyde and p-methoxybenzl (PMB)-protected aldehyde **31** to furnish the aldol adducts **30** and **32**, respectively, which were then assembled to afford a spiroketal **29** in the synthesis of rutamycin B.

The phorbol **33** is a tigliane diterpene whose 12, 13-diesters play a principal role in efforts to understand biological activities such as carcinogenesis and

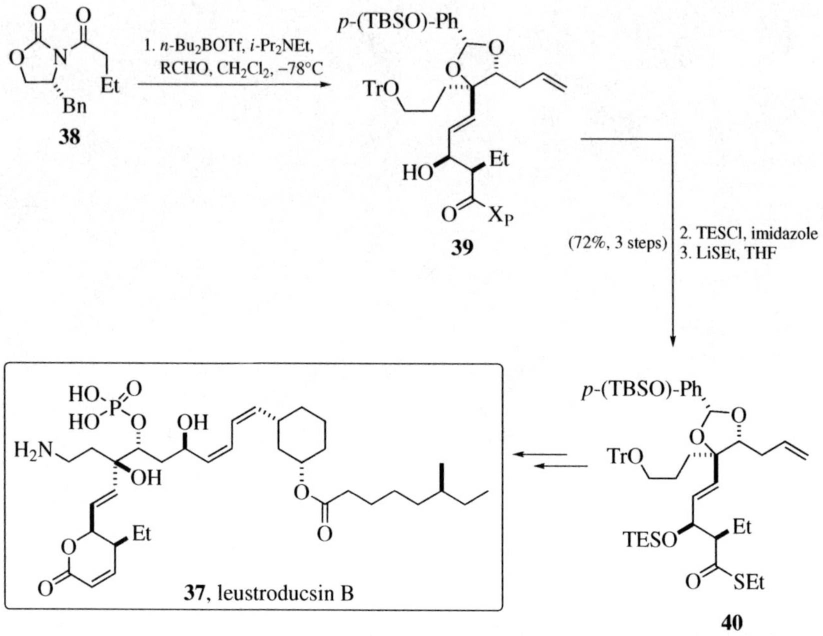

Scheme 2.1n

signal transduction. In the synthesis of **33**, Wender et al. used an Evans aldol reaction protocol to prepare the furan-containing compound **34**[14] (Scheme 2.1m). The highly diastereoselective aldol reaction between N-propionyloxazolidinone (**26S**) and 5-substituted furaldehyde occurred in 96% yield with 98% de stereoselectivity to provide the alcohol **34** as a single diastereomer. Formation of Weinreb amide followed by addition of 3-butenylmagnesium bromide afforded the hydroxy ketone **35**. A highly diastereoselective (30:1) reduction with diisobutylaluminum hydride, consistent with the formation of a six-membered aluminum chelate, provided the diol **36**.

Fukuyama et al. synthesized the alcohol **39** using Evans's chiral auxiliary in the total synthesis of leustroducsin B (**37**), a potent colony-stimulating factor inducer via NF-κB activation at the transcription level[15] (Scheme 2.1n). The asymmetric aldol reaction between **38** and the requisite aldehyde proceeded smoothly to afford **39**. Protection of the secondary alcohol as the TES ether and removal of the chiral auxiliary with LiSEt furnished the thioester **40**.

In an industrial synthesis of the microtuble-stabilizing agent (+)-discodermolide (**41**), Novartis scientists used an Evans asymmetric aldol reaction method to stereoselectively synthesize a $C_{15}-C_{21}$ fragment[16] (Scheme 2.1o). Reaction of the boron enolate of **26R** with the chiral aldehyde **42** gave the corresponding alcohol **43** in 55% yield on a scale of 20 to 25 kg. Transamidation using N, O-dimethylhydroxylamine/triisobutylaluminum complex, DDQ-mediated p-methoxybenzylidene acetal formation, and LiAlH₄ reduction provided the

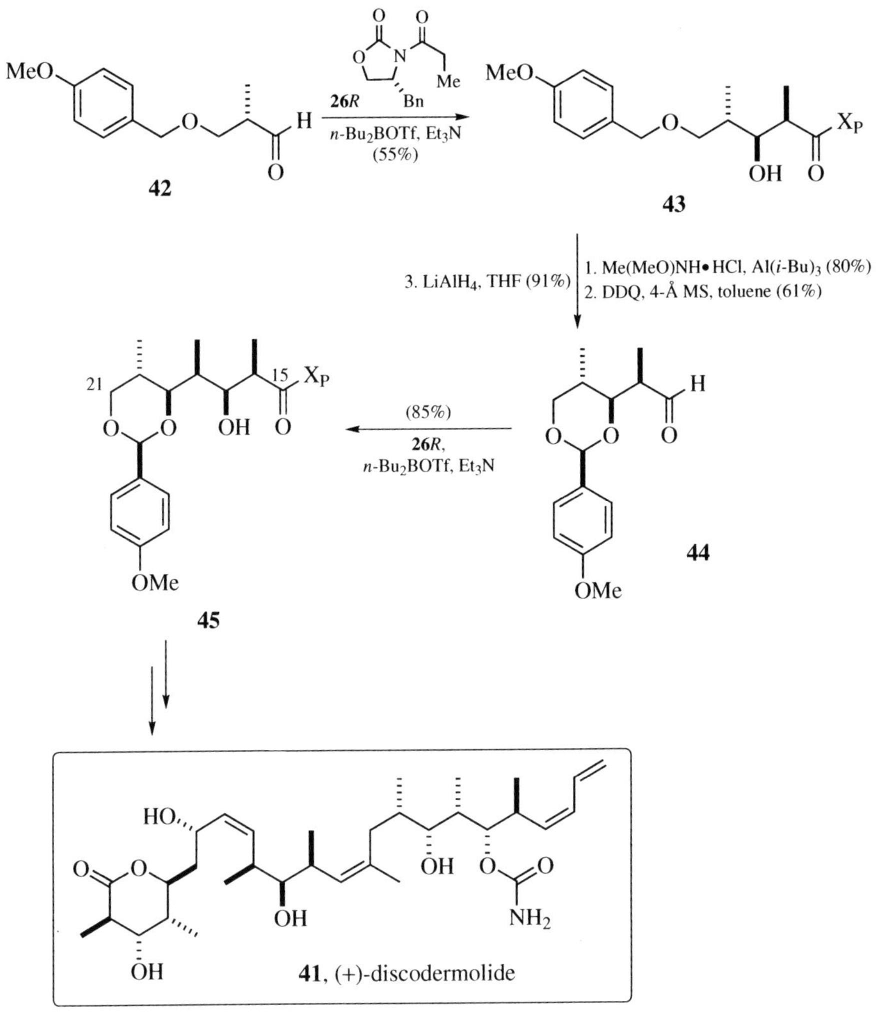

Scheme 2.1o

aldehyde **44**. Another asymmetric aldol reaction between **26R** and **44** afforded the aldol product desired (**45**) in 85% yield with no unwanted diastereoisomers detected.

Oppolzer et al. used (2R)-bornane-10, 2-sultam as an effective chiral auxiliary to achieve a highly enantioselective *syn*-aldol reaction[17] (Scheme 2.1p). Treatment of N-propionylsultam (**46**) with dibutylboron triflate and Hünig's base at −5°C in CH_2Cl_2 followed by addition of aldehydes at −78°C provided, after a simple crystallization, the pure *syn*-aldols **47a**. It is noteworthy that no *anti*-aldol product was observed in the aldol reactions with any of the aldehydes. From the ^{1}H nuclear magnetic resonance (NMR) study, it was confirmed that the boron

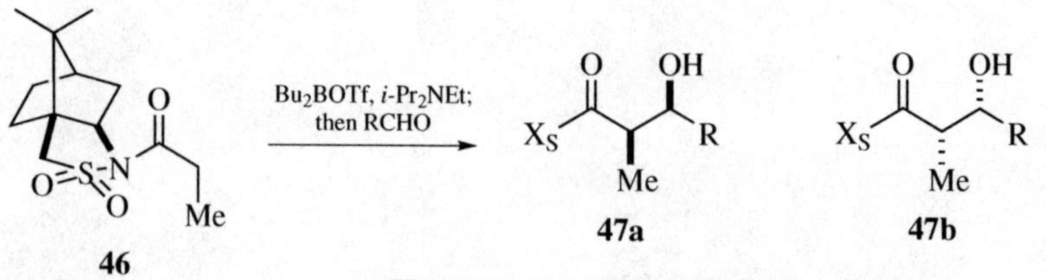

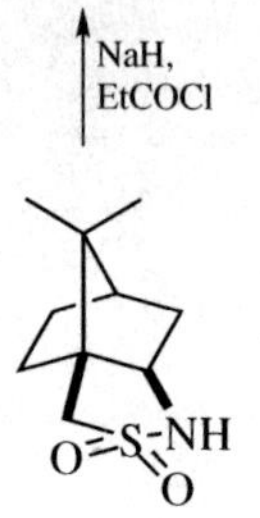

68 ALDOL REACTIONS

Scheme 2.1p

(2R)-bornane-10,2-sultam

R	Yield, %[b]	dr (**47a/47b**)[c]	de[d]
Me	69	>99:<1	>99
Et[a]	80	98:2	>99
i-C$_3$H$_7$	71	97:3	>99
C$_6$H$_5$	80	99:1	>99
(E)-MeCH=CH	54	98:2	>99

[a] Et$_2$BOTf was used.
[b] Yields after crystallization.
[c] Diastereomeric ratio (dr) of the crude products.
[d] Diastereomeric excess (de) after crystallization.

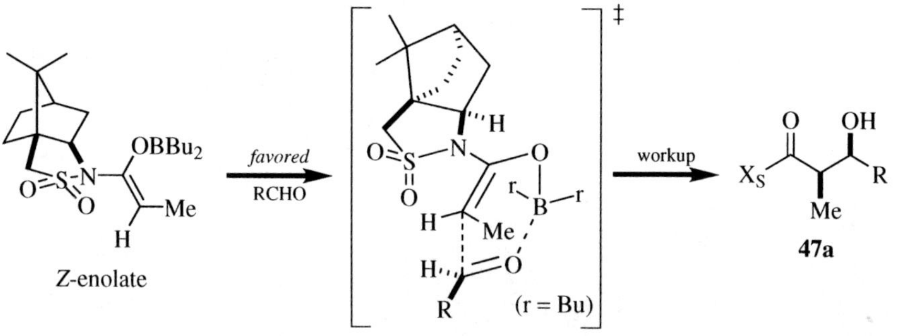

Scheme 2.1q

enolate had a Z-configuration. The aldol reaction is considered to proceed via a Zimmerman–Traxler type of six-membered transition state in which aldehyde approaches from the bottom face of the boron enolate[17] (Scheme 2.1q).

Oppolzer et al. completed an asymmetric synthesis of (−)-denticulatin A (**48**) by using a *syn*-aldol methodology as a key feature[18] (Scheme 2.1r). The diethylboron enolate of *N*-propionylbornanesultam (**46**-*ent*) obtained from diethylboron triflate and Hünig's base underwent a highly stereoselective aldol reaction with the *meso*-dialdehyde **49** to furnish the lactols **50** in 74% yield as a 2:1 epimeric mixture. When the lactols **50** were treated with 1, 2-ethanedithiol in the presence

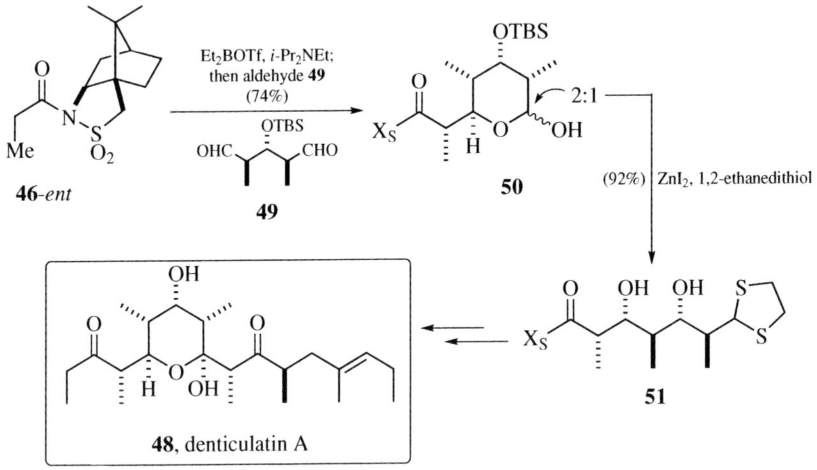

Scheme 2.1r

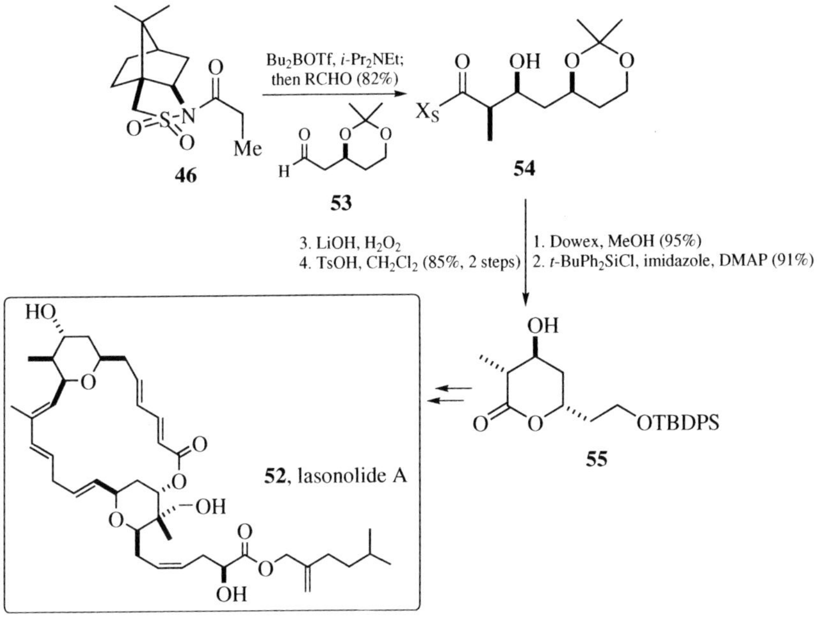

Scheme 2.1s

of ZnI$_2$, the pure dithiolane **51** was obtained in excellent yield via a desilylation/ring-opening sequence.

In an effort toward the enantioselective synthesis of the natural product lasonolide A (**52**), Shishido et al. used Oppolzer et al.'s *syn*-aldol method to access a key intermediate (**55**)[19] (Scheme 2.1s). Aldol reaction of the boron enolate of **46** with

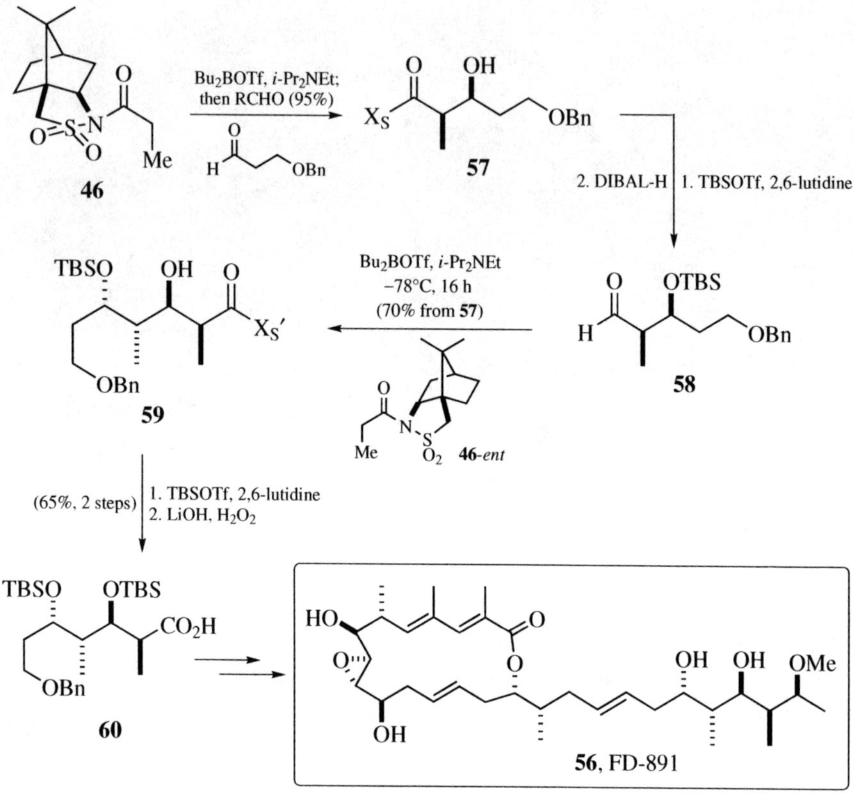

Scheme 2.1t

the aldehyde **53** gave the aldol product desired (**54**) in good yield. Acidic hydrolysis, regioselective protection of the primary alcohol as the *tert*-butyldiphenylsilyl (TBDPS) ether, removal of the chiral auxiliary, and treatment of the resulting acid with *p*-toluenesulfonic acid furnished the lactone **55**. In the stereoselective synthesis of the cytotoxic macrolide FD-891 (**56**), which exhibits antitumor activity, Oppolozer et al.'s asymmetric *syn*-aldol reaction protocol was used to synthesize a key fragment (**60**)[20] (Scheme 2.1t). Thus, the asymmetric aldol reaction of the boron enolate of sultam propionate (**46**) with 3-benzyloxypropanal provided the aldol product **57** as a single isomer in excellent yield. Silylation followed by reductive cleavage of the sultam chiral auxiliary using DIBAL-H afforded the aldehyde **58**, which was subjected to another *syn*-aldol reaction with the reagent **46**-*ent* to give a single aldol product (**59**) in good yield. Silylation of **59** with TBSOTf followed by removal of the chiral auxiliary with LiOOH hydrolysis furnished the carboxylic acid **60**.

Having noticed certain limitations of chlorotitanium aldol reactions on Evans et al.'s chiral auxiliary,[21] in 1997 Crimmins and others developed a brilliant protocol to achieve a highly diastereoselective aldol reaction.[22] Asymmetric aldol

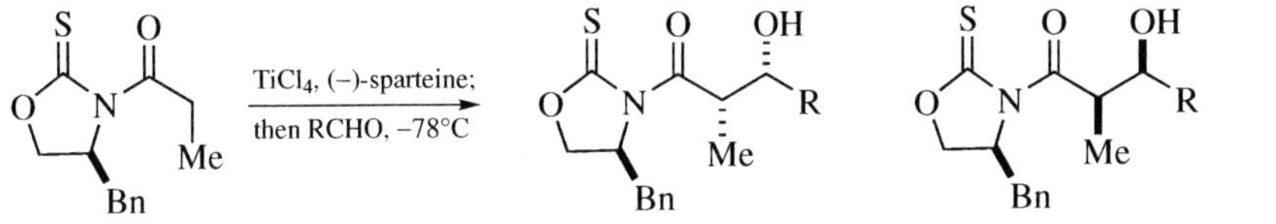

R	62a/62b[a]	Yield, %[a]	62a/62b[b]	Yield, %[b]
i-C$_3$H$_7$	98.8:1.0	70	0:95.5	79
C$_2$H$_5$	97.5:1.0	69	6.4:93.6	72
C$_6$H$_5$	98.7:1.3	89	0.7:97.6[c]	88[c]

[a] 1.0 equiv of TiCl$_4$ and 2.5 equiv of (−)-sparteine were used.
[b] 2.0 equiv of TiCl$_4$ and 1.1 equiv of (−)-sparteine were used.
[c] i-Pr$_2$NEt was used instead of (−)-sparteine.

Scheme 2.1u

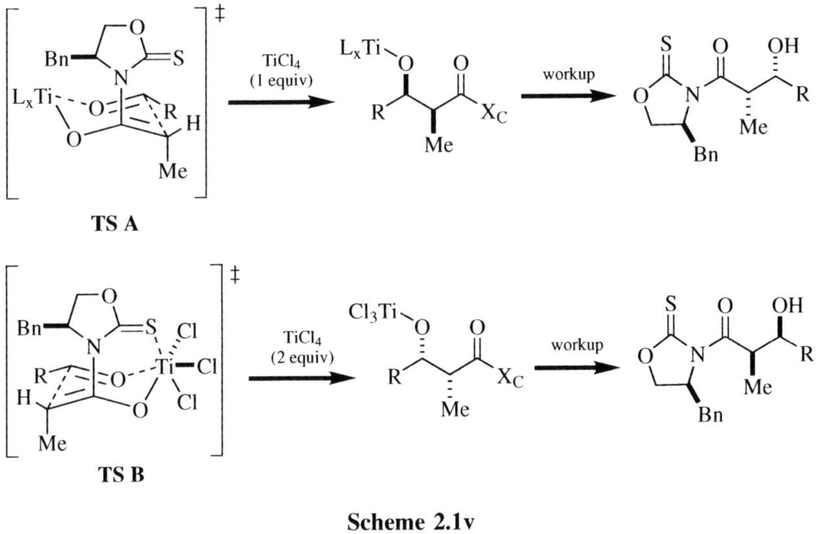

Scheme 2.1v

reactions using chlorotitanium enolates of the *N*-oxazolidinethione propionates **61** proceed with high diastereoselectivity to produce *syn*-aldol products. Whereas the *syn*-aldol **62a** is the major isomer with the use of 1 equivalent of titanium chloride (TiCl$_4$) and 2.5 equivalents of (−)-sparteine, a pseudoenantiomeric (**62b**) becomes a major *syn*-aldol product when 2 equivalents of TiCl$_4$ are employed[22] (Scheme 2.1u). The change in facial selectivity in the aldol additions is proposed to be the result of switching mechanistic pathways between chelating and nonchelating transition states (Scheme 2.1v). In the presence of 2.5 equivalents

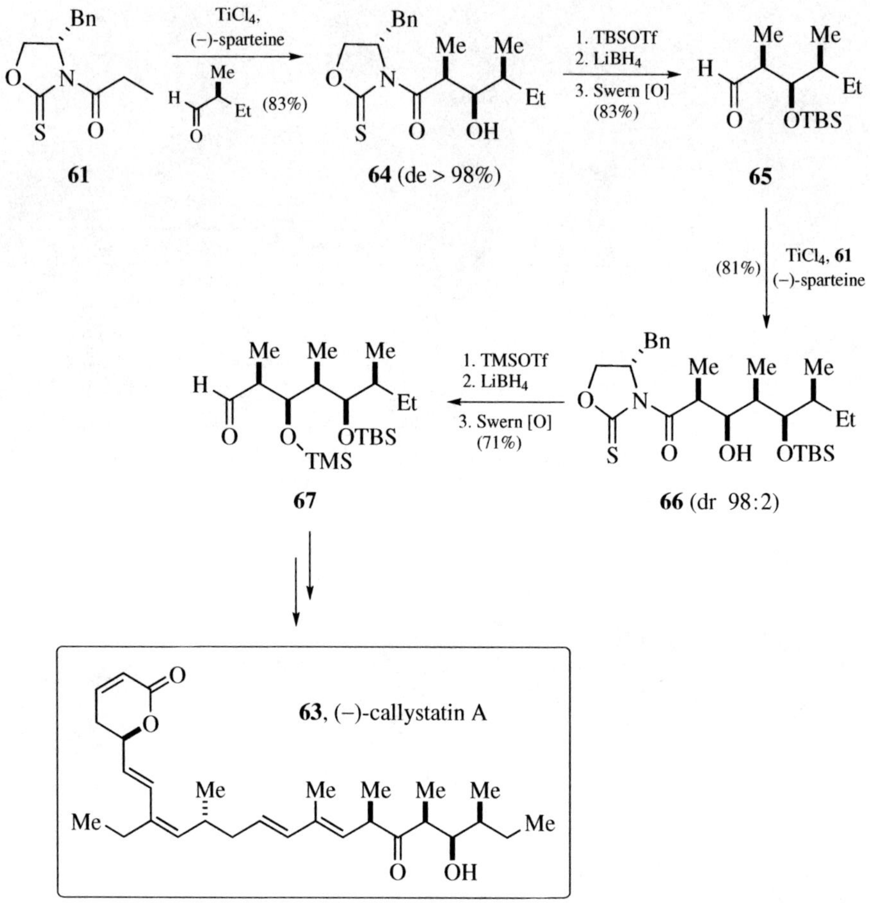

Scheme 2.1w

of (−)-sparteine, nonchelating transition state **A** is operative, possibly due to the coordination of the second equivalent of base to the titanium, resulting in formation of the *syn*-aldol product **62a**. The formation of **62b** can be explained in terms of the highly ordered chelated transition state **B**.[23]

Crimmins and King exploited a highly efficient aldol methodology to complete the asymmetric total synthesis of (−)-callystatin A (**63**), a natural product that exhibits a remarkable in vitro cytotoxicity[24] (Scheme 2.1w). The synthesis of propionate fragment in (−)-callystatin A was achieved by two consecutive asymmetric aldol additions using chlorotitanium enolates of acyloxazolidinethiones. Thus, treatment of *N*-propionyloxazolidinethione (**61**) with titanium chloride and (−)-sparteine followed by addition of (*S*)-2-methylbutanal resulted in formation of the *syn*-aldol adduct **64** in greater than 98% de and 83% yield. For the second aldol reaction, the requisite aldehyde (**65**) was prepared under standard conditions. Execution of the second asymmetric aldol reaction under conditions identical to

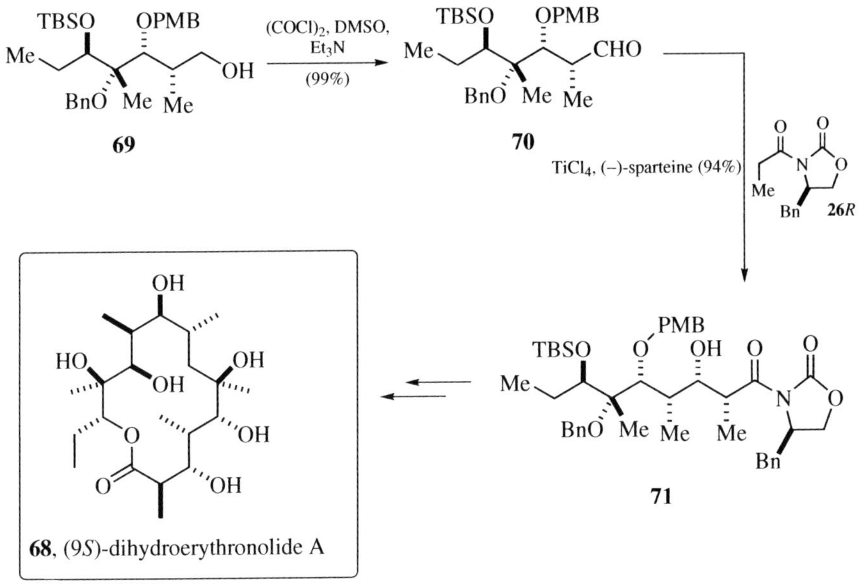

Scheme 2.1x

the first afforded the aldol product **66** in excellent diastereoselectivity. Alcohol protection, removal of chiral auxiliary, and Swern oxidation gave the aldehyde **67**.

Crimmins's $TiCl_4$-mediated asymmetric aldol condensation protocol was used in the enantioselective total synthesis of (9S)-dihydroerythronolide A (**68**)[25] (Scheme 2.1x). Swern oxidation of the primary alcohol **69** provided the aldehyde **70** in almost quantitative yield, which underwent asymmetric aldol condensation with the titanium enolate of (R)-4-benzyl-3-propionyloxazolidin-2-one (**26R**) in the presence of (−)-sparteine to afford the aldol adduct desired (**71**) as a single diastereomer.

Wu and Sun utilized the Crimmins's aldol method in the total synthesis of valilactone (**72**), which shows promising inhibitory activity toward an esterase and pancreas lipase[26] (Scheme 2.1y). The synthesis began with the $TiCl_4$-mediated asymmetric Crimmins aldol condensation between **73** and **74**. The *syn*-aldol **75** was obtained as the only detectable product in 78% yield. The chiral auxiliary in **75** was then removed with concurrent protection of the carboxylic group as a benzyl ester. The thiane protecting group in **76** was removed and the stereoselective reduction of β-hydroxy ketone in **77** using an Evans–Chapman–Carreira reagent[27] (see Chapter 4) generated the 1, 3-*anti*-diol **78**.

In 1995, Boeckman et al. disclosed a highly diastereoselective aldol reaction using the ligand **79** derived from chiral bicyclic lactam[28] (Scheme 2.1z). The imide **80**, readily prepared from bicyclic lactam **79** and propionyl chloride, was converted to the boron Z-enolate, which was then treated with a representative series of aldehydes at −40° C for 48 hours. The levels of diastereoselectivity observed in reactions of boron enolate derived from **80** are comparable to those

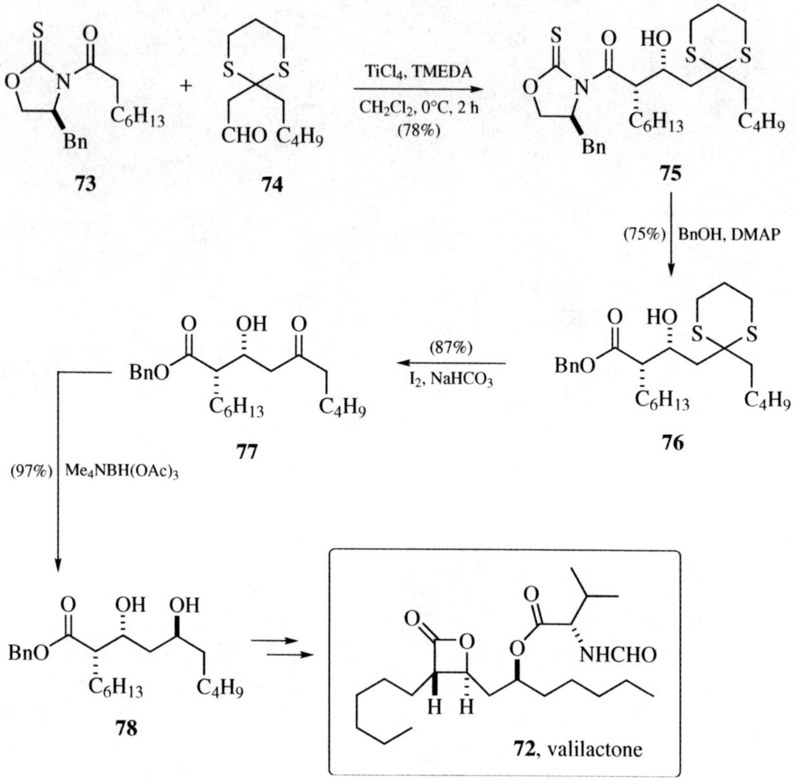

Scheme 2.1y

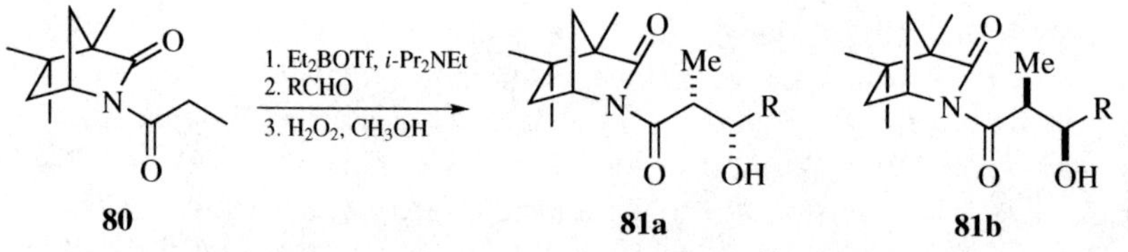

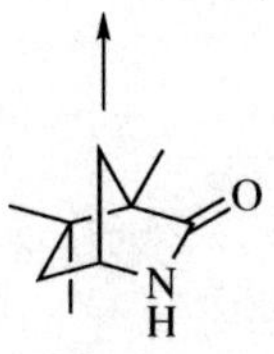

Scheme 2.1z

R	81a/81b	Yield, %
$n\text{-}C_3H_7$	>98:2	80
$i\text{-}C_3H_7$	>98:2	75
$c\text{-}C_6H_{11}$	>98:2	74
$t\text{-}C_4H_9$	>98:2	90
C_6H_5	>98:2	95

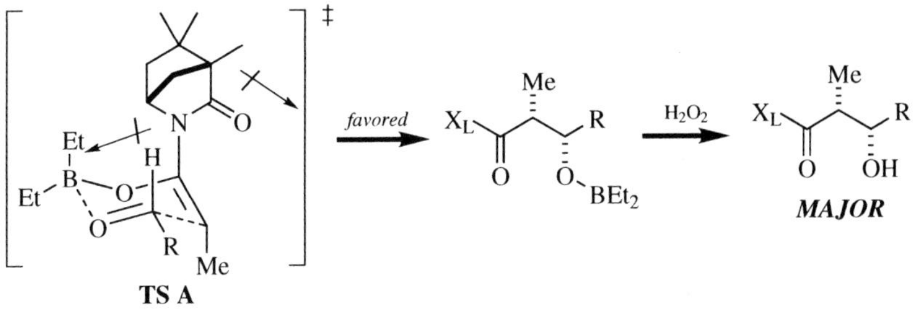

Scheme 2.1aa

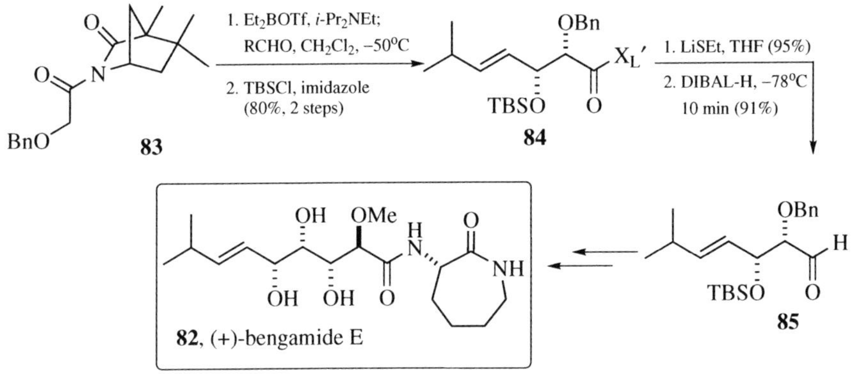

Scheme 2.1bb

obtained by employing the Evans auxiliary. The six-membered transition-state structure **A** is responsible for the high level of diastereoselectivity obtained by the imide **80**. In transition state **A**, the two carbonyl (C=O) dipoles are pointing opposite to each other to minimize the electrostatic repulsion, and the steric congestion is avoided[28] (Scheme 2.1aa).

Using a lactam chiral auxiliary, Boeckman et al. completed the synthesis of (+)-bengamide E (**82**), a member of the natural bengamides family that shows potentially useful antiproliferative activity[29] (Scheme 2.1bb). The imide **83** underwent a highly diastereoselective aldol condensation reaction with (*E*)-4-methyl-2-pentenal to afford the *syn*-aldol adduct expected (**84**) in > 24 : 1 diastereomeric ratio. The chiral auxiliary was removed efficiently with LiSEt, and the resulting thioester was reduced to give the corresponding aldehyde (**85**).

Ghosh et al. reported that the chiral oxazolidinone **87**, derived from (1*S*,2*R*)-*cis*-1-amino-2-indanol (**86**), underwent a highly diastereoselective *syn*-aldol reaction with a variety of aldehydes[30] (Scheme 2.1cc). Reaction of the indanolamine **86** with disuccinyl carbonate in acetonitrile gave the oxazolidinone **87**, which was deprotonated with *n*-BuLi and reacted with propionyl chloride to provide the *N*-propionyl derivative **88**. Reaction of **88** with *n*-Bu₂BOTf and

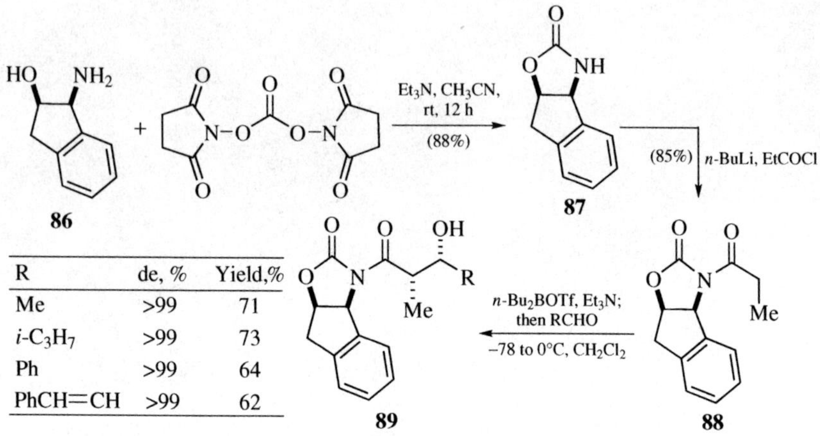

ALDOL REACTIONS

Scheme 2.1cc

R	de, %	Yield,%
Me	>99	71
i-C$_3$H$_7$	>99	73
Ph	>99	64
PhCH=CH	>99	62

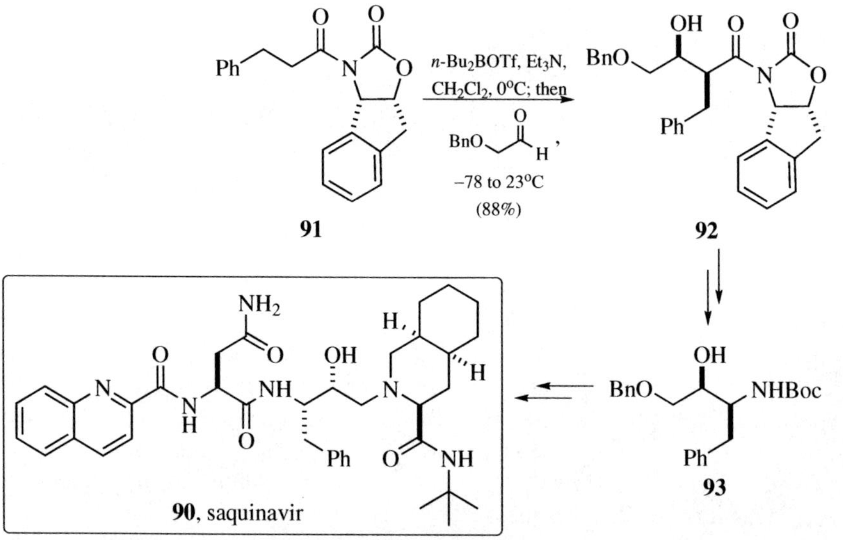

Scheme 2.1dd

triethyl amine at −78°C for 30 minutes and then warming up to 0°C for an hour afforded the boron enolate. Condensation of the enolate with various aldehydes resulted in the formation of a single *syn*-diastereomer (**89**). The exceptionally high diastereoselectivity is probably due to the configurational rigidity of the tricyclic ring system of the chiral auxiliary.[31] The transition state for the foregoing reaction has yet to be published but is assumed to be similar to those for the asymmetric reactions with the Evans and Boeckman chiral auxiliaries.

Ghosh et al. utilized an asymmetric *syn*-aldol reaction methodology to synthesize the core structure of saquinavir (**90**), a protease inhibitor recently approved by the U.S. Food and Drug Administration (FDA) for the treatment of AIDS[32] (Scheme 2.1dd). Aldol reaction of the boron enolate of **91** with benzyloxyacetaldehyde in CH_2Cl_2 at $-78°C$ provided the *syn*-aldol product **92** as a single diastereomer in 88% yield. After removal of the chiral auxiliary and several more manipulations, there was obtained a key intermediate amino alcohol (**93**), from which saquinavir can be synthesized according to known protocol.[33]

REFERENCES

1. (a) Evans, D. A.; Vogel, E.; Nelson, J. V. *J. Am. Chem. Soc.* **1979**, *101*, 6120; (b) Evans, D. A.; Nelson, J. V.; Vogel, E.; Taber, T. R. *J. Am. Chem. Soc.* **1981**, *103*, 3099.
2. Masamune, S.; Choy, W.; Kerdesky, F. A.; Imperiali, B. *J. Am. Chem. Soc.* **1981**, *103*, 1566.
3. Evans, D. A.; Bartroli, J.; Shih, T. L. *J. Am. Chem. Soc.* **1981**, *103*, 2127.
4. Masamune, S.; Hirama, M.; Mori, S.; Ali, S. A.; Garvey, D. S. *J. Am. Chem. Soc.* **1981**, *103*, 1568.
5. Evans, D. A. *Aldrichimica Acta* **1982**, *15*, 23.
6. Evans, D. A.; Takacs, J. M.; McGee, L. R.; Ennis, M. D.; Mathre, D. J.; Bartroli, J. *Pure Appl. Chem.* **1981**, *53*, 1109.
7. Makino, Y.; Iseki, K.; Fujii, K.; Oishi, S.; Hirano, T.; Kobayashi, Y. *Tetrahedron Lett.* **1995**, *36*, 6527.
8. Smith, A. B., III; Barbosa, J.; Wong, W.; Wood, J. L. *J. Am. Chem. Soc.* **1995**, *117*, 10777.
9. Brown, H. C.; Bhat, K. S.; Randad, R.S. *J. Org. Chem.* **1987**, *52*, 320.
10. Fürstner, A.; Ruiz-Caro, J.; Prinz, H.; Waldmann, H. *J. Org. Chem.* **2004**, *69*, 459.
11. Gage, J. R.; Evans, D. A. *Org. Synth. Coll.* Vol. *8*, **1990**, 528.
12. Gage, J. R.; Evans, D. A. *Org. Synth. Coll.* Vol. *8*, **1990**, 339.
13. Evans, D. A.; Ng, H. P.; Rieger, D. L. *J. Am. Chem. Soc.* **1993**, *115*, 11446.
14. Wender, P. A.; Rice, K. D.; Schnute, M. E. *J. Am. Chem. Soc.* **1997**, *119*, 7897.
15. Shimada, K.; Kaburagi, Y.; Fukuyama, T. *J. Am. Chem. Soc.* **2003**, *125*, 4048.
16. (a) Mickel, S. J.; Sedelmeier, G. H.; Niederer, D.; Daeffler, R.; Osmani, A.; Schreiner, K.; Seeger-Weibel, M.; Bérod, B.; Schaer, K.; Gamboni, R.; Chen, S.; Chen, W.; Jagoe, C. T.; Kinder, F. R., Jr.; Loo, M.; Prasad, K.; Repic, O.; Shieh, W.-C.; Wang, R.-M.; Waykole, L.; Xu, D. D.; Xue, S. *Org. Proc. Res. Dev.* **2004**, *8*, 92; (b) Mickel, S. J.; Sedelmeier, G. H.; Niederer, D.; Schuerch, F.; Koch, G.; Kuesters, E.; Daeffler, R.; Osmani, A.; Seeger-Weibel, M.; Schmid, E.; Hirni, A.; Schaer, K.; Gamboni, R.; Bach, A.; Chen, S.; Chen, W.; Geng, P.; Jagoe, C. T.; Kinder, F. R., Jr.; Lee, G. T.; McKenna, J.; Ramsey, T. M.; Repic, O.; Rogers, L.; Shieh, W.-C.; Wang, R.-M.; Waykole, L. *Org. Proc. Res. Dev.* **2004**, *8*, 107.
17. Oppolzer, W.; Blagg, J.; Rodriguez, I.; Walther, E. *J. Am. Chem. Soc.* **1990**, *112*, 2767.
18. Oppolzer, W.; De Brabander, J.; Walther, E.; Bernardinelli, G. *Tetrahedron Lett.* **1995**, *36*, 4413.

19. Deba, T.; Yakushiji, F.; Shindo, M.; Shishido, K. *Synlett* **2003**, 1500.
20. Garcia-Fortanet, J.; Murga, J.; Carda, M.; Marco, J. A. *Org. Lett.* **2006**, *8*, 2695.
21. (a) Evans, D. A.; Rieger, D. L.; Bilodeau, M. T.; Urpi, F. *J. Am. Chem. Soc.* **1991**, *113*, 1047;   (b) Evans, D. A.; Clark, J. S.; Metternich, R.; Novack, V. J.; Sheppard, G. S. *J. Am. Chem. Soc.* **1990**, *112*, 866.
22. (a) Crimmins, M. T.; King, B. W.; Tabet, E. A. *J. Am. Chem. Soc.* **1997**, *119*, 7883; (b) Crimmins, M. T.; King, B. W.; Tabet, E. A.; Chaudhary, K. *J. Org. Chem.* **2001**, *66*, 894;   (c) Crimmins, M. T.; Chaudhary, K. *Org. Lett.* **2000**, *2*, 775.
23. (a) Nagao, Y.; Hagiwara, Y.; Kumagai, T.; Ochiai, M.; Inoue, T.; Hashimoto, K.; Fujita, E. *J. Org. Chem.* **1986**, *51*, 2391;   (b) Evans, D. A.; Rieger, D. L.; Bilodeau, M. T.; Urpi, F. *J. Am. Chem. Soc.* **1991**, *113*, 1047.
24. Crimmins, M. T.; King, B. W. *J. Am. Chem. Soc.* **1998**, *120*, 9084.
25. Peng, Z.-H.; Woerpel, K. A. *J. Am. Chem. Soc.* **2003**, *125*, 6018.
26. Wu, Y.; Sun, Y.-P. *J. Org. Chem.* **2006**, *71*, 5748.
27. Evans, D. A.; Chapman, K. T.; Carreira, E. M. *J. Am. Chem. Soc.* **1988**, *110*, 3560.
28. (a) Boeckman, R. K., Jr.; Connell, B. T. *J. Am. Chem. Soc.* **1995**, *117*, 12368;   (b) Boeckman, R. K., Jr.; Johnson, A. T.; Musselman, R. A. *Tetrahedron Lett.* **1994**, *35*, 8521.
29. Boeckman, R. K., Jr.; Clark, T. J.; Shook, B. C. *Org. Lett.* **2002**, *4*, 2109.
30. Ghosh, A. K.; Duong, T. T.; McKee, S. P. *Chem. Commun.* **1992**, 1673.
31. Senanayake, C. H. *Aldrichimica Acta* **1998**, *31*, 3.
32. Ghosh, A. K.; Hussain, K. A.; Fidanze, S. *J. Org. Chem.* **1997**, *62*, 6080.
33. Parkes, K. E. B.; Bushnell, D. J.; Crackett, P. H.; Dunsdon, S. J.; Freeman, A. C.; Gunn, M. P.; Hopkins, R. A.; Lambert, R. W.; Martin, J. A.; Merrett, J. H.; Redshaw, S.; Spurden, W. C.; Thomas, G. J. *J. Org. Chem.* **1994**, *59*, 3656.

2.2. Asymmetric *Anti*-Aldol Reaction

The synthetic methodology for the efficient construction of *anti*-aldol has also been explored. In 1986, Masamune and others reported a new asymmetric aldol reaction that afforded an aldol product in greater than 98% de favoring *anti*-isomer[1] (Scheme 2.2a). The chiral *E*-enolates **1**, derived from *S*-3-(3-ethyl)pentylpropanethioate and (2*S*,5*S*)-dimethylborolane trifloromethanesulfonate, underwent aldol reactions smoothly at −78°C, and *anti*-aldol products were obtained for a variety of aldehydes with greater than 30:1 *anti/syn* stereoselectivity. More important, the *anti*-aldol products were of greater than 97% optical purity. In explaining the results, Masamune et al. invoked a six-membered chairlike transition-state model (Scheme 2.2b). In transition state **A**, the methyl group in the enolate steers the 3-ethyl-3-pentanethiol group toward the borolane moiety, thus forcing the enolate to attack aldehyde on its *si*-face.

In 1997, Masamune et al. disclosed another *anti*-selective aldol reaction method, using a readily available chiral ligand, norephedrine[2] (Scheme 2.2c). The ester **3** was treated with 2 equivalents of dicyclohexylboron triflate (Cy$_2$BOTf) and 2.4 equivalents of triethylamine at −78°C for 2 hours. When aldehyde was

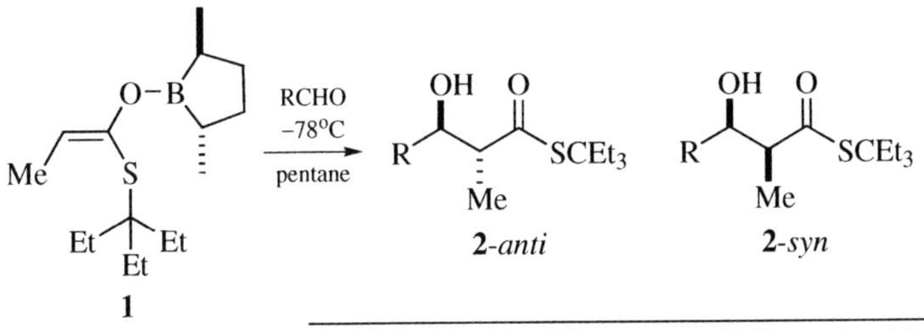

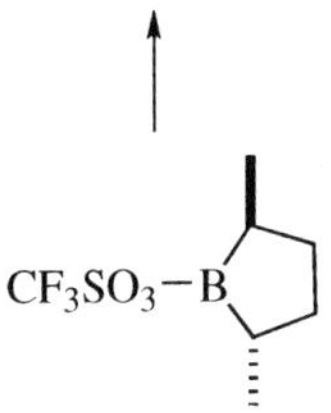

Scheme 2.2a

R	anti/syn	ee, % (2-anti)	Yield, %
n-C$_3$H$_7$	33:1	97.9	91
i-C$_3$H$_7$	30:1	99.5	85
t-C$_4$H$_9$	30:1	99.9	95
c-C$_6$H$_{11}$	32:1	98.0	82
C$_6$H$_5$	33:1	99.8	71
BnO(CH$_2$)$_2$	>30:1	97.1	93

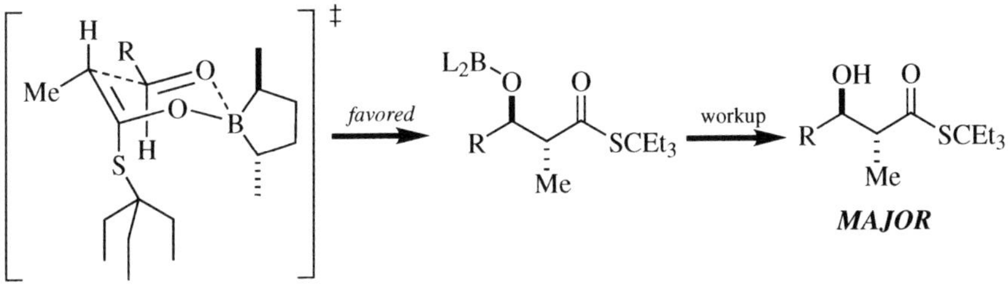

Scheme 2.2b

added and the aldol reaction was conducted at $-78°C$ for 1 hour and then $0°C$ for another hour, the *anti*-aldol product **4-*anti*** was obtained in high yield with superb *anti*-selectivity (*anti*/*syn* = 98:2) and high diastereofacial selectivity (>95:5) for all of the aliphatic, aromatic, and α,β-unsaturated aldehydes.

As the boron enolate from dicyclohexylboron triflate and triethylamine at $-78°C$ was determined to be an *E*-isomer, the predominant formation of *anti*-aldol product can be explained via the classic Zimmerman–Traxler six-membered chairlike transition state[3] (Scheme 2.2d). In transition state **A**, the norephedrine unit of the boron enolate presumably arranges itself in such a way that the phenyl group directs the approach of the aldehyde.[4]

Masamune et al. applied the newly developed enantioselective *anti*-aldol reaction to the syntheses of two key fragments of miyakolide **5**, a bryostatin-like marine metabolite[5] (Scheme 2.2e). The readily synthesized aldehyde **8** was treated with chiral enol borinate generated from the ester **3** to give the aldol **9** in 85%

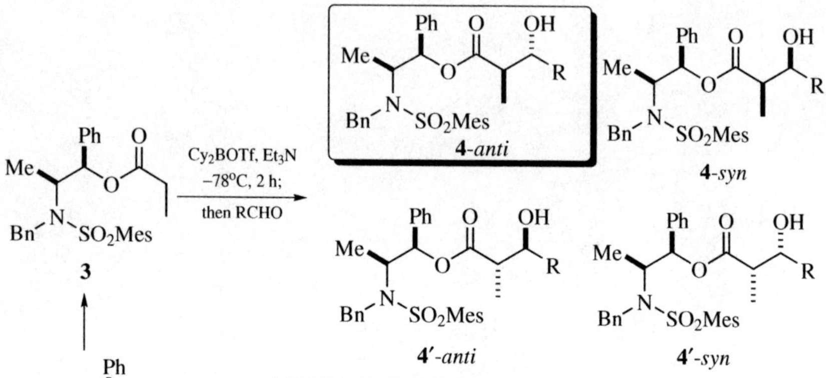

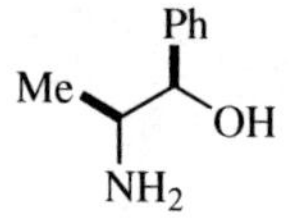

Scheme 2.2c

R	Yield, %[a]	dr (**4**-anti / **4'**-anti)
Et	90	96.1:3.9
n-C$_3$H$_7$	95	95.2:4.8
i-C$_3$H$_7$	98	97.7:2.3
c-C$_6$H$_{11}$	91	95.2:4.8
t-C$_4$H$_9$	96	99.4:0.6
C$_6$H$_5$	93	94.7:5.3
(E)-MeCH=CH	96	98.0:2.0
BnO(CH$_2$)$_2$	94	94.8:5.2

[a] Yields of *syn*-isomers were less than 2%.

yield with high selectivity (15 : 1), eventually furnishing the C$_6$–C$_{13}$ fragment **6**. The synthesis of the fragment **7** started with the chiral aldehyde **10** and featured a double asymmetric *anti*-aldol reaction. Thus, **10** reacted with enol borinate from the ester **3**-*ent* to provide the aldol **11** as the major isomer in the ratio 15 : 1.

The norephedrine-derived Masamune asymmetric aldol reaction was utilized in the total synthesis of (+)-testudinariol A (**12**), a triterpene marine natural product that possesses a highly functionalized cyclopentanol framework with four contiguous stereocenters appended to a central 3-alkylidene tetrahydropyran[6] (Scheme 2.2f). The norephedrine-derived ester **13** was enolized with dicyclohexylboron triflate and triethylamine in dichloromethane and then treated with 3-benzyloxypropanal to afford the aldol adduct (**14**) as a 97 : 3 mixture of *anti*/*syn* diastereomers in 72% yield. Diastereoselectivity within the *anti*-manifold was 90 : 10. Protection of alcohol as the methoxyethoxymethyl (MEM) ether followed by conversion of the ester to an aldehyde by LiAlH$_4$ reduction and subsequent Swern oxidation gave the aldehyde **16** in 64% yield over three steps.

Masamune's norephedrine-based *anti*-aldol methodology was again employed successfully in the total synthesis of the antitumor macrolide natural product rhizoxin D (**17**)[7] (Scheme 2.2g). The synthesis began with an *anti*-aldol addition of boron enolate of **3**-*ent* to the aldehyde **18**. The addition proceeded with

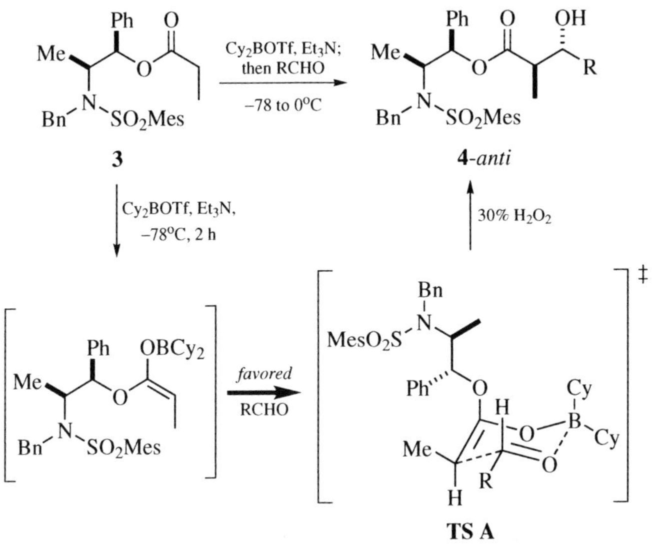

Scheme 2.2d

excellent diastereoselectivity (90% de) and in good yield to afford the *anti*-aldol **19**. Silylation and reductive cleavage of the chiral auxiliary then provided the alcohol **20**.

Perhaps Walker and Heathcock were the first to accomplish an asymmetric *anti*-aldol reaction using acyloxazolidinone chiral auxiliary[8] (Scheme 2.2h). An aldehyde was precomplexed with Et_2AlCl in CH_2Cl_2 at $-78°C$ and the dibutylboron enolate was added to the cold solution. The *anti/syn* selectivity is satisfactory for aliphatic aldehydes, with the *anti/syn* ratios ranging from 86:14 to 95:5. With benzaldehyde, however, the ratio is only 74:26, leaving room for improvement. As the reaction was mediated by Lewis acid and the predominant product was an *anti*-diastereomer, Heathcock proposed open transition states to explain the stereochemical outcome of the aldol reactions.[9]

Two decades after the discovery of a powerful *syn*-diastereoselective aldol reaction, Evans and colleagues reported a highly diastereoselective *anti*-aldol reaction with chiral acyloxazolidinones promoted by catalytic amounts of magnesium chloride ($MgCl_2$) in the presence of triethylamine and chlorotrimethylsilane[10] (Scheme 2.2i). Ethyl acetate (EtOAc) is the optimal solvent in promoting the reaction of **23S** with benzaldehyde to afford aldol product in 91% yield and with 32:1 diastereoselectivity. Variation of the aldehyde reaction component results in large changes in diastereoselectivity, providing moderate-to-high selectivity of *anti*-aldol products. One limitation to this aldol procedure is that aliphatic aldehydes such as hydrocinnamaldehyde are relatively unreactive, resulting in low conversion.

Because the major product is an *anti*-diastereomer and the Z-enolate is involved, it is unlikely that a traditional Zimmerman–Traxler transition state

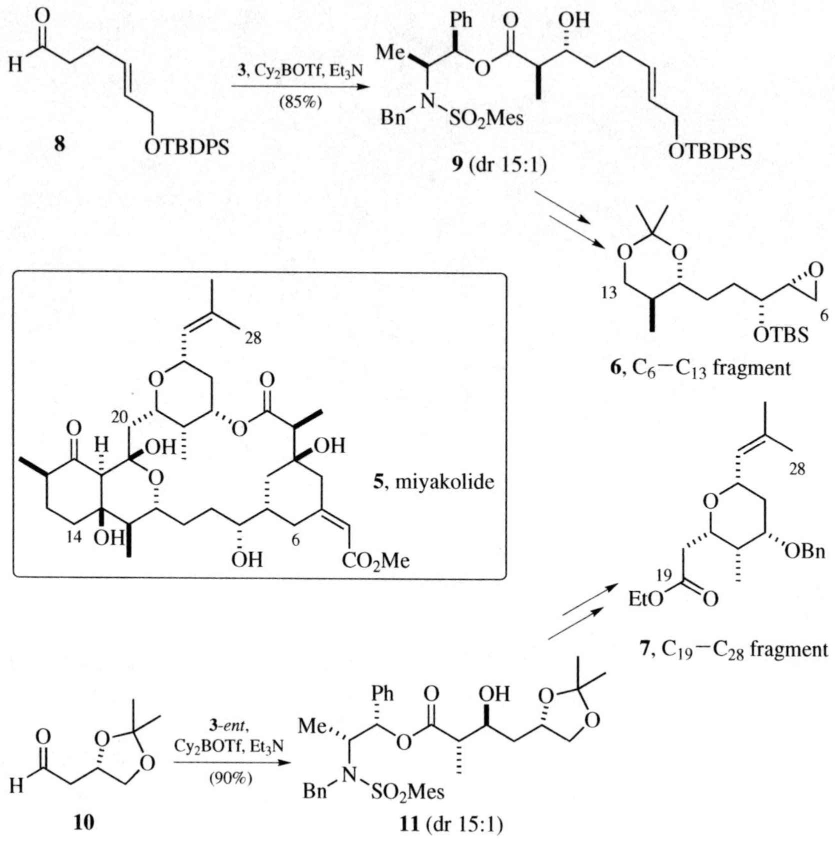

Scheme 2.2e

is operating here. To gain solid ground to explain the unusual result, semiempirical calculations were carried out on the two possible transition states where the magnesium metals are six-coordinated. This study suggests that in this case, twist-boat transition state **A** is more stable than corresponding chair transition state **B** by 2.5 kcal/mol[11] (Scheme 2.2j).

Chiral acylthiazolidinethiones such as **26** can readily be prepared from commercially available amino acids in three steps[12] (Scheme 2.2k). They have been employed as a synthetically useful auxiliary in diastereoselective aldol reactions.[13] The magnesium-catalyzed aldol reaction of the thiazolidinethione **26S** with cinnamaldehyde afforded **27** as a major diastereomer in 87% yield. Interestingly, compound **27** is the opposite *anti*-aldol diastereomer to that seen with the oxazolidinone **23S**.

Because the acylthiazolidinethione-derived magnesium-enolate exhibits the opposite face selection to that of its oxazolidinone counterpart, it is necessary to assume that the thione C=S moiety is not coordinating to the Mg center in

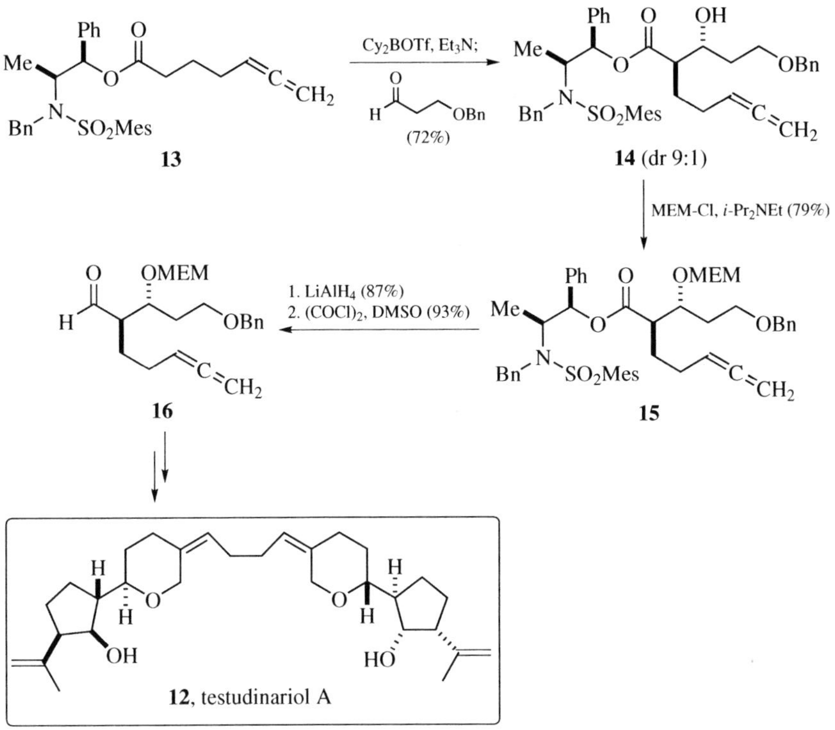

Scheme 2.2f

the aldol reaction transition state[11] (Scheme 2.2l). The enolate and benzaldehyde would then be arranged so as to minimize the electrostatic interactions of the two dipoles, C=O and C=S. Between the two possible transition states, calculations found that a boatlike transition state is favored over a chairlike six-membered transition state, in good agreement with the experimental result.

In the total synthesis of the migrastatin **28**, an *anti*-selective aldol reaction of *N*-propionyl oxazolidinone (**23R**) with the aldehyde **29** was employed successfully to furnish the key intermediate (**30**) as a single isomer[14] (Scheme 2.2m). Protection of the alcohol as a TES ether and reductive cleavage of the chiral auxiliary with LiBH$_4$ gave the alcohol **31**. Evans et al. employed *anti*-aldol methodology during synthesis of the C$_8$–C$_{18}$ ketone fragment **32** of the (−)-aflastatin A C$_9$–C$_{27}$ degradation polyol[15] (Scheme 2.2n). The synthesis began with MgCl$_2$-catalyzed directed aldol addition to provide the *anti*-aldol adduct **33** (dr > 20:1). The imide **33** was converted into the Weinreb amide **34**, protected as a *p*-methoxybenzyl ether and then reduced to afford the C$_8$–C$_{11}$ aldehyde **35** in high yield.

(−)-Stemoamide (**36**), isolated from the roots and rhizomes of *Stemona tuberosa* in 1992, is one of the structurally simplest members of the *Stemona* family and has shown many biological activities. Olivo and others used a highly

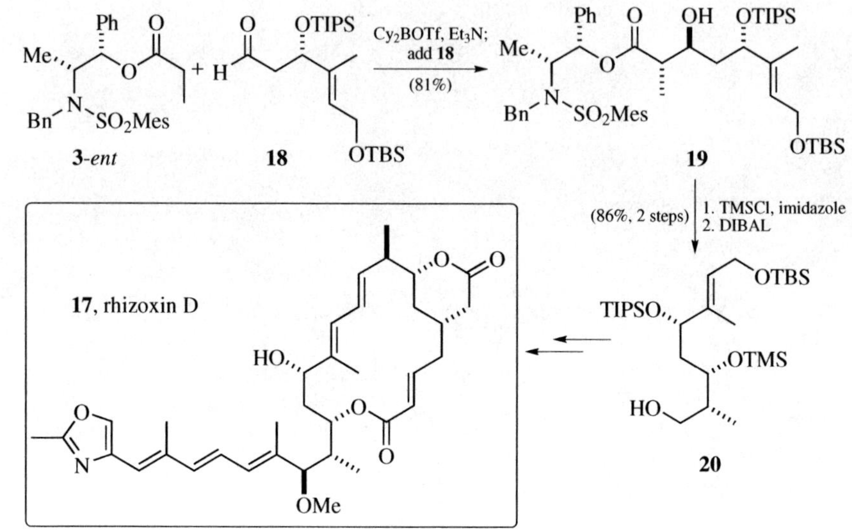

84 ALDOL REACTIONS

Scheme 2.2g

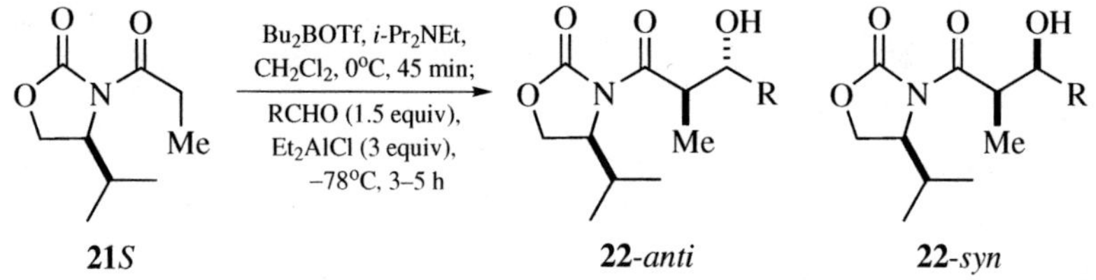

Scheme 2.2h

RCHO	anti/syn	Yield, %
i-BuCHO	86:14	86
C$_2$H$_5$CHO	88:12	81
i-PrCHO	95:5	63
t-BuCHO	95:5	65
methacrolein	90:10	67
C$_6$H$_5$CHO	74:26	62

diastereoselective *anti*-aldol reaction to prepare the aldol product **38**[16] (Scheme 2.2o). When the thiazolidinethione **37**, derived from (*R*)-phenylglycine, was subjected to an *anti*-aldol reaction with cinnamaldehyde in the presence of a catalytic amount of magnesium bromide, the aldol product expected (**38**) was obtained in 74% yield with the stereochemistry desired for the synthesis of (−)-stemoamide. The chiral auxiliary was removed by NaBH$_4$ reduction and the resulting alcohol was oxidized using Ley's reagent[17] to furnish the aldehyde **39**.

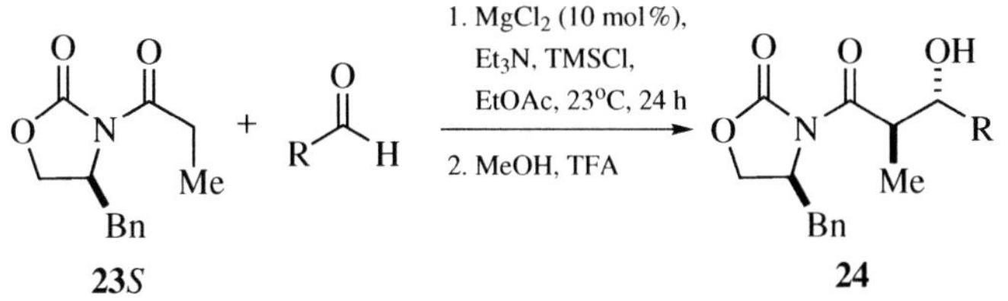

RCHO	dr[a]	Yield, %
C_6H_5CHO	32:1	91
$p\text{-MeC}_6H_4CHO$	24:1	—
$p\text{-OMeC}_6H_4CHO$	32:1	91
$p\text{-NO}_2C_6H_4CHO$	7:1	71
cinnamaldehyde	21:1	92
α-naphthaldehyde	14:1	91
furfural	6:1	80
methacrolein	16:1	77

[a] dr, diastereomeric ratio reported as major isomer/sum of other diastereomers.

Scheme 2.2i

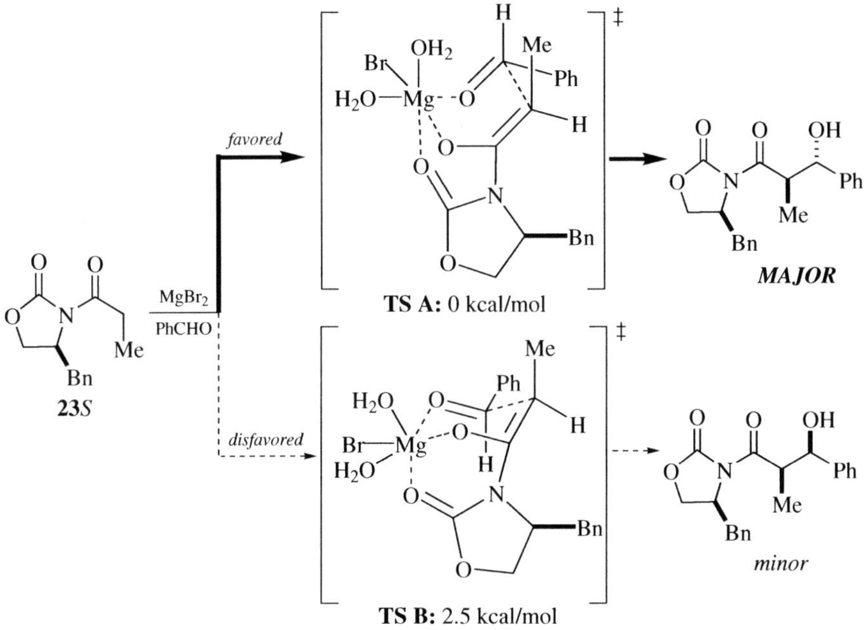

Scheme 2.2j

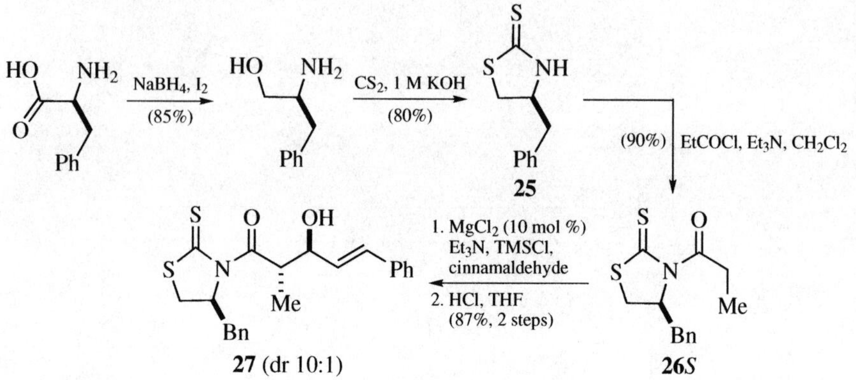

Scheme 2.2k

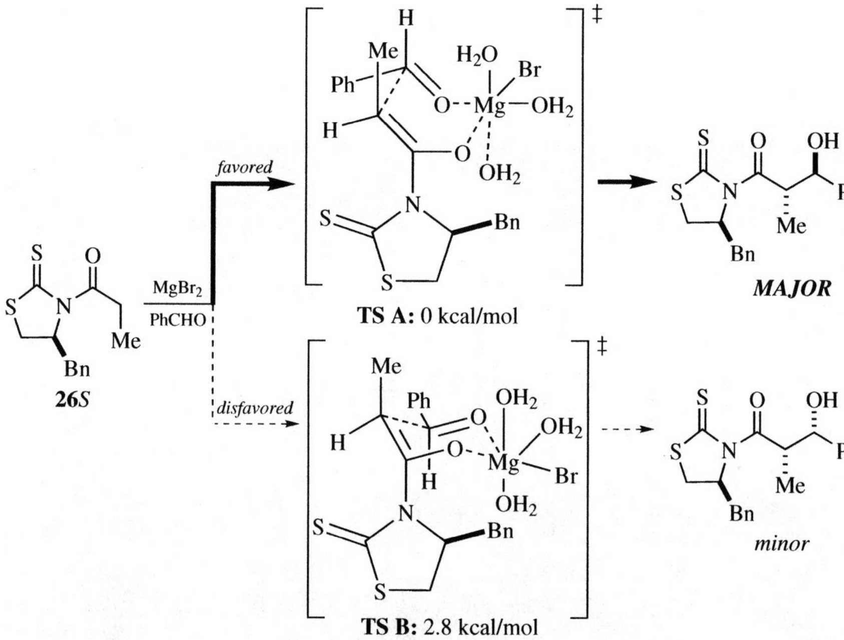

Scheme 2.2l

(−)-Talaumidin (**40**) exhibits significant neurite outgrowth-promoting and neuroprotective activities in the biological assays and is a promising agent for the treatment of neurodegenerative diseases such as Alzheimer's and Parkinson's. In the synthesis of (−)-talaumidin (**40**), Fukuyama et al. used the recently disclosed Evans magnesium-catalyzed *anti*-aldol methodology[18] (Scheme 2.2p).

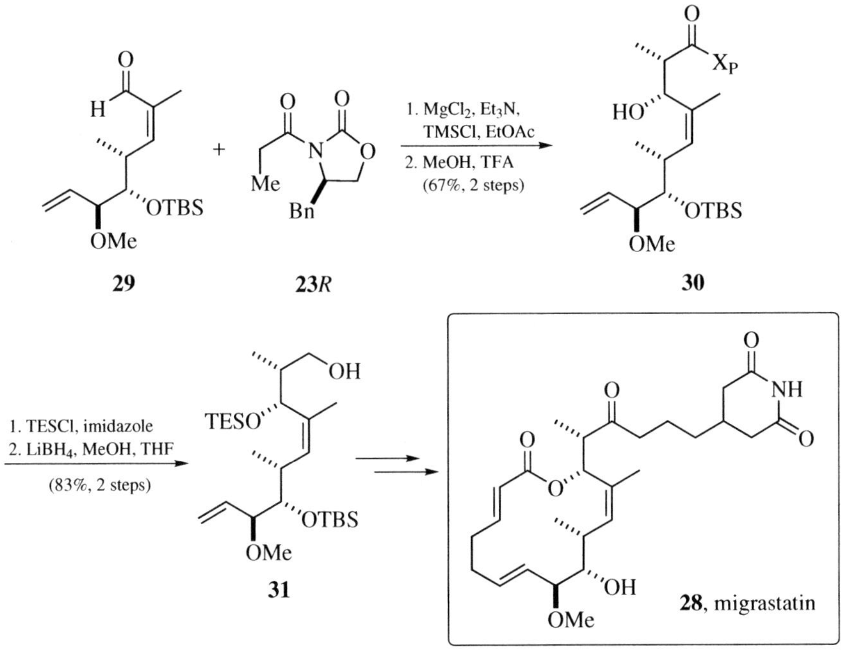

Scheme 2.2m

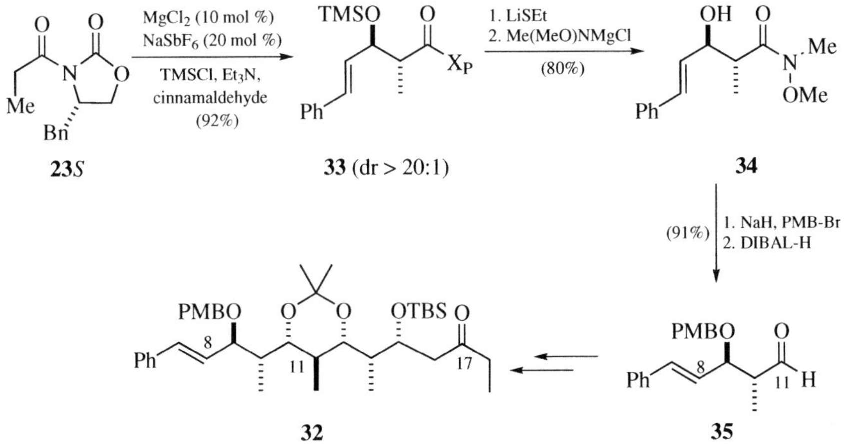

Scheme 2.2n

The reaction of 4-benzyloxy-3-methoxybenzaldehyde (**41**) with *N*-propionyl oxazolidinone (**23S**) in the presence of chlorotrimethylsilane along with a catalytic amount of MgCl$_2$ provided the aldol adduct **42** in modest yield with a high

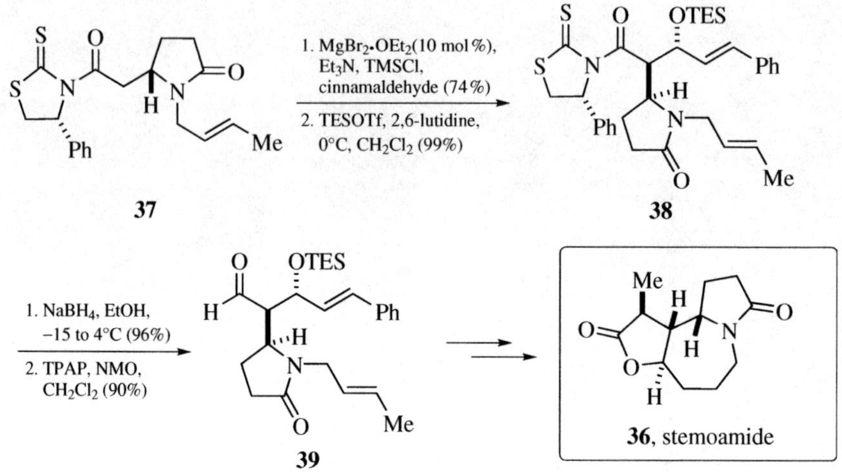

Scheme 2.2o

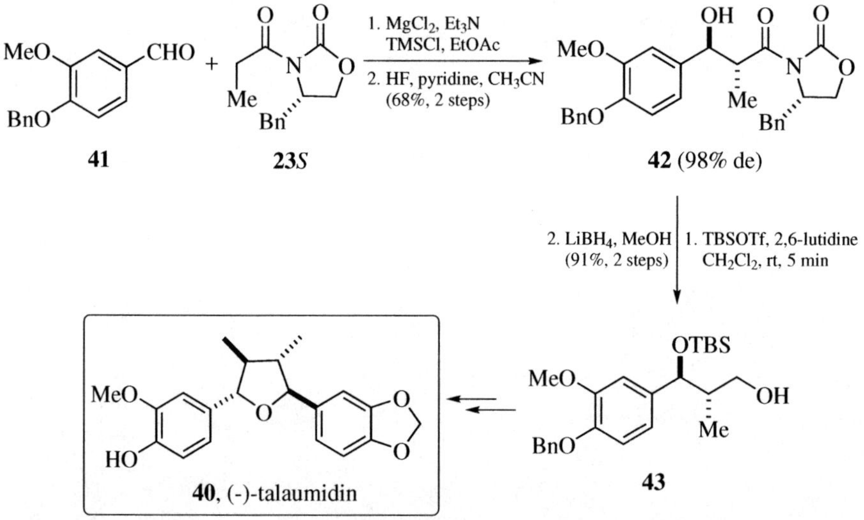

Scheme 2.2p

diastereoselectivity of 98% de. Protection of the hydroxyl group as a TBS ether and removal of the chiral auxiliary using LiBH$_4$ efficiently furnished the primary alcohol **43**.

The chiral sulfonamide **45**, which can be prepared in two steps from commercially available (1R,2S)-cis-1-amino-2-indanol (**44**), was introduced by Ghosh and Onishi for the synthesis of enantiomerically pure *anti*-aldol products via titanium enolate[19a] (Scheme 2.2q).

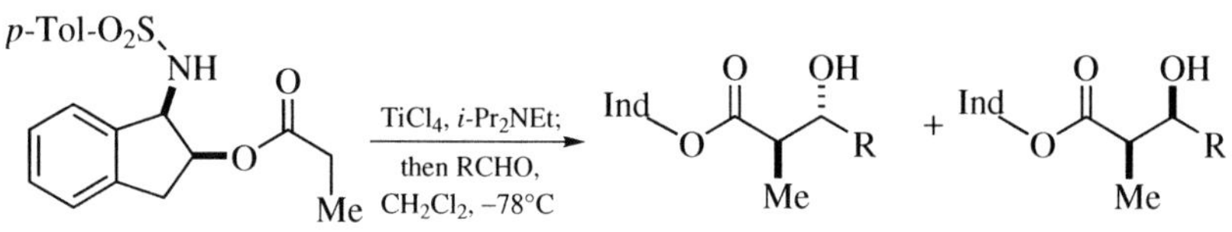

Scheme 2.2q

R	anti/syn	Yield, %
CH_3	85 : 15	50
C_2H_5	85 : 15	50
i-C_3H_7	85 : 15	91
i-C_4H_9	>99 : 1	97
C_6H_5	45 : 55	85

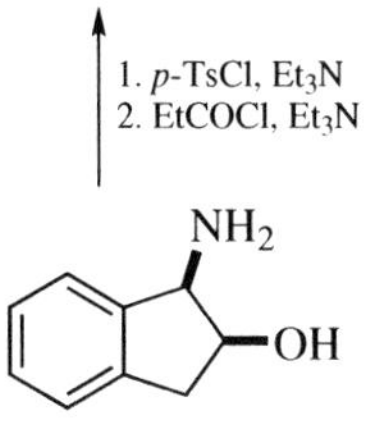

Scheme 2.2r

The aldol reactions of the titanium Z-enolates proceeded smoothly with various aldehydes precomplexed with titanium chloride at $-78°$ C. The diastereoselectivity is high to excellent, with the single exception of benzaldehyde. The high degree of diastereoselection associated with this current asymmetric *anti*-aldol process can be rationalized by a Zimmerman–Traxler type of six-membered chairlike transition state **A**[19a] (Scheme 2.2r). The model is based on the assumptions that the titanium enolate is a seven-membered metallocycle with a chairlike conformation, and a second titanium metal is involved in the transition state, where it is chelated to indanolyloxy oxygen as well as to the aldehyde carbonyl in a six-membered chairlike transition-state structure.

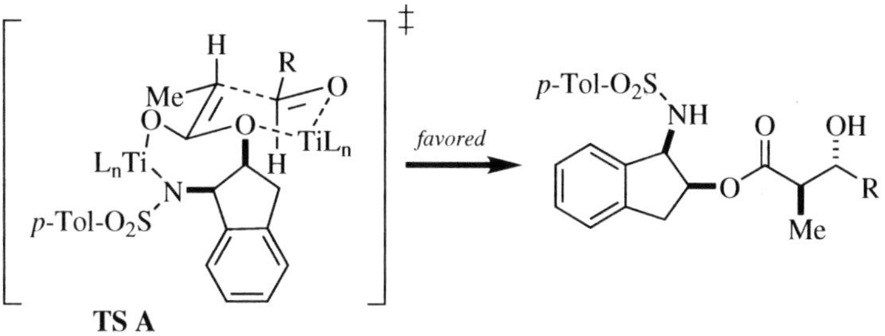

Ghosh and Kim recently disclosed a new chiral auxiliary (**47**) that exhibits an enhanced diastereoselectivity over the previously utilized auxiliary (**44**) in the *anti*-aldol reaction[19b] (Scheme 2.2s). Based on this *anti*-aldol strategy, Ghosh and Fidanze achieved an asymmetric synthesis of (−)-tetrahydrolipstatin (**50**), which was isolated from *Streptomyces toxytricini*[20] (Scheme 2.2t). (−)-Tetrahydrolipstatin is a potent inhibitor of pancreatic protease and has been

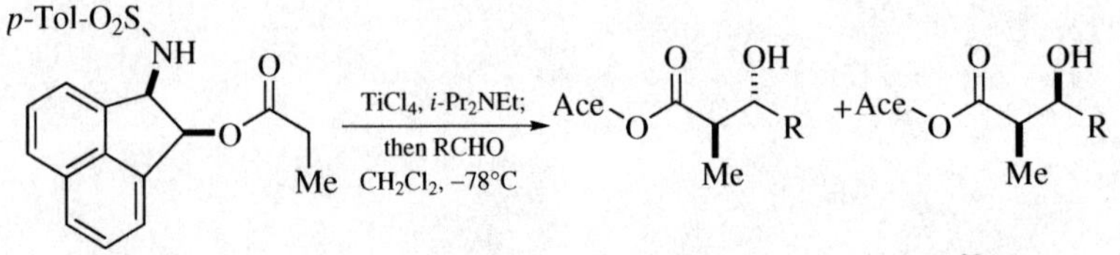

48 49-*anti* 49-*syn*

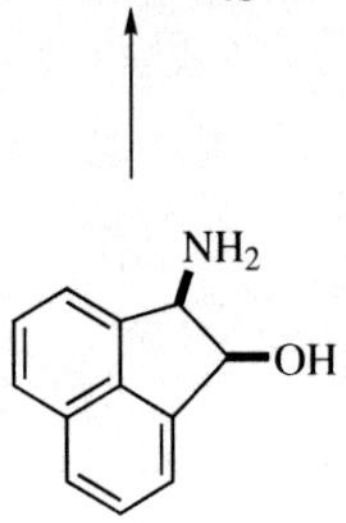

47

R	*anti/syn*	Yield, %
CH_3	80 : 20	71
C_2H_5	92 : 8	92
$BnCH_2CH_2$	95 : 5	97
i-C_3H_7	96 : 4	95
c-C_6H_{11}	99 : 1	84
C_6H_5	93 : 7	93

Scheme 2.2s

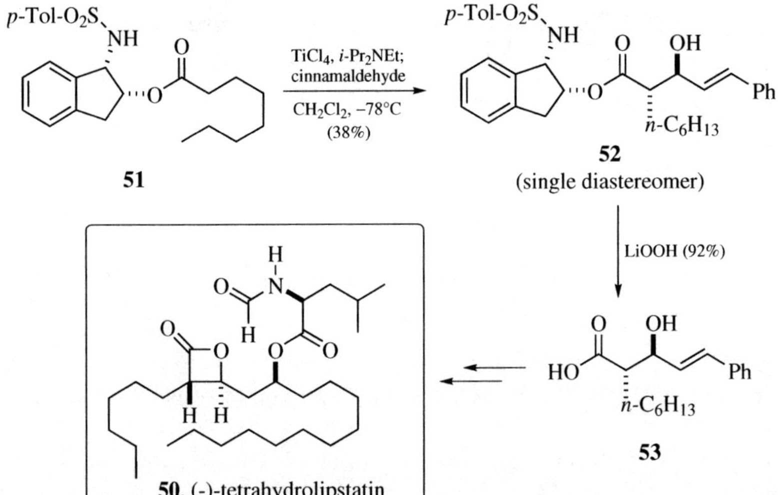

51

52 (single diastereomer)

53

50, (-)-tetrahydrolipstatin

Scheme 2.2t

marketed in several countries as an antiobesity drug under the trade name Xenical. The key step of the synthesis is the diastereoselective *anti*-aldol reaction of the titanium enolate of the ester **51**. When the titanium enolate of **51** was reacted with *trans*-cinnamaldehyde, an *anti*-aldol product (**52**) was obtained as a single diastereomer in 38% yield. Hydrolytic cleavage of the chiral auxiliary with LiOOH then provided the acid **53**.

REFERENCES

1. Masamune, S.; Sato, T.; Kim, B.; Wollmann, T. A. *J. Am. Chem. Soc.* **1986**, *108*, 8279.
2. Abiko, A.; Liu, J.-F.; Masamune, S. *J. Am. Chem. Soc.* **1997**, *119*, 2586.
3. Andrus, M. B.; Sekhar, B. B. V. S.; Turner, T. M.; Meredith, E. L. *Tetrahedron Lett.* **2001**, *42*, 7197.
4. (a) Myers, A. G.; Yang, B. H.; Chen, H.; McKinstry, L.; Kopecky, D. J.; Gleason, J. L. *J. Am. Chem. Soc.* **1997**, *119*, 6496; (b) Myers, A. G.; Yang, B. H.; Chen, H.; Gleason, J. L. *J. Am. Chem. Soc.* **1994**, *116*, 9361.
5. Yoshimitsu, T.; Song, J. J.; Wang, G.-Q.; Masamune, S. *J. Org. Chem.* **1997**, *62*, 8978.
6. Amarasinghe, K. K. D.; Montgomery, J. *J. Am. Chem. Soc.* **2002**, *124*, 9366.
7. Lafontaine, J. A.; Provencal, D. P.; Gardelli, C.; Leahy, J. W. *Tetrahedron Lett.* **1999**, *40*, 4145.
8. Walker, M. A.; Heathcock, C. H. *J. Org. Chem.* **1991**, *56*, 5747.
9. (a) Mahrwald, R. *Chem. Rev.* **1999**, *99*, 1095; (b) Gennari, C. in *Stereoselectivities in Lewis Acid Promoted Reactions*; Schinzer, D., Ed.; Kluwer Academic Publishers: Dordrecht, the Netherlands, **1989**; Chap. 4; (c) Murata, S.; Suzuki, M.; Noyori, R. *J. Am. Chem. Soc.* **1980**, *102*, 3248; (d) Mulzer, J.; Bruntrup, G.; Finke, J.; Zippel, M. *J. Am. Chem. Soc.* **1979**, *101*, 7723.
10. Evans, D. A.; Tedrow, J. S.; Shaw, J. T.; Downey, C. W. *J. Am. Chem. Soc.* **2002**, *124*, 392.
11. Evans, D. A.; Downey, C. W.; Shaw, J. T.; Tedrow, J. S. *Org. Lett.* **2002**, *4*, 1127.
12. (a) Nagao, Y.; Yamada, S,; Kumagai, T.; Ochiai, M.; Fujita, E. *J. Chem. Soc., Chem. Commun.* **1985**, 1418; (b) Crimmins, M. T.; Chaudhary, K. *Org. Lett.* **2000**, *2*, 775.
13. Crimmins, M. T.; King, B. W.; Tabet, E. A. *J. Am. Chem. Soc.* **1997**, *119*, 7883.
14. Gaul, C.; Njardarson, J. T.; Danishefsky, S. J. *J. Am. Chem. Soc.* **2003**, *125*, 6042.
15. Evans, D. A.; Trenkle, W. C.; Zhang, J.; Burch, J. D. *Org. Lett.* **2005**, *7*, 3335.
16. Olivo, H. F.; Tovar-Miranda, R.; Barragan, E. *J. Org. Chem.* **2006**, *71*, 3287.
17. Griffith, W. P.; Ley, S. V.; Whitcombe, G. P.; White, A. D. *J. Chem. Soc., Chem. Commun.* **1987**, 1625.
18. Esumi, T.; Hojyo, D.; Zhai, H.; Fukuyama, Y. *Tetrahedron Lett.* **2006**, *47*, 3979.
19. (a) Ghosh, A. K.; Onishi, M. *J. Am. Chem. Soc.* **1996**, *118*, 2527; (b) Ghosh, A. K.; Kim, J.-H. *Org. Lett.* **2003**, *5*, 1063.
20. Ghosh, A. K.; Fidanze, S. *Org. Lett.* **2000**, *2*, 2405.

2.3. Proline-Catalyzed Asymmetric Aldol Reaction

An approach has recently been made in which asymmetric aldol reactions are performed without the need for preformed metal enolates.[1] In 2000, List and co-workers reported that the cyclic amino acid L-proline is an effective catalyst for the asymmetric aldol reaction of acetone with a variety of aromatic and aliphatic aldehydes[2] (Scheme 2.3a). When L-proline was mixed with acetone

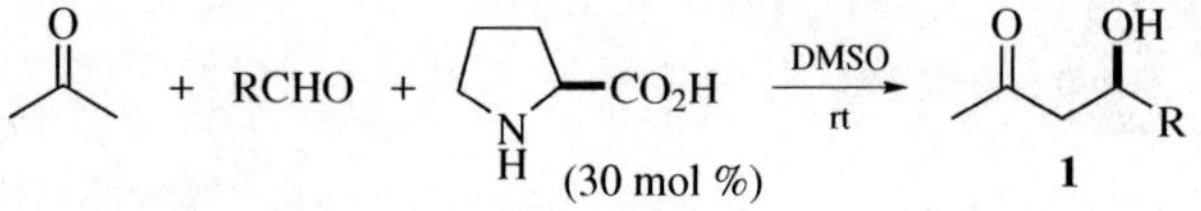

ALDOL REACTIONS

Scheme 2.3a

Product	Yield, %	ee, %	Product	Yield, %	ee, %
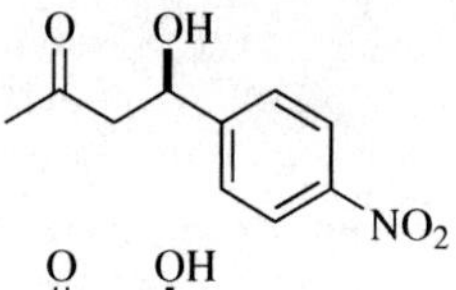 4-NO$_2$-C$_6$H$_4$ aldol	68	76	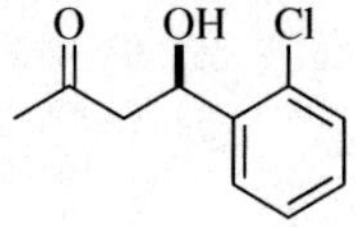 2-Cl-C$_6$H$_4$ aldol	94	69
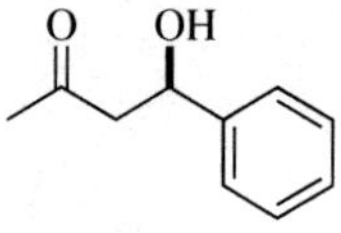 C$_6$H$_5$ aldol	62	60	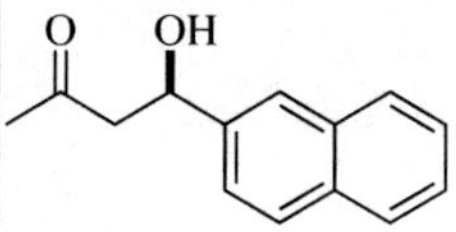 2-naphthyl aldol	54	77
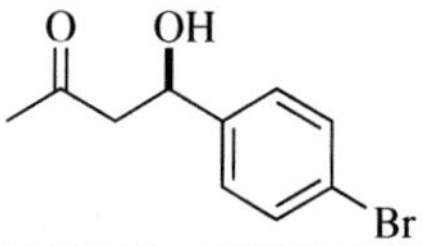 4-Br-C$_6$H$_4$ aldol	74	65	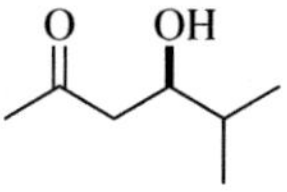 isopropyl aldol	97	96

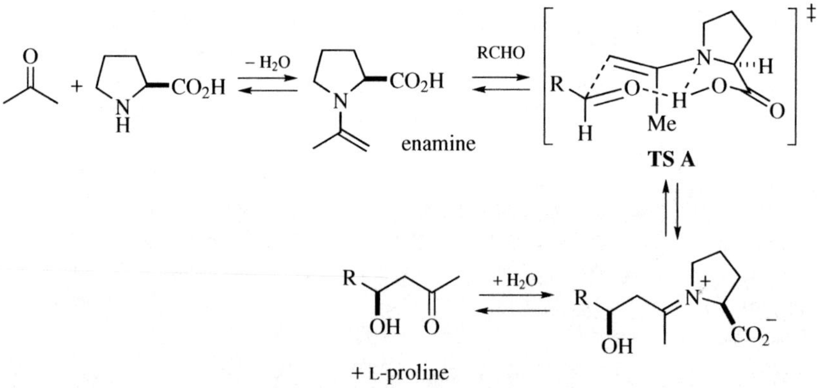

Scheme 2.3b

and 4-nitrobenzaldehyde in anhydrous dimethylsulfoxide (DMSO) solvent, the aldol product was obtained in 68% yield and 76% ee.

Other aromatic aldehydes provided products with similar enantiomeric excess. Although α-unbranched aldehydes such as pentanal did not yield any significant amount of the desired aldol products, the reaction of isobutyraldehyde gave the corresponding aldol product in 97% yield and 96% ee. The reaction is considered to proceed via an enamine mechanism. The enantioselectivity of the reaction can be explained in terms of a metal-free version of a six-membered transition state

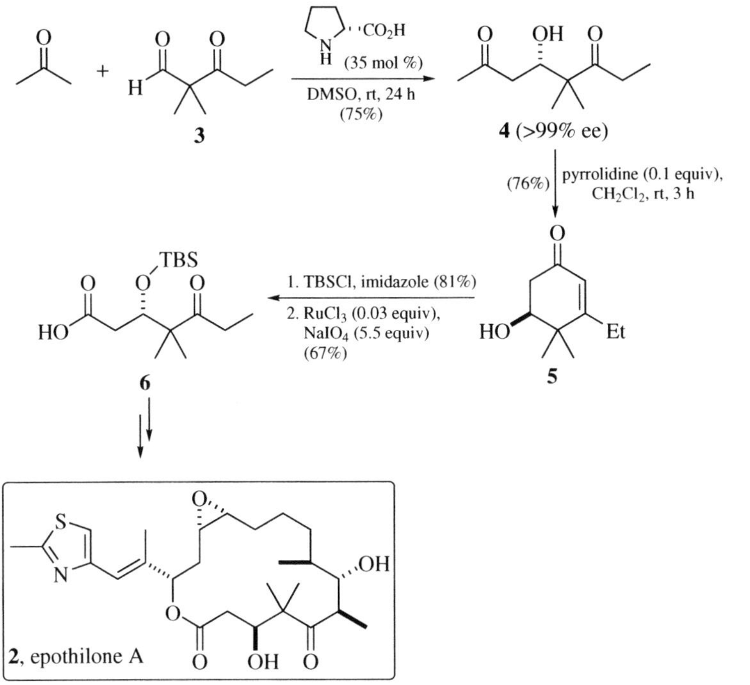

Scheme 2.3c

in which the tricyclic hydrogen-bonded framework provides a solid ground for enantiofacial selectivity[3] (Scheme 2.3b).

Epothilone A (**2**) is a natural product that exhibits taxoterelike anticancer activity. A new synthesis of the ketoacid **6**, a common C_1-C_6 fragment used in the total synthesis of epothilone A, was accomplished by directed aldol reaction of acetone with the aldehyde **3**[4] (Scheme 2.3c). The aldol reaction of acetone with the aldehyde **3** in the presence of D-proline proceeded smoothly to furnish the expected aldol product (**4**) in 75% yield and with greater than 99% ee. Intramolecular aldol reaction of the hydroxy ketone **4** in the presence of pyrrolidine gave the cyclohexenone **5** in good yield. Protection of the alcohol as a TBS ether followed by oxidation of the alkene then produced the desired ketoacid (**6**).

Not only does acetone undergo a highly enantioselective aldol reaction, but hydroxy acetone exhibits excellent stereoselectivity to produce the *anti*-aldol products **7**[5] (Scheme 2.3d). For example, L-proline catalyzed the aldol reaction between hydroxy acetone and cyclohexanecarbaldehyde to furnish the *anti*-diol in 60% yield with a greater than 20 : 1 diastereomeric ratio. The enantiofacial selectivity of the *anti*-isomer was higher than > 99%. Diastereoselectivities are very high with α-substituted aldehydes, whereas low selectivities are recorded in reactions with aromatic aldehydes and with α-unsubstituted aliphatic aldehydes. It is noteworthy that the levels of enantiofacial selectivity for the *anti*-aldol products

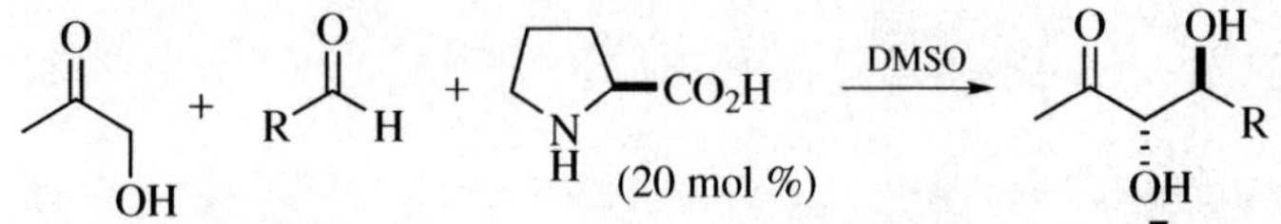

Scheme 2.3d

Product	ee, % (anti/syn)	Yield, %	Product	ee, % (anti/syn)	Yield, %
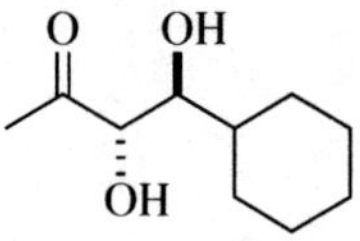 cyclohexyl anti-diol	>99 (>20:1)	60	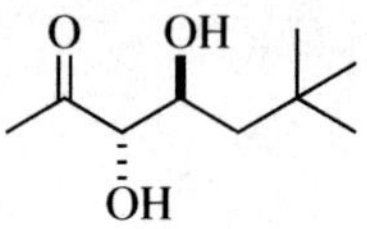 tert-butyl anti-diol	>97 (1.7:1)	38
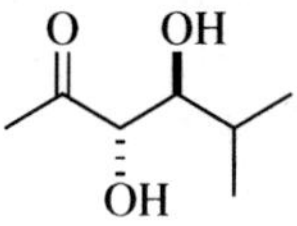 isopropyl anti-diol	>99 (>20:1)	62	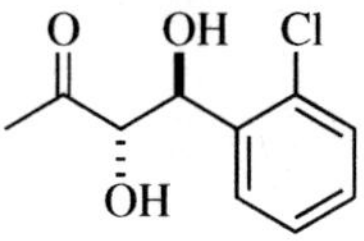 2-chlorophenyl anti-diol	67 (1.5:1)	95
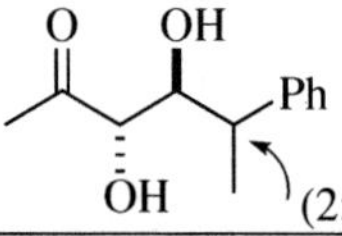 benzyl anti-diol	>95 (>20:1) (2:1)	51	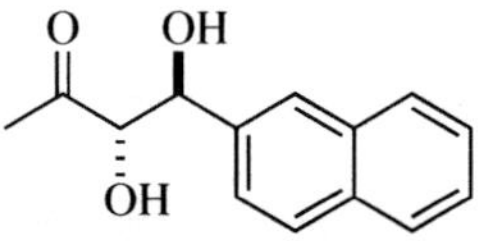 2-naphthyl anti-diol	79 (3:1)	62

Scheme 2.3d

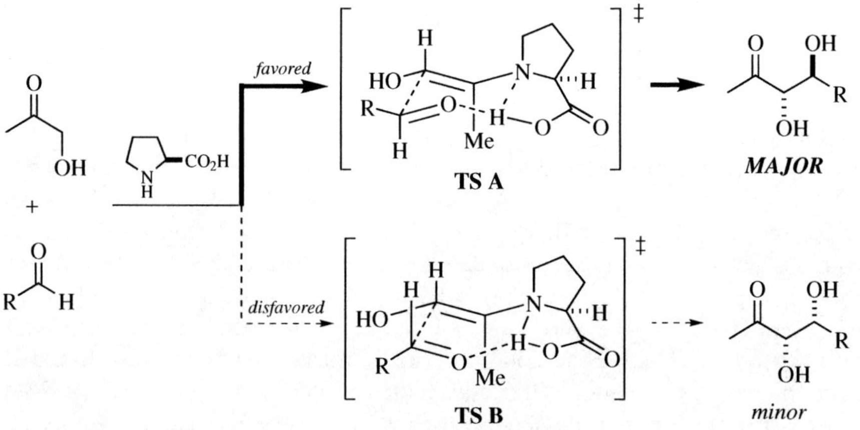

Scheme 2.3e

from both α-substituted and α-unsubstituted aliphatic aldehydes are exceptionally high.

Based on the presumption that the enamine double bond would possess an (E)-configuration, the diastereofacial selectivity can be explained by comparing two potential transition states **A** and **B**[5] (Scheme 2.3e). Thus, *anti*-diol products

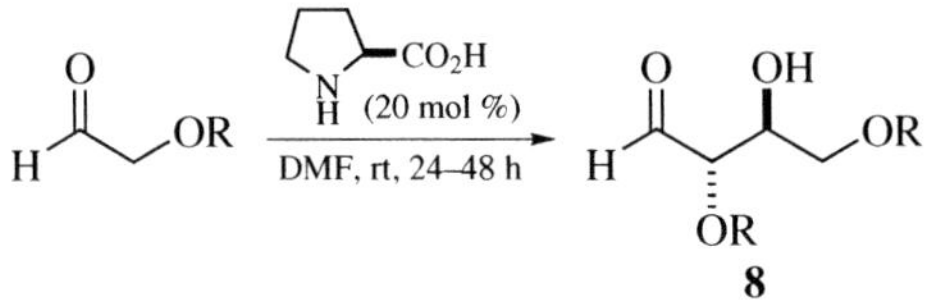

Scheme 2.3f

Product	ee, % (anti/syn)	Yield, %
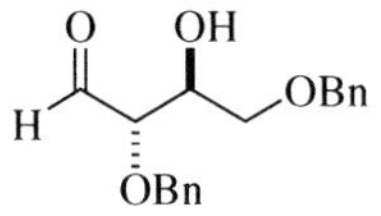 (H-CO-CH(OBn)-CH(OH)-CH2-OBn)	98 (4:1)	73
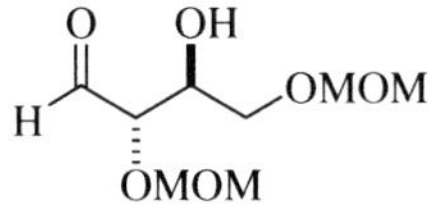 (H-CO-CH(OMOM)-CH(OH)-CH2-OMOM)	96 (4:1)	42
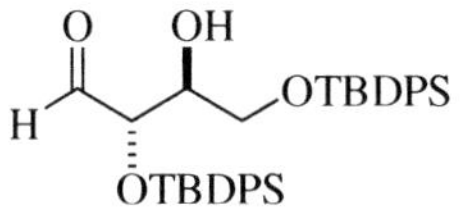 (H-CO-CH(OTBDPS)-CH(OH)-CH2-OTBDPS)	96 (9:1)	61

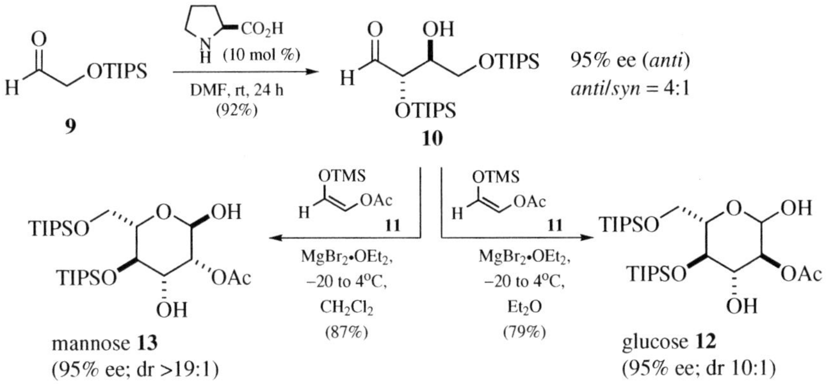

Scheme 2.3g

are formed via a six-membered chairlike transition state **A** in which the alkyl (R) group occupies an equatorial position. The formation of *syn*-diol products can be rationalized by a boatlike transition state **B**, where the facial selectivity of the aldehyde is reversed.

When α-alkoxyaldehyde substrates were subjected to organocatalytic conditions, a highly enantioselective aldol dimerization reaction occurred[6] (Scheme 2.3f). Substrates bearing relatively electron-rich alkoxy groups provide dimers with synthetically useful levels of enantioselectivity and reactivity. Moreover, the aldehyde with bulky α-silyloxy substituent can readily be utilized to produce *anti*-diol with moderate levels of diastereoselectivity.

Northrup and MacMillan reported an elegant two-step carbohydrate synthesis using the method described above[7] (Scheme 2.3g). The first step of the synthesis was an L-proline-catalyzed enantioselective dimerization of triisopropylsilyl (TIPS)-protected α-hydroxyaldehyde (**9**), which afforded α,γ-oxy-protected L-erythrose (**10**) in excellent yield and with high enantioselectivity. A Mukaiyama aldol reaction in ether between the TIPS-protected aldehyde **10** and the α-acetoxy enolsilane **11** in the presence of $MgBr_2 \cdot OEt_2$ afforded the glucose **12**, whereas the analogous reaction in dichloromethane provided the mannose **13** with high selectivity.

REFERENCES

1. For reviews, see (a) Gröger, H.; Wilken, J. *Angew. Chem. Int. Ed*. **2001**, *40*, 529. (b) List, B. *Chem. Commun*., **2006**, 819.
2. List, B.; Lerner, R. A.; Barbas, C. F., III. *J. Am. Chem. Soc*. **2000**, *122*, 2395.
3. Bahmanyar, S.; Houk, K. N. *J. Am. Chem. Soc*. **2001**, *123*, 11273.
4. Zheng, Y.; Avery, M. A. *Tetrahedron* **2004**, *60*, 2091.
5. Sakthivel, K.; Notz, W.; Bui, T.; Barbas, C. F., III. *J. Am. Chem. Soc*. **2001**, *123*, 5260.
6. Northrup, A. B.; Mangion, I. K.; Hettche, F.; MacMillan, D. W. C. *Angew. Chem. Int. Ed*. **2004**, *43*, 2152.
7. Northrup, A. B.; MacMillan, D. W. C. *Science* **2004**, *305*, 1752.

3 Metal Allylation Reactions

GENERAL CONSIDERATIONS

The addition of allylic organometallic reagents to carbonyl compounds gives homoallylic alcohols[1] (Scheme 3.I). The homoallylic alcohol products are often obtained with high stereoselectivity and can be further manipulated to access synthetically valuable intermediates such as β-hydroxy carbonyl compounds. As a result, the addition reaction of allylic metal reagents to carbonyl compounds has been a subject of great interest to synthetic organic chemists. Over the last three decades, numerous allylic boron and silicon reagents have been invented that participate in highly stereoselective syntheses of homoallylic alcohols.[2] The boron and silicon reagents are popular not only because they afford products with high stereoselectivity, but also because the stereochemical outcome of the addition reaction is predictable. Due to the high Lewis acidity of boron, the allylic boron reagents tend to undergo addition to aldehydes and ketones through a six-membered cyclic transition state. Unlike their boron counterparts, only certain types of allylic silicon reagents can have a closed six-membered chairlike transition state, due to the low Lewis acidity of silicon.

Allylic boron reagents have allowed organic chemists a great deal of control over stereochemistry in carbon–carbon bond-forming reactions. Hoffmann et al. discovered that crotylboronates and their derivatives add to aldehydes with high diastereoselectivity.[3] For example, when the crotylboronates **1E** or **1Z** were treated with an equimolar amount of aldehydes at −78°C, the homoallylic alcohols **2** were obtained in almost quantitative yields after workup. More significantly, the diastereoselectivity of the reaction was always greater than 93:7, with *E*-boronate **1E** affording *anti*-isomer and *Z*-boronate **1Z** yielding *syn*-isomer[4] (Scheme 3.II).

In a related study, Hoffmann and Kemper also synthesized the γ-methoxyallylboronates **3E** and **3Z** and performed addition reactions into a variety of aldehydes to explore the possibility of a stereoselective synthesis of vicinal diols[5] (Scheme 3.III). Once again, the reaction is highly diastereoselective, affording an *anti*-isomer from the *E*-boronate and a *syn*-isomer from the *Z*-boronate. The nearly perfect diastereoselectivity observed in the reaction of crotylboronates

---

Six-Membered Transition States in Organic Synthesis, By Jaemoon Yang
Copyright © 2008 John Wiley & Sons, Inc.

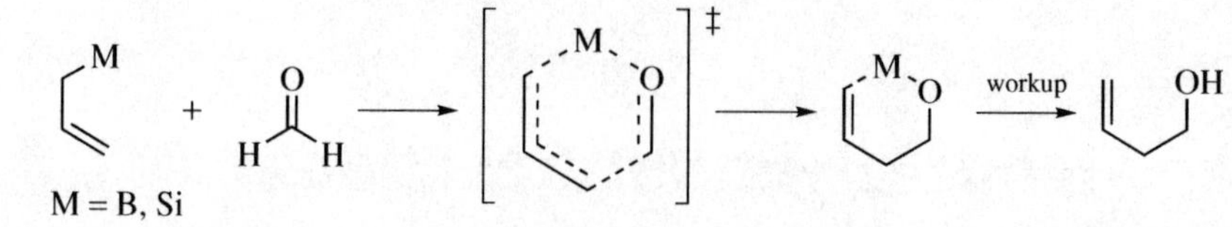

Scheme 3.I

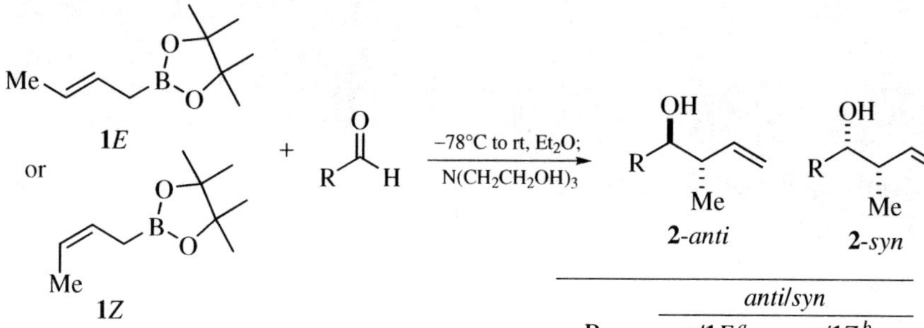

Scheme 3.II

R	anti/syn w/1E [a]	anti/syn w/1Z [b]
Ph	94:6	4:96
Me	93:7	3:97
Et	93:7	3:97
i-Pr	96:4	6:94

[a] E/Z = 93:7.
[b] E/Z = <5:>95.

with aldehydes is explained in terms of a six-membered chairlike transition state[4] (Scheme 3.IV). In the favored six-membered chairlike transition state **A**, the residue R of the aldehyde prefers to occupy an equatorial position.

To explain the stereochemical outcome of the reaction of allylic boron reagents with carbonyl compounds, Houk and Li carried out calculations on the transition structures of the model reaction of formaldehyde and allylboronic acid[6] (Scheme 3.V). The bimolecular complex formed initially between allylboronic acid and formaldehyde would rearrange via a six-membered transition state to form an intermediate. Calculations show that chair transition state **A** is 8.2 kcal/mol more stable than twist-boat transition structure **B**, clearly confirming that the six-membered chairlike transition-state model is a legitimate scheme to predict the stereochemical outcome of the boron allylation reaction.

Houk and Li also performed calculations on the reaction between allylboronic acid and acetaldehyde[6] (Scheme 3.VI). Transition state **A**, in which the methyl group of acetaldehyde occupies an equatorial position is more stable than transition state **B** by 5.5 kcal/mol. Thus, the theoretical studies support the transition-state models in Scheme 3.IV proposed by Hoffmann and others.

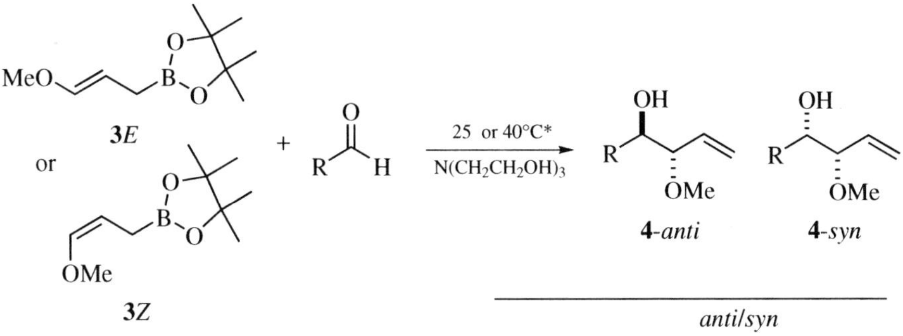

GENERAL CONSIDERATIONS

Scheme 3.III

R	anti/syn w/3E[a]	anti/syn w/3Z[b]
Ph	95:5	<5:>95
Me	95:5	7:93
Et	93:7	8:92
i-Pr	95:5	11:89

[a] $E/Z = 89:11$.
[b] $E/Z = <5:>95$.

* 25°C with **3E** and 40°C with **3Z**.

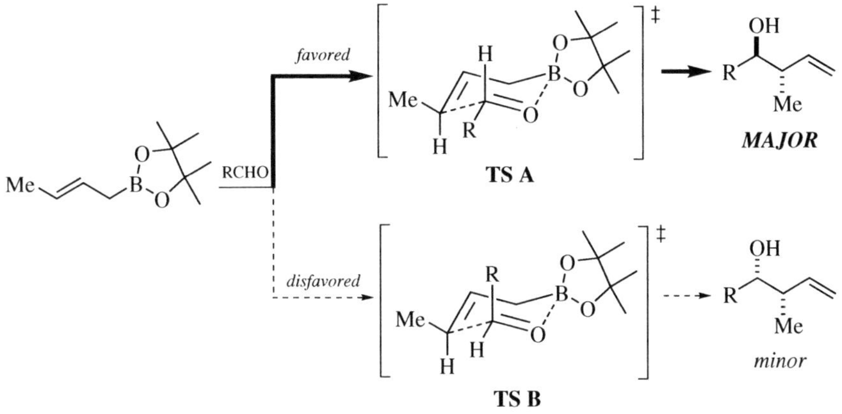

Scheme 3.IV

An interesting diastereoselectivity pattern was observed when α-halogen-substituted allylboronates were added to aldehydes. In this reaction, (Z)-alkenes were obtained as the major products[7] (Scheme 3.VII). Hoffmann and Landmann explained the results by examining two competing six-membered chairlike transition states (Scheme 3.VIII). Among the possible factors that favor the transition state **A**, they pointed out that dipole–dipole interactions could play a dominant

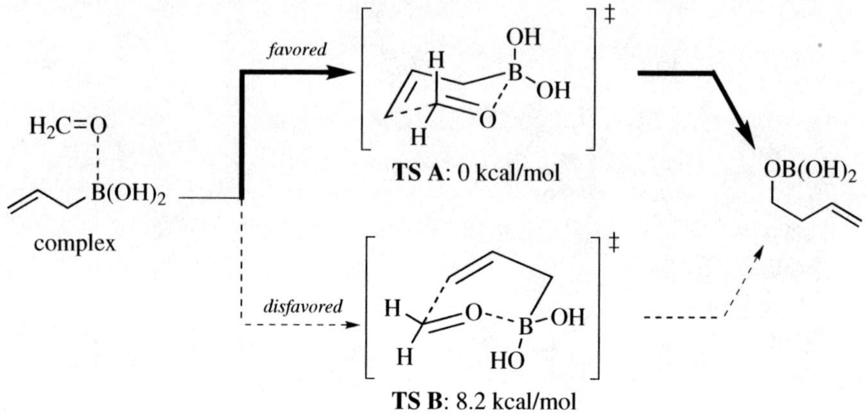

Scheme 3.V

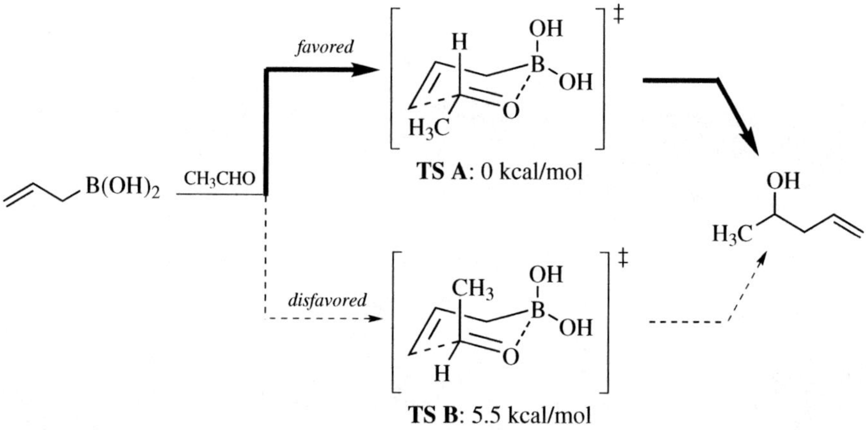

Scheme 3.VI

role: The dipoles of the C–X and the B–O bonds in transition state **A** run opposite to each other, to render it a minimum net dipole moment.[8]

Theoretical support was obtained to explain the experimental results observed with α-chloro- or α-bromo-substituted pinacolboronates (**5**)[6] (Scheme 3.IX). When calculations were performed on the reaction between the α-fluoro-substituted allylboronic acid and formaldehyde, transition state **A**, in which fluorine atom occupies an axial position, was found to be more stable than transition state **B** by 3.5 kcal/mol.

In the next two sections we examine a number of stereoselective addition reactions of boron and silicon allylic reagents to carbonyl compounds.

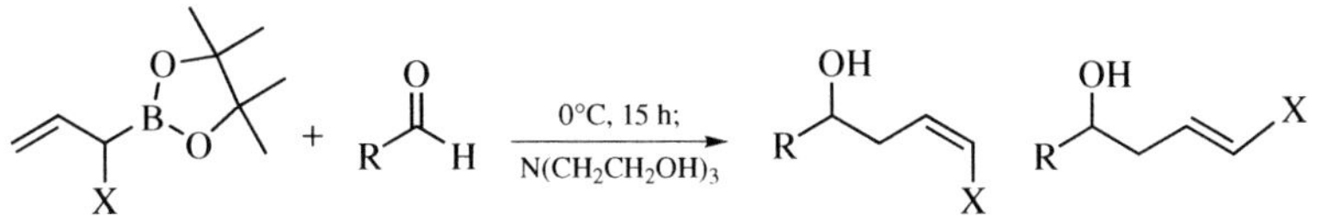

GENERAL CONSIDERATIONS 101

Scheme 3.VII

	X = Cl		X = Br	
R	6a/6b	Yield, %	6a/6b	Yield, %
Ph	95:5	82	97:3	80
Me	93:7	63	93:7	78
Et	94:6	86	94:6	82
i-Pr	96:4	83	96:4	83

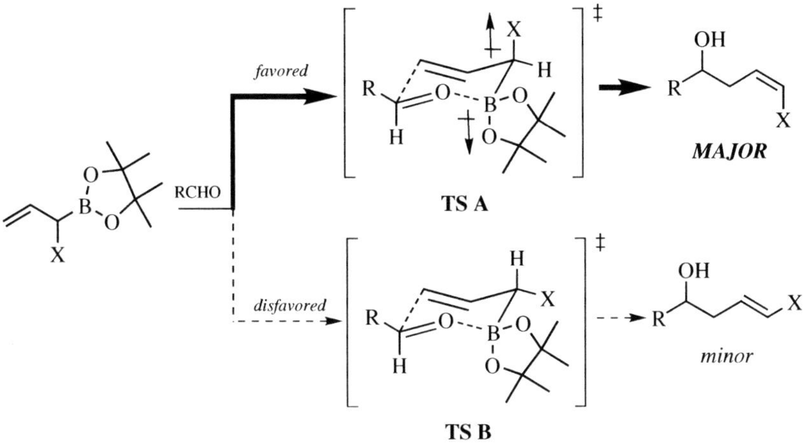

Scheme 3.VIII

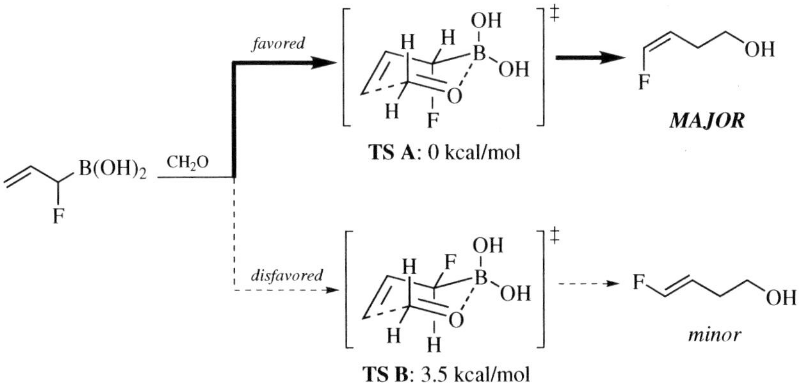

Scheme 3.IX

REFERENCES

1. (a) Bartlett, P. A. *Tetrahedron* **1980**, *36*, 3; (b) Hoffmann, R. W. *Angew. Chem. Int. Ed*. **1982**, *21*, 555; (c) Yamamoto, Y.; Asao, N. *Chem. Rev*. **1993**, *93*, 2207.
2. Kennedy, J. W. J.; Hall, D. G. *Angew. Chem. Int. Ed*. **2003**, *42*, 4732.
3. Hoffmann, R. W.; Niel, G.; Schlapbach, A. *Pure Appl. Chem*. **1990**, *62*, 1993.
4. (a) Hoffmann, R. W.; Zeiβ, H.-J. *J. Org. Chem*. **1981**, *46*, 1309; (b) Hoffmann, R. W.; Kemper, B. *Tetrahedron* **1984**, *40*, 2219.
5. (a) Hoffmann, R. W.; Kemper, B. *Tetrahedron Lett*. **1981**, *22*, 5263; (b) Hoffmann, R. W.; Kemper, B. *Tetrahedron Lett*. **1982**, *23*, 845.
6. Houk, K. N.; Li, Y. *J. Am. Chem. Soc*. **1989**, *111*, 1236.
7. Hoffmann, R. W.; Landmann, B. *Tetrahedron Lett*. **1983**, *24*, 3209.
8. Hoffmann, R. W.; Landmann, B. *Chem. Ber*. **1986**, *119*, 1039.

REACTIONS

3.1. Boron Allylation Reaction

In 1978, Herold and Hoffmann reported their finding that the chiral allylboronate **1**, derived from (+)-camphor, added to a variety of aldehydes to give the homoallylic alcohol **2** with moderate to good levels of enantioselectivity[1] (Scheme 3.1a). Presumably, the reaction proceeds via a six-membered chairlike transition structure[2] (Scheme 3.1b). The major enantiomer would be formed through transition state **A**, where the allyl group is delivered on the *si*-face of the aldehyde, which is activated by an internal coordination of boron atom to the carbonyl oxygen. Transition state **B** would be disfavored over **A**, due to the possible steric

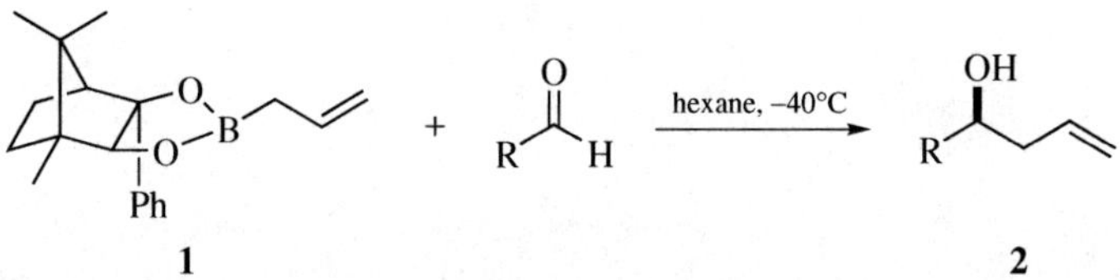

R	ee, %	Yield, %
CH_3	65	92
C_2H_5	77	91
$n\text{-}C_3H_7$	72	93
$i\text{-}C_3H_7$	70	88
$t\text{-}C_4H_9$	45	85
C_6H_5	36	90

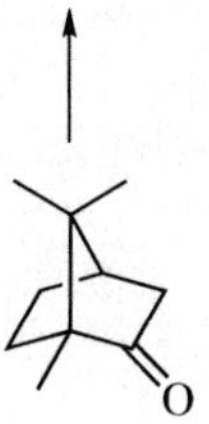

(+)-camphor

Scheme 3.1a

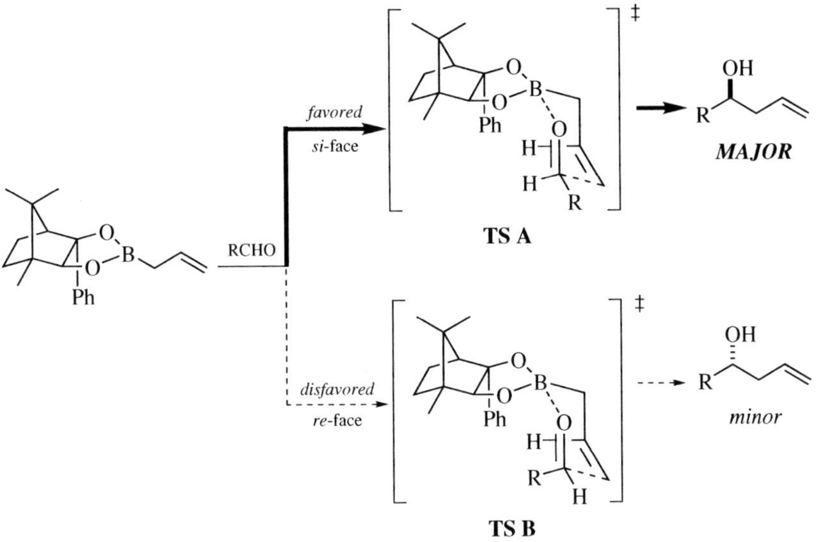

Scheme 3.1b

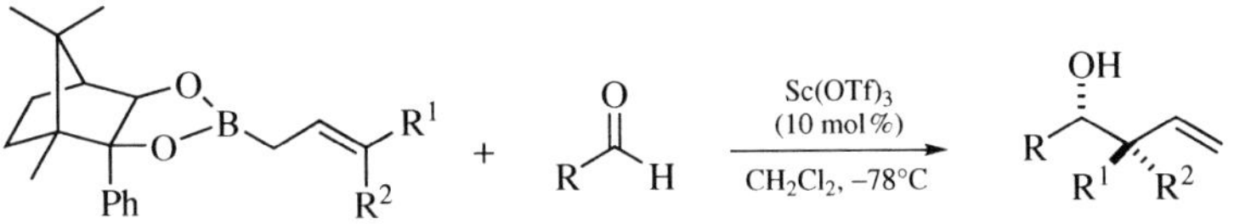

Scheme 3.1c

R^1, R^2	R	ee, %	Yield, %
H, H	PhCH$_2$CH$_2$	97	97
	TBDMSOCH$_2$	90	76
	C$_6$H$_5$	92	85
Me, H	PhCH$_2$CH$_2$	96	71
	TBDMSOCH$_2$	95	74
	C$_6$H$_5$	97	60
H, Me	PhCH$_2$CH$_2$	96	52
	TBDMSOCH$_2$	96	57
	C$_6$H$_5$	59	53

interactions between the alkyl (R) group of aldehyde and the hydrogen and phenyl substituents of the camphor frame. Computational study to support the foregoing hypothesis has yet to be done.

In 2003, Hall et al. made an impressive improvement in enantioselectivity by employing a scandium catalyst[3] (Scheme 3.1c). When a catalytic amount

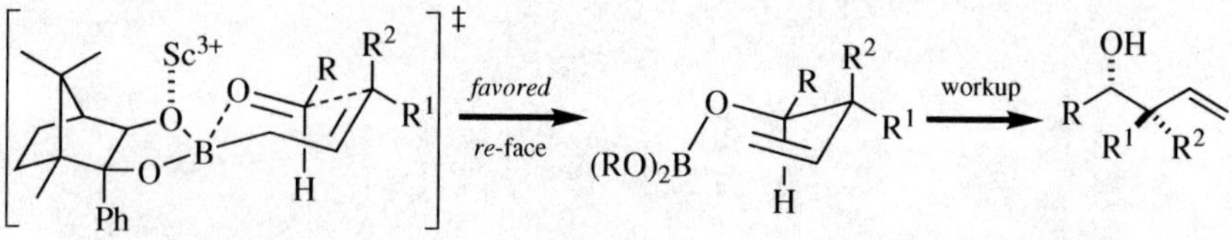
Sc3+
R2
R
R1
H
Ph
favored
re-face
(RO)2B
workup
OH

TS A

Scheme 3.1d

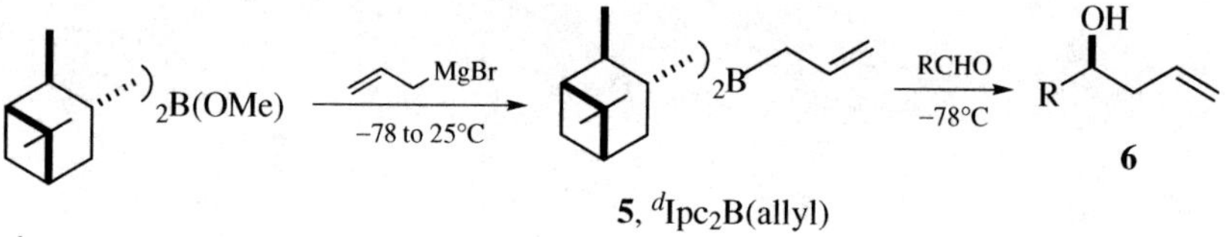
2B(OMe)
MgBr
−78 to 25°C
2B
RCHO
−78°C
OH
R
6
5, dIpc2B(allyl)

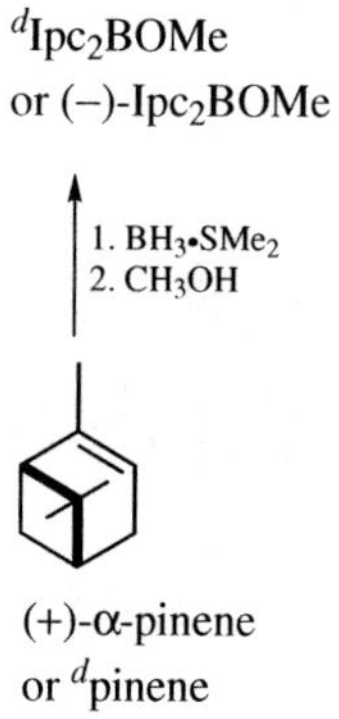
dIpc2BOMe
or (−)-Ipc2BOMe
1. BH3•SMe2
2. CH3OH
(+)-α-pinene
or dpinene

R	ee, %[a]	Yield, %
CH_3	93 (>99)	74
C_2H_5	86 (—)	71
n-C_3H_7	87 (96)	72
i-C_3H_7	90 (96)	86
t-C_4H_9	83 (>99)	88
C_6H_5	96 (96)	81

[a] The %ee values in parentheses were obtained when allylboration reaction was conducted. at −100°C after the removal of MgBr(OMe) salt.

Scheme 3.1e

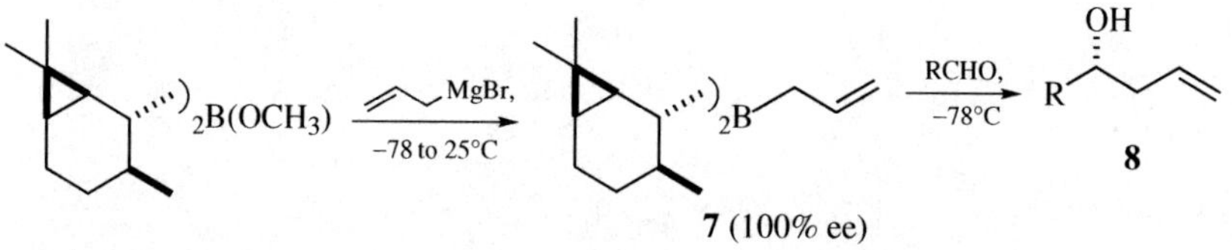
2B(OCH3)
MgBr,
−78 to 25°C
2B
RCHO,
−78°C
OH
R
8
7 (100% ee)

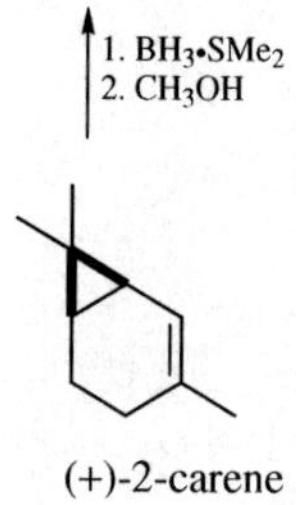
1. BH3•SMe2
2. CH3OH
(+)-2-carene

R	ee, %
CH_3	98
C_2H_5	94
n-C_3H_7	94
i-C_3H_7	94
t-C_4H_9	99
C_6H_5	95

Scheme 3.1f

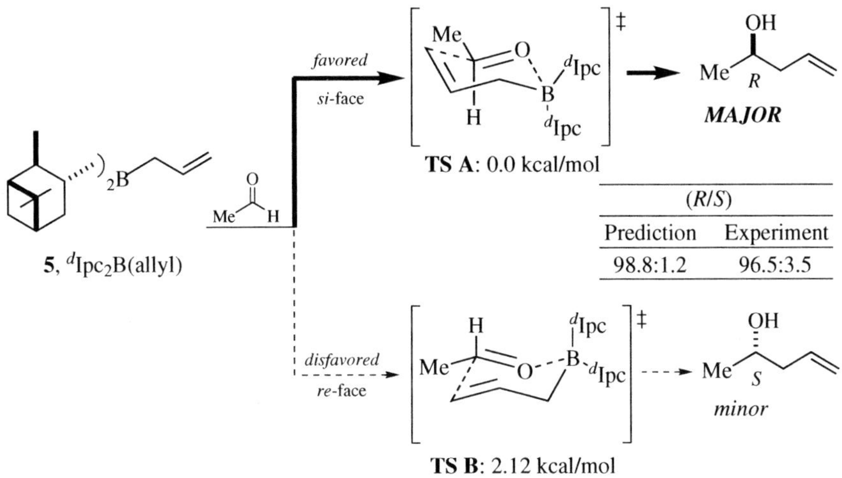

Scheme 3.1g

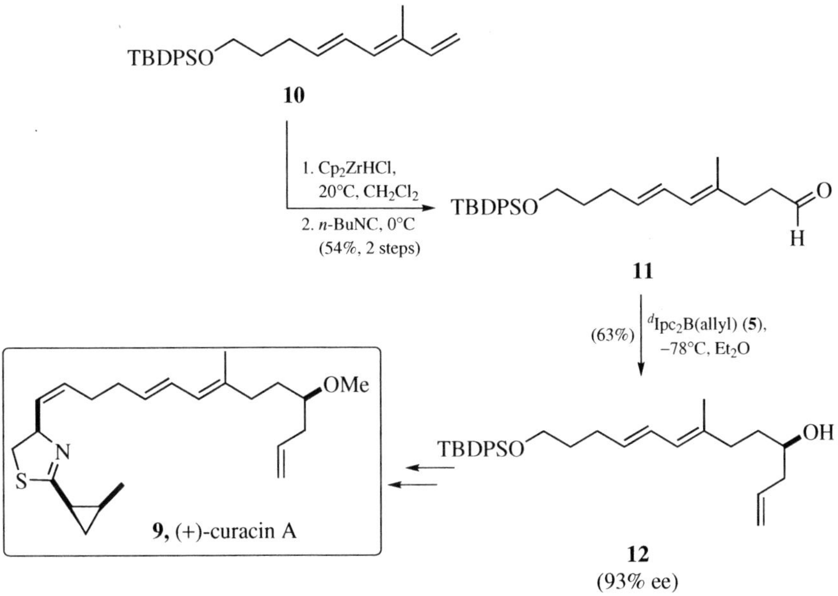

	(R/S)	
	Prediction	Experiment
	98.8:1.2	96.5:3.5

Scheme 3.1h

of the Lewis acid Sc(OTf)$_3$ is used in the allylboration reaction with Hoffmann's camphor-based reagent (**3**), there is not only a reaction rate increase but also a dramatic improvement in enantioselectivity. Both aliphatic and aromatic aldehydes are excellent substrates for allylation with allyl- and (*E*)-crotylboronates to afford

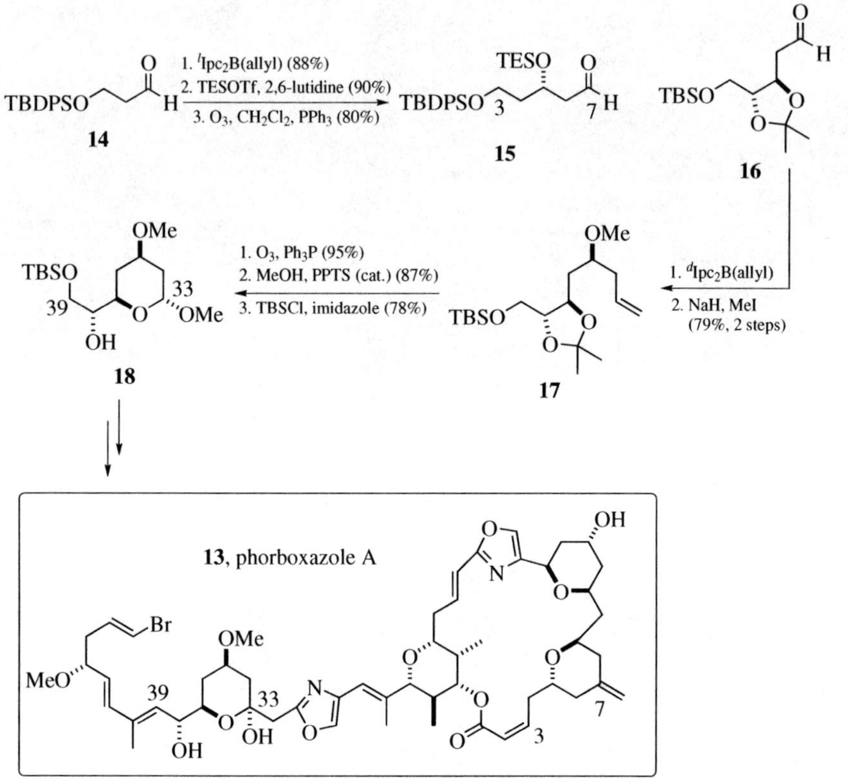

Scheme 3.1i

corresponding homoallylic alcohols of synthetically useful levels of enantioselectivity. The only system yet to be improved is the combination of Z-crotylboronate with benzaldehyde. The allylboration is proposed to proceed via a six-membered chairlike transition state in which the Lewis acidic scandium metal coordinates to the boronate oxygen[4] (Scheme 3.1d).

Five years after Hoffmann's disclosure of the enantioselective allylation reaction, Brown reported in 1983 a new chiral allylborane reagent, B-allyldiisopinocampheylborane [**5**; ^{d}Ipc$_2$B(allyl) or (+)-Ipc$_2$B(allyl)]. ^{d}Ipc$_2$B(allyl) is readily prepared in three steps from commercially available (+)-α-pinene (Scheme 3.1e). Thus, hydroboration of (+)-α-pinene (^{d}pinene) with borane–dimethyl sulfide followed by methanolysis of the Ipc$_2$BH gives B-methoxydiisopinocampheylborane [^{d}Ipc$_2$BOMe or (−)-Ipc$_2$BOMe]. The reaction of Ipc$_2$BOMe with allylmagnesium bromide then generates the allylborane reagent **5**. ^{d}Ipc$_2$B(allyl) has been utilized successfully for asymmetric carbon–carbon bond formation reaction to furnish secondary homoallylic alcohols (**6**) with optical purities in the range 83 to 96%[5] (Scheme 3.1e).

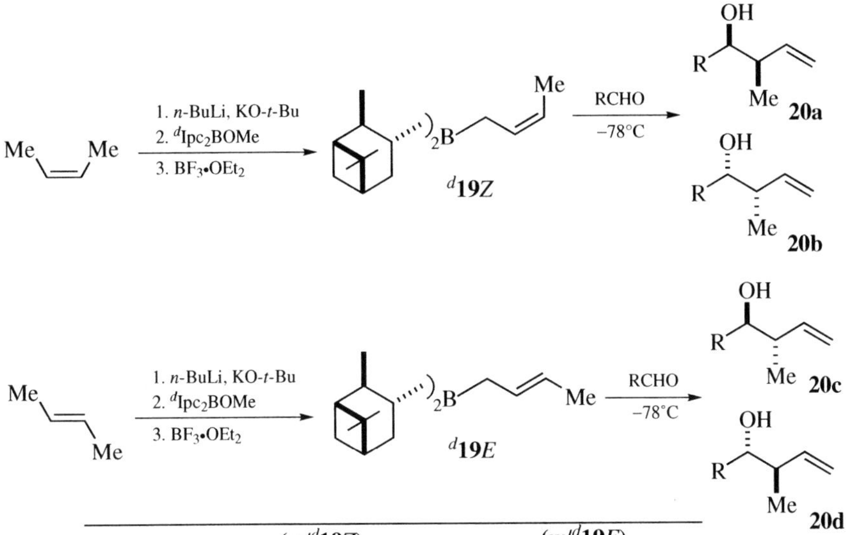

	(w/^{d}19Z)		(w/^{d}19E)	
R	20a/20b	Yield, %	20c/20d	Yield, %
CH$_3$	95:5	75	95:5	78
C$_2$H$_5$	95:5	70	95:5	70
CH$_2$=CH	95:5	63	95:5	65
C$_6$H$_5$	94:6	72	94:6	79

Scheme 3.1j

The reaction provides a uniformly high level of enantioselectivity regardless of the nature of the aldehydes used. The new chiral reagent, B-allyldiisopinocampheylborane, therefore has one significant advantage over Hoffmann's reagent, as even aromatic aldehydes are good substrates to access homoallylic alcohols with high enantioselectivity. Later, Racherla and Brown discovered that the enantioselectivity dramatically improved when the reaction was conducted at −100 °C after the removal of MgBr(OMe) salt.[6] An even more efficient reagent than B-allyldiisopinocampheylborane (**5**) is B-allylbis(2-isocaranyl)borane [**7**; 2-^{d}Icr$_2$B(allyl)], which undergoes a highly efficient asymmetric allylboration with a variety of aldehydes to afford the homoallylic alcohol **8** in 94 to 99% ee[7] (Scheme 3.1f).

Although Brown and co-workers proposed a six-membered transition state for the asymmetric allylboration reaction in which the aldedyde oxygen initially coordinates to boron followed by an internal transfer of the allyl group from boron to the carbonyl carbon,[8] a quantitative analysis to explain the enantioselectivity was not available until 1993, when Gennari et al. conducted a computational study to rationalize the enantiofacial selectivity of Brown allylation[9] (Scheme 3.1g). Calculation predicts that transition state **A**, in which the allyl group attacks the *si*-face of the aldehyde, is favored over transition state **B** by 2.12 kcal/mol.

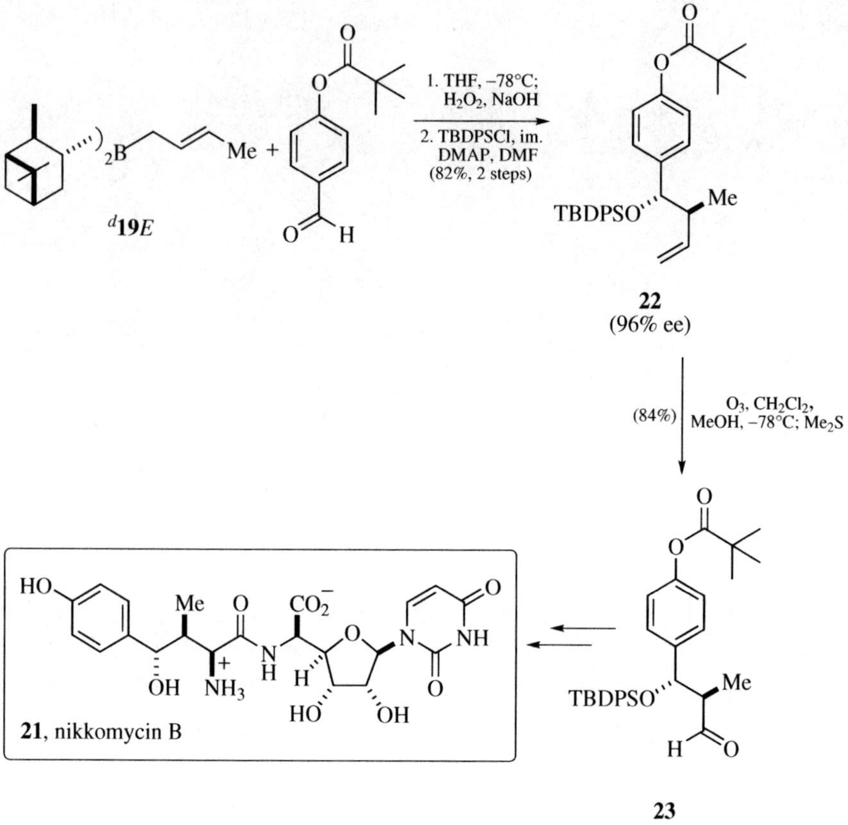

Scheme 3.1k

The enantioselectivity calculated from this energy difference agrees well with the experimental value.

The Brown allylation protocol was used successfully in the total synthesis of (+)-curacin A (**9**), a structurally novel antimitotic agent[10] (Scheme 3.1h). Hydrozirconation[11] of the triene **10** at the terminal vinyl group followed by quenching the zirconium intermediate with *n*-butyl isocyanide gave the aldehyde **11** after acidic aqueous workup. Asymmetric Brown allylation of **11** with ^{d}Ipc$_2$B(allyl) (**5**) gave the homoallylic alcohol **12** in 93% ee. In the total synthesis of phorboxazole A (**13**), one of the most potent cytotoxic natural products discovered to date, the Brown asymmetric allylation method, was employed repeatedly to access a variety of homoallylic alcohols[12] (Scheme 3.1i). Construction of the C_3–C_7 fragment began with Brown allylation of the aldehyde **14** with ^{l}Ipc$_2$B(allyl) to acquire the corresponding homoallylic alcohol, which was subsequently protected as a TES ether. Upon ozonolysis of the alkene, the aldehyde **15** was obtained in good yield. Another asymmetric Brown allylation was performed to construct the C_{33}–C_{39} subunit of phorboxazole A. Treatment

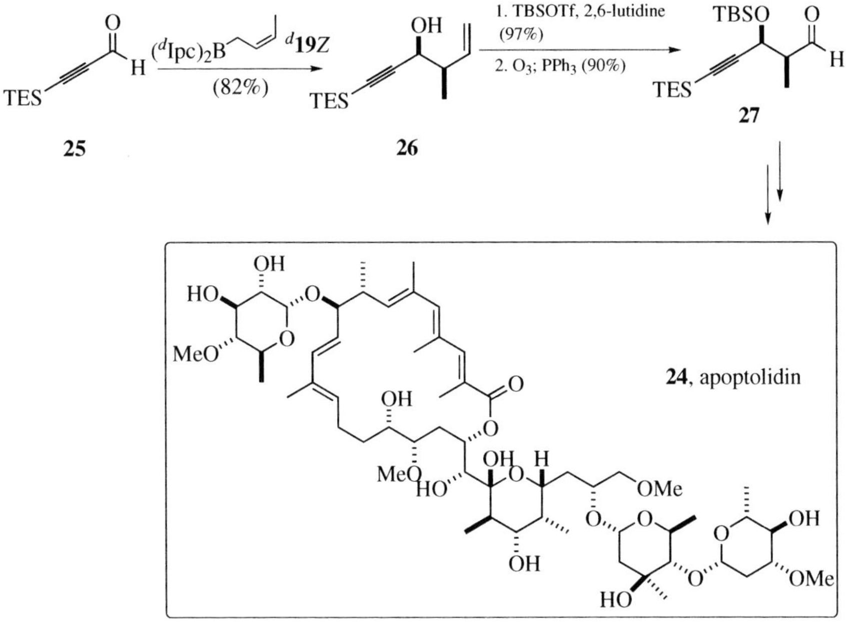

Scheme 3.1l

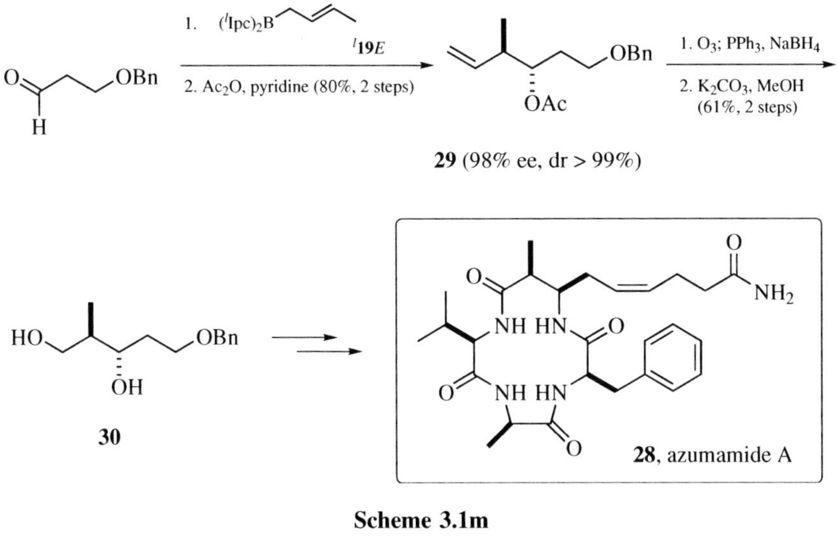

Scheme 3.1m

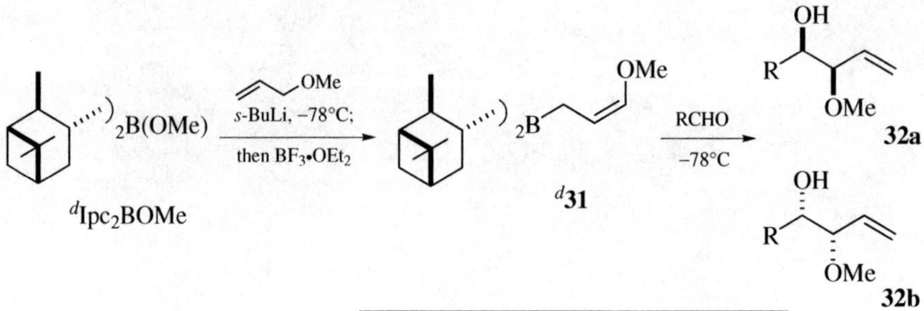

R	32a/32b	Yield, %
CH$_3$	95:5	57
C$_2$H$_5$	94:6	65
(CH$_3$)$_2$CH	94:6	57
C$_6$H$_5$	95:5	72
CH$_2$=CH	94:6	63

Scheme 3.1n

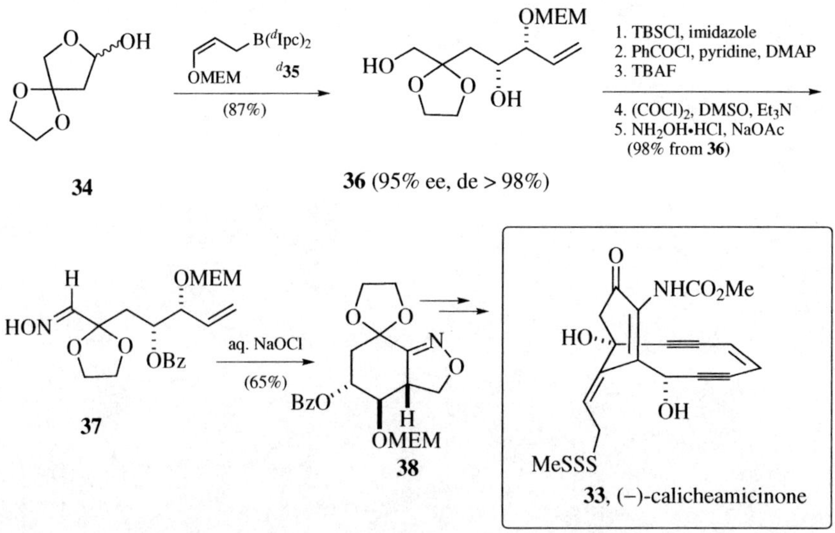

Scheme 3.1o

of the aldehyde **16** with ^{d}Ipc$_2$B(allyl) resulted in the corresponding homoallylic alcohol, which was then converted to its methyl ether (**17**). Cleavage of the double bond via ozonolysis, acidic methanolysis to generate the cyclic acetal, and reprotection of the primary hydroxyl group as a TBS ether afforded **18**.

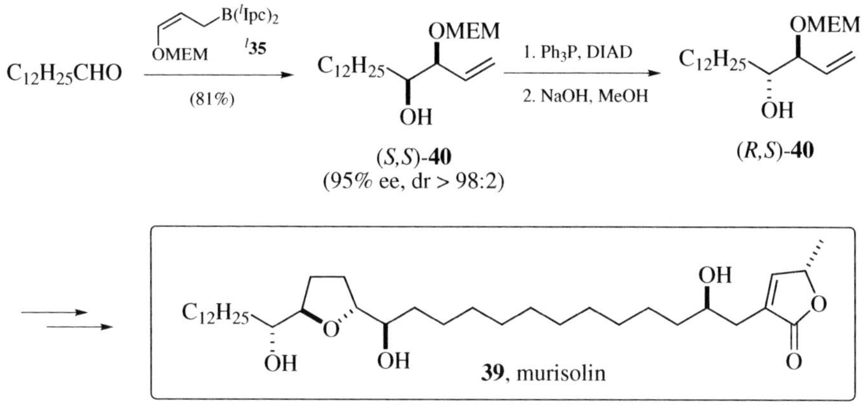

Scheme 3.1p

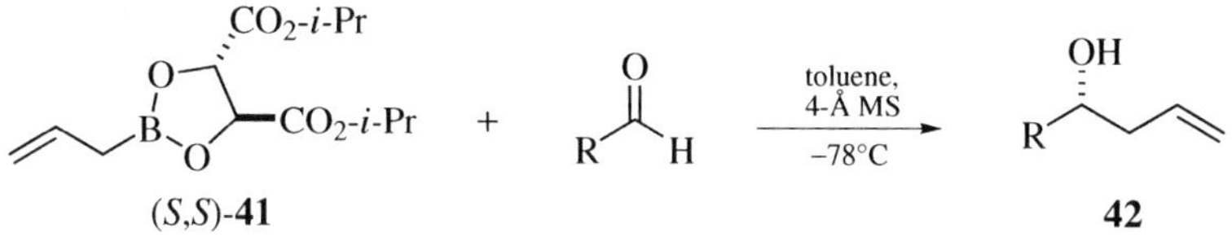

Scheme 3.1q

R	ee, %	Yield, %
$CH_3(CH_2)_8$	79	86
$c\text{-}C_6H_{11}$	87	72
$t\text{-}C_4H_9$	82	—
C_6H_5	71	78

Brown and Bhat also developed highly stereoselective crotylation reactions using Z- and E-crotyldiisopinocampheylborane reagents[8a,13] (Scheme 3.1j). The reagents are prepared from cis- and trans-2-butene, respectively. The 2-butenes are metalated with potassium tert-butoxide and n-butyllithium in THF at −45°C. Treatment of the resulting crotylpotassiums with B-methoxydiisopinocampheylborane at −78°C followed by boron trifluoride etherate affords the crotylborane reagents ^{d}19Z and ^{d}19E, respectively. Reaction of ^{d}19Z and ^{d}19E with aliphatic and aromatic aldehydes provides β-methylhomoallylic alcohols in good yields and with high enantioselectivity.

Barrett and Lebold used the Brown asymmetric crotylation to prepare the homoallylic alcohol 22 in the total synthesis of nikkomycin B 21, a natural product that exhibits fungicidal, insecticidal, and acaricidal activities[14] (Scheme 3.1k).

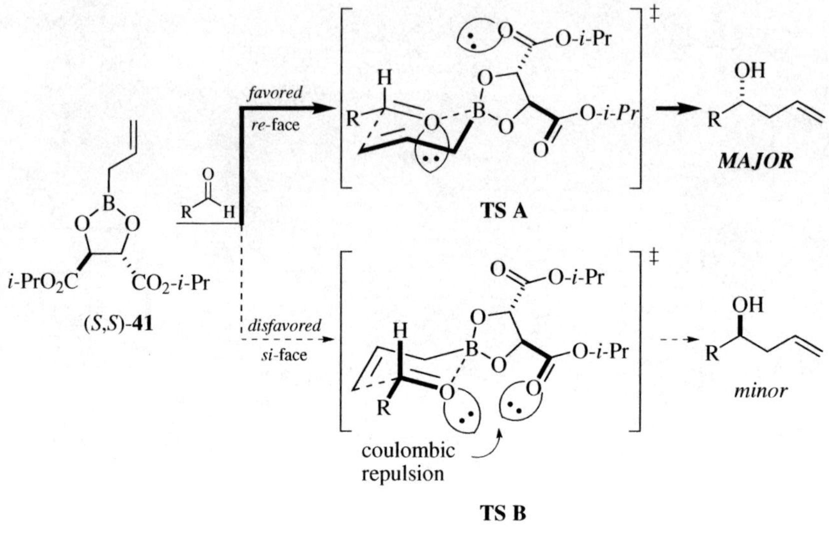

Scheme 3.1r

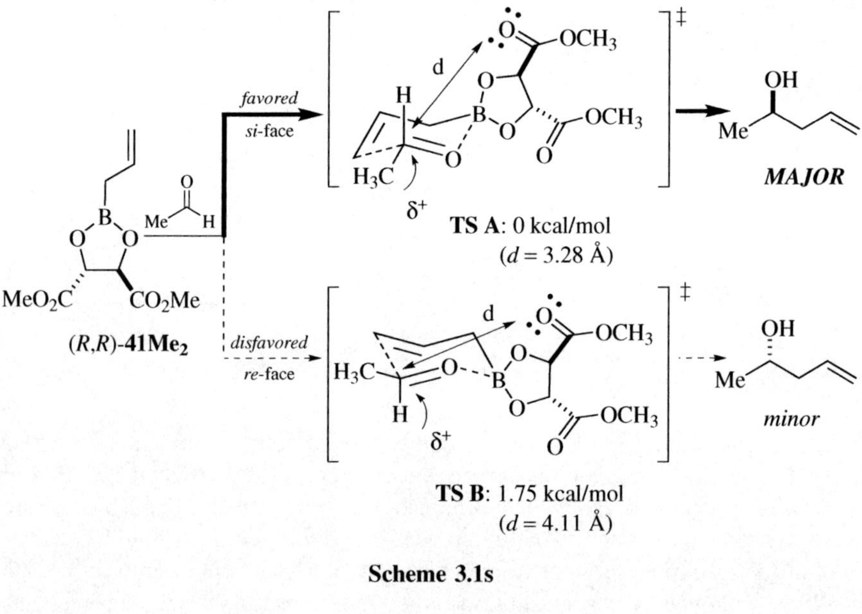

Scheme 3.1s

Reaction of 4-(pivaloyloxy)benzaldehyde with the E-crotyldiisopinocampheylborane ^{d}**19E** gave the corresponding homoallylic alcohol **22** in a 98 : 2 enantiomeric ratio. After protection of the alcohol as a *tert*-butyldiphenylsilyl (TBDPS) ether, the alkene was subjected to ozonolysis to provide the β-hydroxy aldehyde **23**.

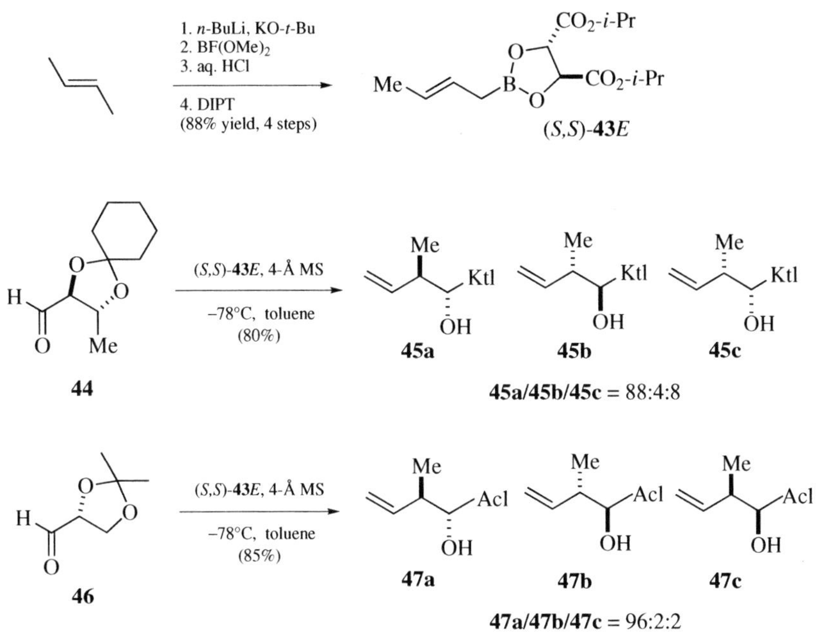

Scheme 3.1t

The Brown asymmetric crotylation method was utilized in the total synthesis of the apoptolidin **24**, an attractive synthetic target with many unique biological activities, including the selective induction of apoptosis in rat glia cells[15] (Scheme 3.1l). In this synthesis, the acetylenic aldehyde **25** was treated with (Z)-crotyl-^{d}Ipc$_2$borane (^{d}**19Z**) to afford the alcohol **26** selectively and in 82% yield. Protection of **26** as a TBS ether, followed by ozonolysis of the olefinic bond, gave the aldehyde **27** in high yield.

Brown's crotylboration protocol was used effectively in the synthesis of azumamide A **28**. Azumamides are unusual cyclic peptides that show potent inhibitory activity on histone deacetylase enzymes. A highly diastereo- and enantioselective (dr > 99%; 98% ee) crotylation of 3-benzyloxypropanal with the chiral reagent (E)-crotyl-^{l}Ipc$_2$borane (^{l}**19E**) afforded the homoallylic alcohol **29**. Subsequent reductive ozonolysis and K$_2$CO$_3$-mediated hydrolysis of the acetate furnished the diol **30**[16] (Scheme 3.1m).

Asymmetric allylboration has also been applied to γ-methoxyallyl derivatives. Isomerically pure (Z)-γ-methoxyallyldiisopinocampheylborane (^{d}**31**), prepared from ^{d}Ipc$_2$BOMe and the lithium anion of allyl methyl ether, reacts with various aldehydes to afford the syn-β-methoxyhomoallylic alcohol (**32a**) in a highly regio- and stereoselective manner[17] (Scheme 3.1n). This one-pot synthesis of enantiomerically pure 1,2-diol derivatives went as smoothly as the asymmetric Brown crotylation, affording products with uniformly high diastereoselectivity.

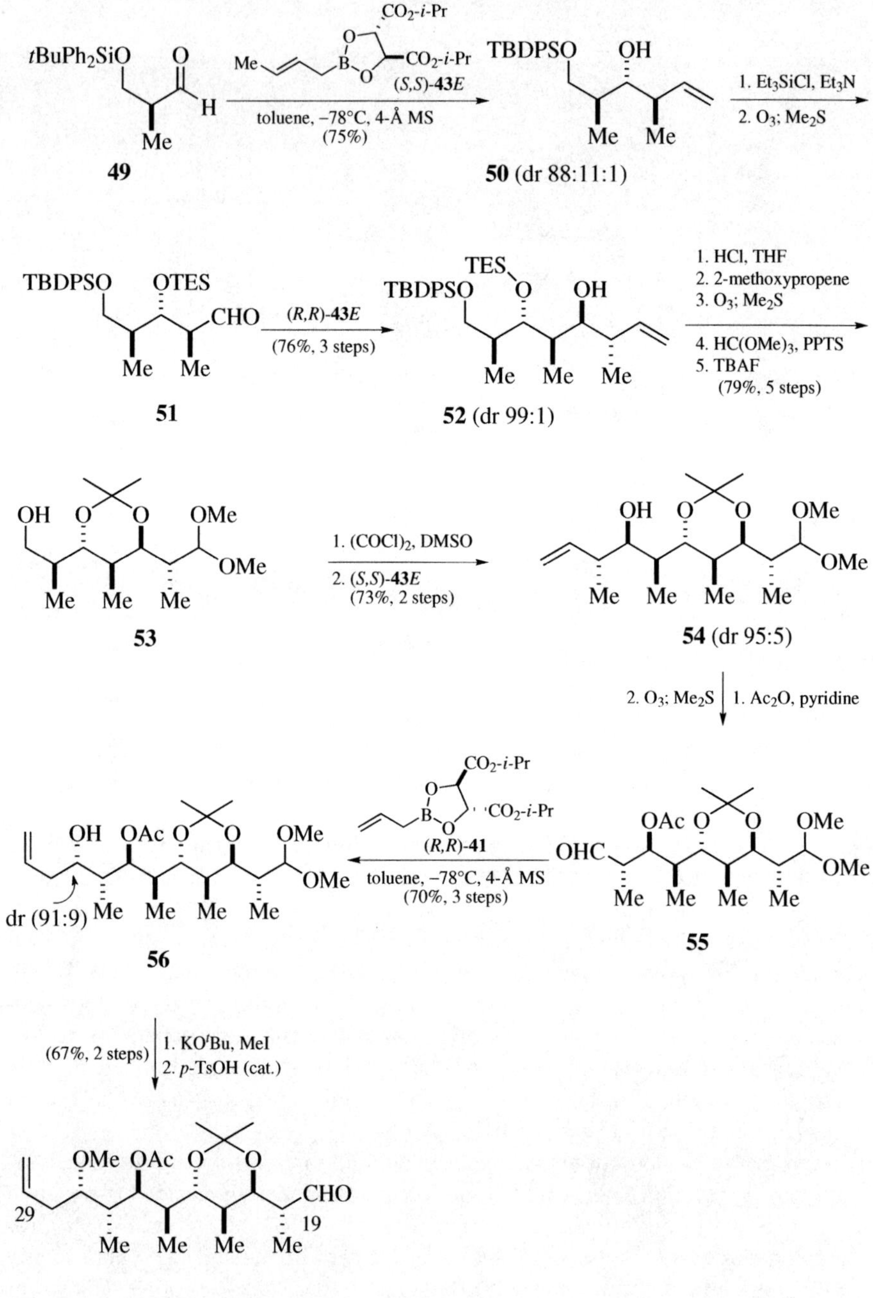

114 METAL ALLYLATION REACTIONS

48, C$_{19}$—C$_{29}$ segment of rifamycin S

Scheme 3.1u

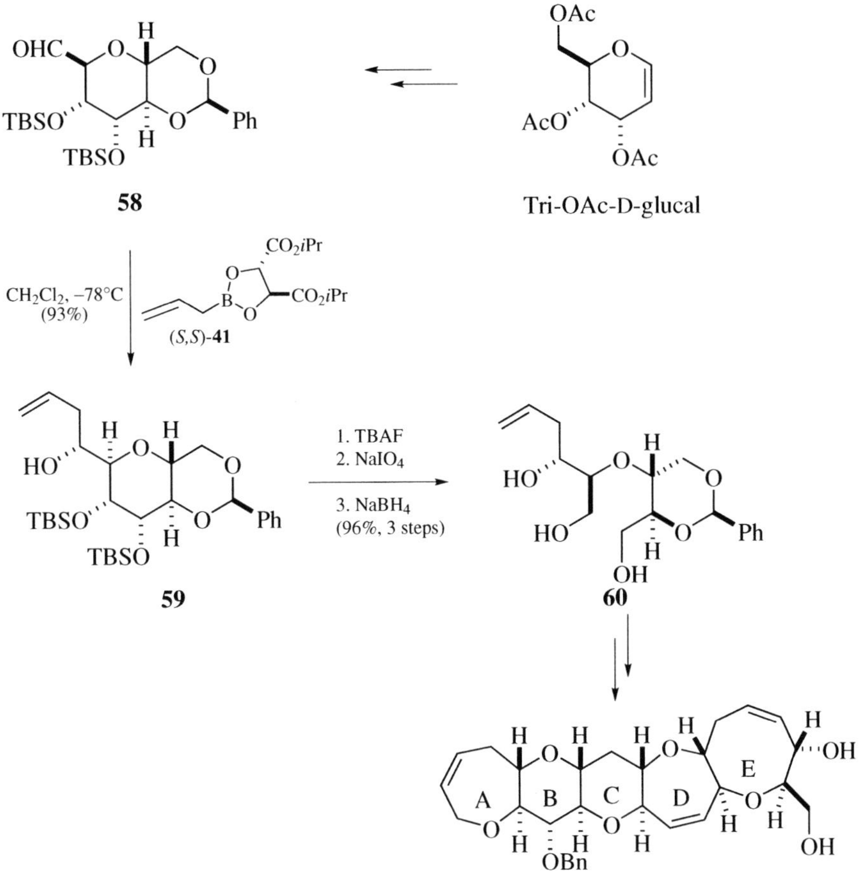

57, ABCDE-ring part of CTX3C

Scheme 3.1v

The Brown allylboration was used in the enantioselective total synthesis of (−)-calicheamicinone **33**[18] (Scheme 3.1o). Thus the lactol **34**, readily prepared from tetronic acid, was treated with the allylborane ^{d}**35** to give **36** in a highly stereoselective manner (95% ee, > 98% de). Compound **36** was converted to the aldoxime **37** by standard chemistry. Generation of the nitrile oxide with aqueous sodium hypochlorite was accompanied by spontaneous [3 + 2]-dipolar cycloaddition to afford **38** in 65% yield.

In the synthesis of a library of (+)-murisolin (**39**) and its 15 other stereoisomers, Curran and others used Brown allylation strategy to obtain the four diastereoisomers of the homoallylic alcohol **40**[19] (Scheme 3.1p). Thus, the homoallylic alcohol (*S,S*)-**40** was prepared in 95% ee from allylborane reagent ^{l}**35** and corresponding aldehyde. By using the Mitsunobu reaction, (*R,S*)-**40** was obtained. Similarly, (*R,R*)-**40** was synthesized via the enantiomeric borane reagent ^{d}**35**

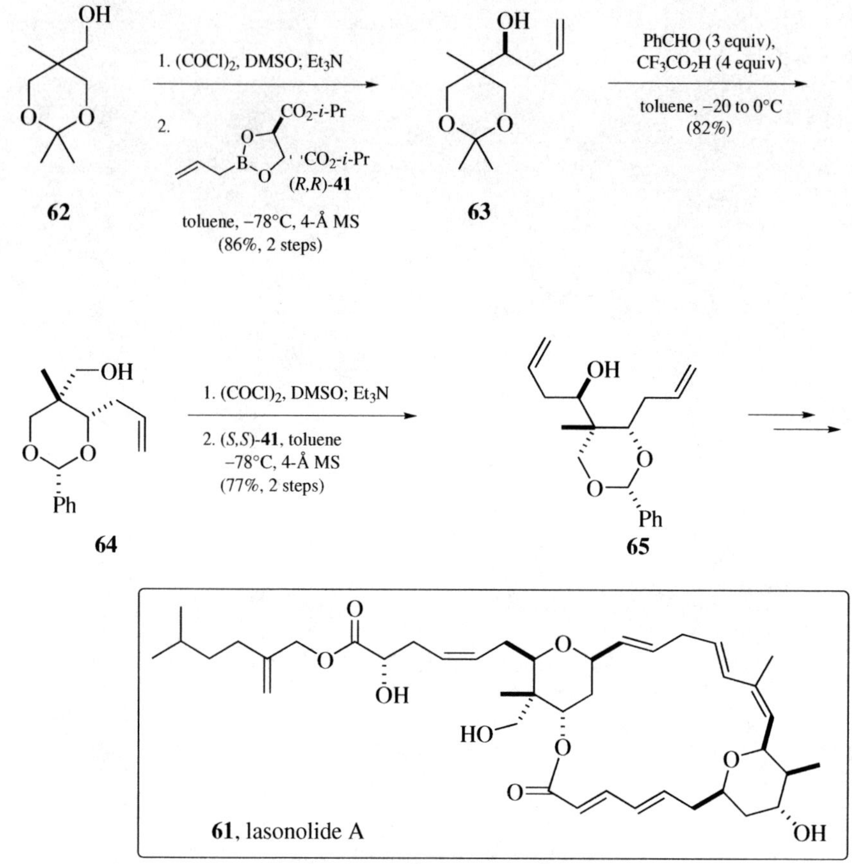

Scheme 3.1w

derived from ^{d}Ipc$_2$BOMe, and (S,R)-**40** was obtained by the Mitsunobu reaction of (R,R)-**40**. The four isomers thus secured were used in the total synthesis of (+)-murisolin and of its 15 isomers.

Roush et al. discovered that the tartrate ester–modified allylboronates, such as diisopropyl tartrate allylboronate (S,S)-**41**, react with achiral aldehydes to give the homoallylic alcohols **42** in good yields and high levels of enantioselectivity of up to 87% ee when the reaction is carried out in toluene in the presence of 4-Å molecular sieves[20] (Scheme 3.1q). To rationalize the asymmetric induction realized by **41**, two six-membered transition states were compared (Scheme 3.1r). It was reasoned that transition state **A** was favored over transition state **B** due mainly to the nonbonded electronic repulsive interactions of the lone-pair electrons of the aldehyde oxygen and the carbonyl oxygen of the tartrate ester.

In an effort to elucidate the electronic effects in the stereochemistry-determining transition states, Gung and co-workers conducted a computational study in 2002.[21] The calculations carried out on the allylation reaction between

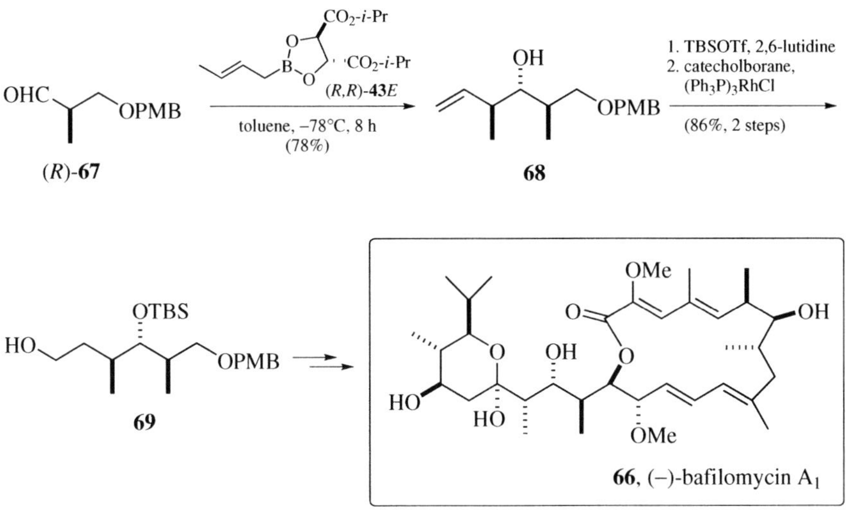

Scheme 3.1x

(R,R)-dimethyl tartrate allylboronate and acetaldehyde showed that transition state **A** is more stable than **B** by 1.75 kcal/mol (Scheme 3.1s). The major force for the energy difference is an attractive Coulomb interaction between the ester oxygen and the boron-complexed aldehyde carbonyl group: The distance between the two interacting charges is shorter in **A** (3.28 Å) than in **B** (4.11 Å). The authors concluded that the repulsive *n/n* interaction proposed initially might play a lesser role than speculated previously.

The tartrate-based *E*-crotylboronate (*S*,*S*)-**43E**, which can readily be prepared from *E*-2-butene, underwent highly enantioselective crotylation reactions with the chiral aldehydes **44** and **46**[22] (Scheme 3.1t). Best results in both cases were obtained in reactions performed at −78 °C in toluene in the presence of 4-Å molecular sieves. Under these conditions, the reaction of the L-deoxythreose ketal **44** was highly selective to generate **45a** in 22:1 diastereofacial selectivity. Similarly, the reaction of the D-glyceraldehyde acetonide **46** with (*S*,*S*)-**43E** showed exceptionally high 48:1 selectivity.

In order to apply tartrate ester–modified allyl- and crotylboronates to synthetic problems,[23] Roush and Palkowitz undertook the stereoselective synthesis of the C_{19}–C_{29} fragment **48** of rifamycin S, a well-known member of the ansamycin antibiotic group[24] (Scheme 3.1u). The synthesis started with the reaction of (*S*,*S*)-**43E** and the chiral aldehyde (*S*)-**49**. This crotylboration provided the homoallylic alcohol **50** as the major component of an 88:11:1 mixture. Compound **50** was transformed smoothly into the aldehyde **51**, which served as the substrate for the second crotylboration reaction. The alcohol **52** was obtained in 71% yield and with 98% diastereoselectivity. After a series of standard functional group manipulations, the alcohol **53** was oxidized to the corresponding aldehyde and underwent the third crotylboronate addition, which resulted in a 95:5 mixture

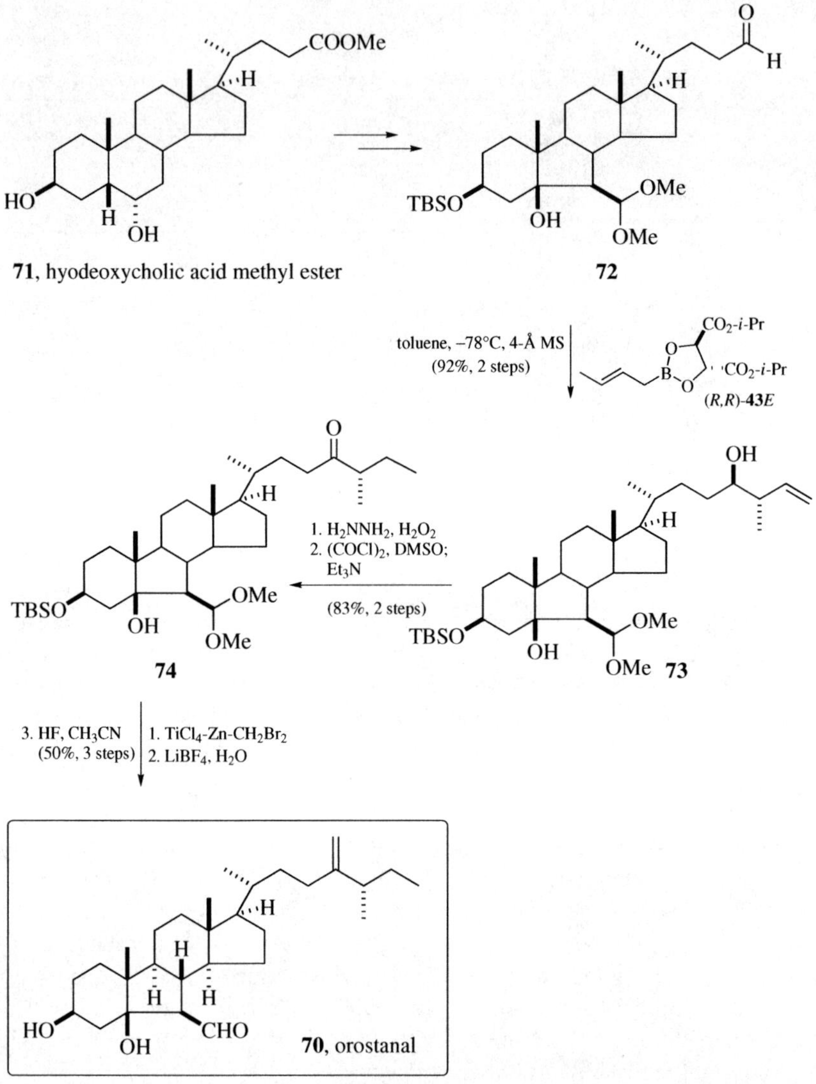
COOMe
H
HO
H
OH
71, hyodeoxycholic acid methyl ester
O
H
H
TBSO
OH
OMe
OMe
72
toluene, –78°C, 4-Å MS
(92%, 2 steps)
CO_2-i-Pr
CO_2-i-Pr
(R,R)-43E
O
H
TBSO
OH
OMe
OMe
74
1. H_2NNH_2, H_2O_2
2. $(COCl)_2$, DMSO; Et_3N
(83%, 2 steps)
OH
H
TBSO
OH
OMe
OMe
73
3. HF, CH_3CN
(50%, 3 steps)
1. $TiCl_4$-Zn-CH_2Br_2
2. $LiBF_4$, H_2O
H
H
H
H
HO
OH
CHO
70, orostanal

Scheme 3.1y

of **54** and its isomer. Acylation of **54** followed by ozonolysis provided **55**, which was treated with allylboronate (*R,R*)-**41** under standard conditions. The allylation product **56** was obtained as a 91 : 9 mixture.

The Roush allylboration method was also used in the synthesis of the ABCDE-ring part of ciguatoxin CTX3C **57**[25] (Scheme 3.1v). The aldehyde **58**, which was prepared from tri-*O*-acetylD-glucal in six steps, was treated with the chiral allylboronate reagent (*S,S*)-**41** to give the homoallylic alcohol **59** as the sole product. Removal of the TBS groups followed by oxidative cleavage of the

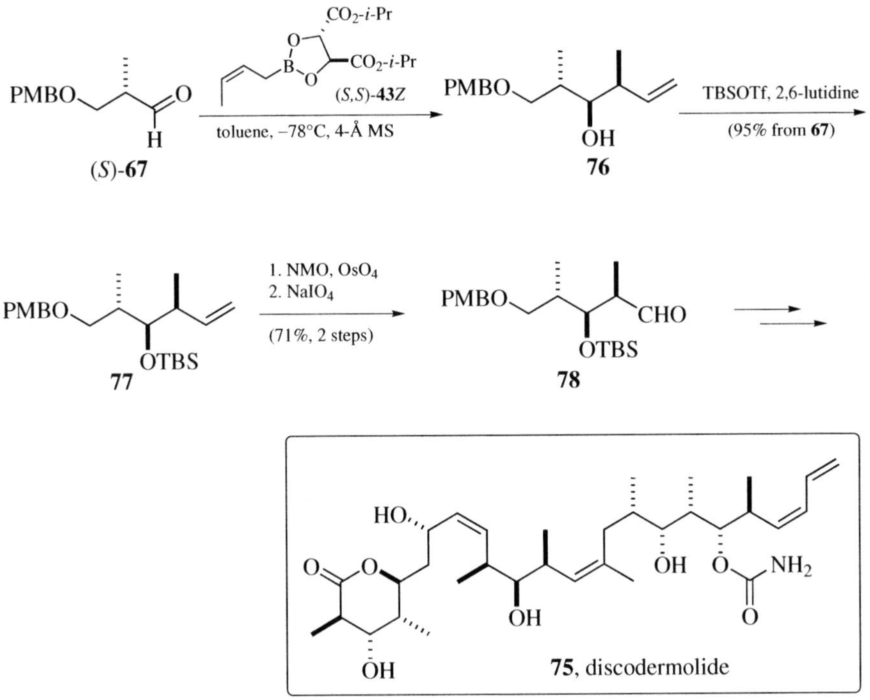

Scheme 3.1z

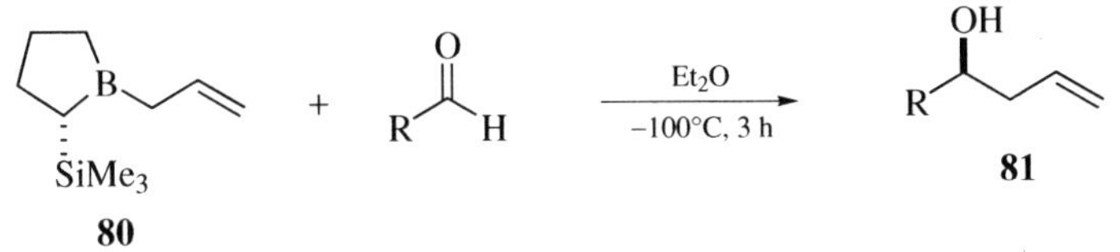

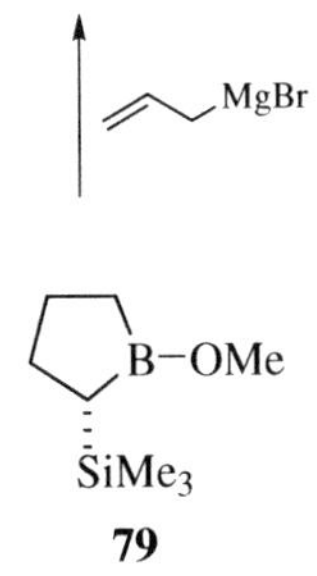

Scheme 3.1aa

R	ee, %	Yield, %
C_2H_5	96	80
$i\text{-}C_3H_7$	96	85
$t\text{-}C_4H_9$	97	90
$CH_3CH=CH$	97	85
$BnOCH_2CH_2$	92	84

Scheme 3.1bb

R	ee, %[a]
$n\text{-}C_5H_{11}$	90 (95)
$c\text{-}C_6H_{11}$	93 (97)
C_6H_5	94 (95)
$(E)\text{-}C_6H_5CH\!=\!CH$	98 (97)

[a] The % ee values in parentheses are for reactions run in toluene.

Scheme 3.1cc

resulting 1,2-diol and the subsequent reduction afforded the triol **60** in 96% overall yield.

The tartrate-based allylboronates were also used in the total synthesis of (+)-lasonolide A (**61**), which displays antitumor activity by inhibiting the invitro proliferation of A-549 human lung carcinoma cells[26] (Scheme 3.1w). In the synthesis, the alcohol **62** was oxidized and then treated with the allylboronate (R,R)-**41** to give the homoallylic alcohol **63** with 78% ee. Compound **63** was then

Scheme 3.1dd

treated with benzaldehyde in the presence of trifluoroacetic acid to furnish a separable 5 : 1 mixture favoring the benzylidene desired (**64**). Swern oxidation of **64** followed by a second asymmetric allylation, this time with (*S,S*)-**41**, yielded the alcohol **65** with an enhanced enantioselectivity of 91% ee.

Roush et al. applied the diastereoselective crotylboration methodology in the total synthesis of bafilomycin A_1 (**66**), a potent vacuolar ATPase inhibitor that displays broad antibiotic activity[27] (Scheme 3.1x). In the synthesis, the known aldehyde (*R*)-**67** was treated with (*E*)-crotylboronate (*R,R*)-**43E** to provide an 85 : 15 mixture of the homoallylic alcohol **68** and the undesired 3,4-*anti*-4,5-*syn* diastereomer with an isolated 78% yield of **68**. Alcohol protection as a TBS ether followed by hydroboration mediated by Wilkinson's catalyst efficiently provided the primary alcohol **69**.

Liu and Zhou applied Roush's crotylboration to the stereoselective synthesis of the orostanal **70**, a novel sterol that induces apoptosis in human acute promyelotic leukemia cells[28] (Scheme 3.1y). The aldehyde **72**, prepared from hyodeoxycholic acid methyl ester, underwent asymmetric reaction with crotylboronate (*R,R*)-**43E** to furnish **73**. Hydrogenation of the terminal alkene followed by Swern oxidation gave the ketone **74**. Methylenation of the ketone and removal of the protective groups afforded orostanal in 50% yield.

Another synthetic application of Roush's crotylboration methodology using a (*Z*)-crotylboronate can be found in the formal synthesis of (+)-discodermolide (**75**)[29] (Scheme 3.1z). The aldehyde (*S*)-**67**, which was prepared from the Roche ester, reacted with (*Z*)-crotylboronate (*S,S*)-**43Z** to give the *syn*-homoallylic alcohol **76**. Silylation of alcohol and oxidative cleavage of the alkene **77** provided the aldehyde **78**, from which the final product (**75**) can be synthesized according to a known procedure.[30]

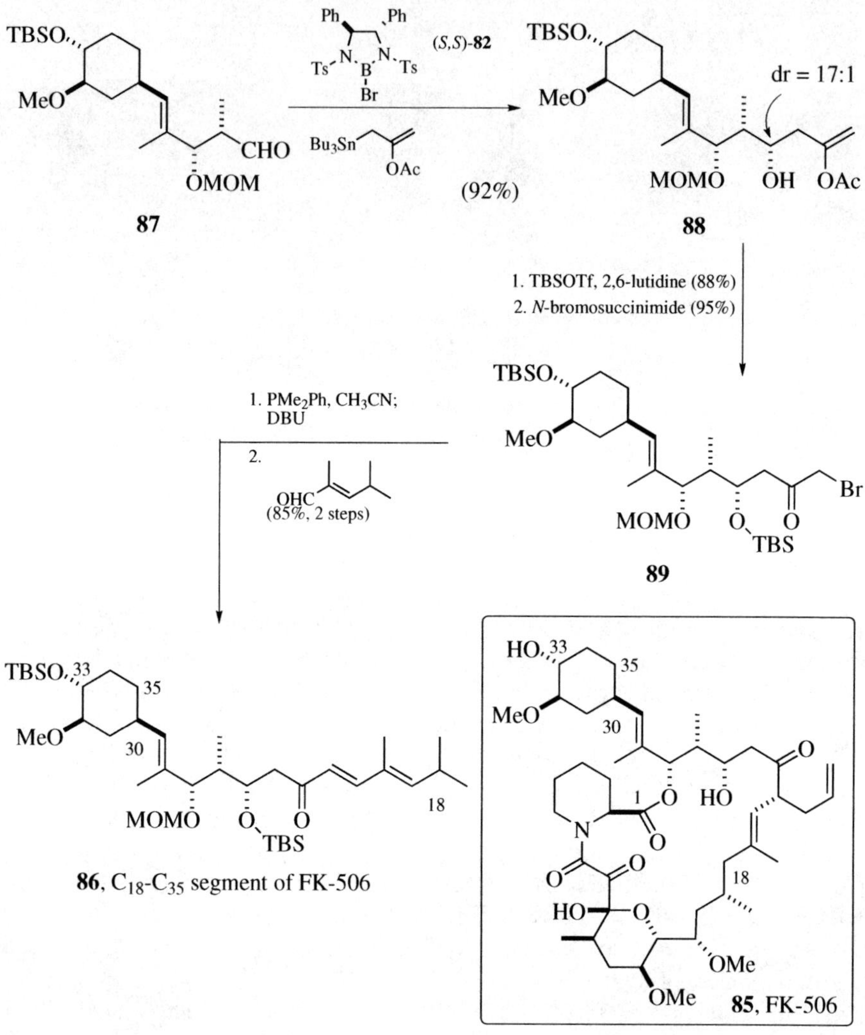

Scheme 3.1ee

Short and Masamune reported that the monosubstituted C_1-symmetric borolane derivative **80**, prepared by addition of allylmagnesium bromide to a solution of (S)-B-methoxy-2-(trimethylsilyl)borolane (**79**), was a highly efficient reagent for asymmetric allylboration [31] (Scheme 3.1aa). The borolane reagent shows satisfactory reactivity and high stereoselectivity toward a variety of aldehydes at $-100°$ C. To explain the stereochemical outcome of the reaction, a six-membered chairlike transition state was proposed[31] (Scheme 3.1bb). Transition state **A**, in which the allyl group attacks the si-face of the aldehyde is presumably operating here based on steric interactions. Transition state **B** would be of higher energy than transition state **A**, due to the severe nonbonded repulsions between the

Scheme 3.1ff

trimethylsilyl group and the two hydrogens of the allyl group. It is noteworthy that the allylation reaction of isobutyraldehyde with the chiral reagent bearing a butyl group on the borolane ring gave product in much reduced enantioselectivity of 72% ee. Thus, the Masamune asymmetric allylation reaction was an elegant demonstration of the effective steric role played by a trimethylsilyl group.[32]

In 1989, Corey et al. reported a highly enantioselective allylation of aldehydes utilizing (R,R)-1,2-diamino-1,2-diphenylethane (stilbenediamine, or stien) as an efficient chiral auxiliary[33] (Scheme 3.1cc). Reaction of the bis-*p*-toluenesulfonyl derivative of (R,R)-stien[34] in CH_2Cl_2 with 1 equivalent of BBr_3 followed by treatment of the resulting bromoborane (**82**) with allyltributyltin generated the chiral allylborane **83**. Reaction of **83** with aliphatic and aromatic aldehydes in CH_2Cl_2 or toluene at −78 °C produced the corresponding homoallylic alcohol **84** in excellent optical purities and greater than 90% yield.

The absolute configuration observed for the homoallylic alcohol **84** derived from the (R,R)-allylborane reagent **83** can be rationalized on the basis of a chair-like six-membered transition state[33] (Scheme 3.1dd). The major enantiomer is formed via transition state **A**, in which the boron allyl group attacks the *re*-face

Scheme 3.1gg

of the aldehye. Transition state **B** would be disfavored, due to the nonbonded interaction between the sulfonyl group and the pseudoaxial aldehydic hydrogen.

Corey and Huang used this allylation method in construction of the C_{18}–C_{35} subunit **86** of the macrocyclic immunosuppressant FK-506 (**85**)[35] (Scheme 3.1ee). The aldehyde **87** underwent asymmetric allylation with the cyclic borane reagent generated in situ by reaction of bromoborane (S,S)-**82** and 2-acetoxyallyltributyl-stannane. The allylation proceeded with high 17 : 1 diastereoselectivity to furnish the homoallylic alcohol **88** in 92% yield after column chromatography. Silylation of **88** followed by treatment with N-bromosuccinimide yielded the bromomethyl ketone **89**, which was further elaborated via a Wittig reaction to afford the C_{18}–C_{35} fragment **86**.

The highly enantioselective allylation method developed by Corey was utilized iteratively during the total synthesis of phorboxazole A (**13**), a natural product that exhibits unprecedented cytostatic activity against all 60 cell lines of the National Cancer Institute human cancer test panel[36] (Scheme 3.1ff). The stannane **90** underwent effective transmetalation from $0\,°C$ to room temperature over 12 hours with the (R,R)-bromoborane **82**. The asymmetric allylation of **91** gave the homoallylic alcohol **92** with excellent diastereoselectivity. After further manipulations, the aldehyde **93** was generated, setting the stage for the second asymmetric allylation with the stannane **90** and the (S,S)-bromoborane **82** to produce the polyol derivative **94** in a 92 : 8 diastereomeric ratio.

R/S	
Prediction	Experiment
85:15	71:29

Scheme 3.1hh

Corey's asymmetric allylation methodology was utilized in the total synthesis of amphidinolide T3 (**95**), a marine natural product that exhibits significant antitumor properties[37] (Scheme 3.1gg). The asymmetric allylation of the aldehyde **96** was carried out successfully with chiral allylborane reagent generated in situ from allyltributylstannane and (R,R)-**82** to furnish the homoallylic alcohol desired (**97**) in 85% yield with excellent diastereoselectivity. Subsequent conversion of the alcohol to the tosylate ester followed by treatment with potassium hydroxide resulted in formation of the trisubstituted tetrahydrofuran **98**.

To rapidly generate a structurally diverse set of chiral ligands for asymmetric reactions, a new research program using computer-aided design has recently been launched. Using this technique, Kozlowski et al. identified a structurally unique diol (**100**) for enantioselective boron allylation reaction. Compound **100** was synthesized from the known dioxatetracyclic compound **99**, and the validity of the computational prediction was evaluated for an asymmetric boron allylation reaction[38] (Scheme 3.1hh).

The allyl boronate **101**, prepared by treatment of the *cis*-decalin diol **100** with trisallylborane, underwent allylation with dihydrocinnamaldehyde to afford the corresponding homoallylic alcohols favoring the *R*-enantiomer, which was the major enantiomer predicted based on transition-state calculations. Although the level of enantioselectivity realized with **100** is low, further refinements of the computational parameters should lead to the discovery of more efficient chiral ligands.[39]

REFERENCES

1. Herold, T.; Hoffmann, R. W. *Angew. Chem. Int. Ed.* **1978**, *17*, 768.
2. Herold, T.; Schrott, U.; Hoffmann, R. W.; Schnelle, v. G. E.; Ladner, W.; Steinbach, K. *Chem. Ber.* **1981**, *114*, 359.
3. Lachance, H.; Lu, X.; Gravel, M.; Hall, D. G. *J. Am. Chem. Soc.* **2003**, *125*, 10160.
4. Rauniyar, V.; Hall, D. G. *J. Am. Chem. Soc.* **2004**, *126*, 4518.
5. (a) Brown, H. C.; Jadhav, P. K. *J. Am. Chem. Soc.* **1983**, *105*, 2092; (b) Brown, H. C.; Jadhav, P. K.; Bhat, K. S.; Perumal, T. *J. Org. Chem.* **1986**, *51*, 432.
6. Racherla, U. S.; Brown, H. C. *J. Org. Chem.* **1991**, *56*, 401.
7. Brown, H. C.; Randad, R. S.; Bhat, K. S.; Zaidlewicz, M.; Racherla, U. S. *J. Am. Chem. Soc.* **1990**, *112*, 2389.
8. (a) Brown, H. C.; Bhat, K. S. *J. Am. Chem. Soc.* **1986**, *108*, 5919; (b) Brown, H. C.; Racherla, U. S.; Pellechia, P. J. *J. Org. Chem.* **1990**, *55*, 1868.
9. Vulpetti, A.; Gardner, M.; Gennari, C.; Bernardi, A.; Goodman, J. M.; Paterson, I. *J. Org. Chem.* **1993**, *58*, 1711.
10. (a) Wipf, P.; Xu, W. *J. Org. Chem.* **1996**, *61*, 6556; (b) Xu, W. Ph.D. dissertation, University of Pittsburgh, Pittsburgh, PA, **1997**.
11. (a) Wipf, P.; Xu, W. *J. Org. Chem.* **1993**, *58*, 825; (b) Wipf, P.; Xu, W. *Tetrahedron Lett.* **1994**, *35*, 5197.
12. White, J. D.; Kuntiyong, P.; Lee, T. H. *Org. Lett.* **2006**, *8*, 6039.
13. Brown, H. C.; Bhat, K. S. *J. Am. Chem. Soc.* **1986**, *108*, 293.
14. Barrett, A. G. M.; Lebold, S. A. *J. Org. Chem.* **1991**, *56*, 4875.
15. Nicolaou, K. C.; Li, Y.; Fylaktakidou, K. C.; Mitchell, H. J.; Wei, H.-X.; Weyershausen, B. *Angew. Chem. Int. Ed.* **2001**, *40*, 3849.
16. Izzo, I.; Maulucci, N.; Bifulco, G.; De Riccardis, F. *Angew. Chem. Int. Ed.* **2006**, *45*, 7557.
17. Brown, H. C.; Jadhav, P. K.; Bhat, K. *J. Am. Chem. Soc.* **1988**, *110*, 1535.
18. Smith, A. L.; Hwang, C.-K.; Pitsinos, E.; Scarlato, G. R.; Nicolaou, K. C. *J. Am. Chem. Soc.* **1992**, *114*, 3134.
19. (a) Zhang, Q.; Lu, H.; Richard, C.; Curran, D. P. *J. Am. Chem. Soc.* **2004**, *126*, 36; (b) Curran, D. O.; Zhang, Q.; Richard, C.; Lu, H.; Gudipati, V.; Wilcox, C. S. *J. Am. Chem. Soc.* **2006**, *128*, 9561.
20. Roush, W. R.; Walts, A. E.; Hoong, L. K. *J. Am. Chem. Soc.* **1985**, *107*, 8186.
21. Gung, B. W.; Xue, X.; Roush, W. *J. Am. Chem. Soc.* **2002**, *124*, 10692.
22. Roush, W. R.; Halterman, R. L. *J. Am. Chem. Soc.* **1986**, *108*, 294.

23. For a comprehensive review of synthetic applications of asymmetric boron allylation reactions, see Chemler, S. R.; Roush, W. R. in Modern Carbonyl Chemistry; Otera, J., Ed.; Wiley-VCH: Weinheim, Germany, 2000; Chapt. 11.
24. Roush, W. R.; Palkowitz, A. D. *J. Am. Chem. Soc.* **1987**, *109*, 953.
25. Fujiwara, K.; Goto, A.; Sato, D.; Ohtaniuchi, Y.; Tanaka, H.; Murai, A.; Kawai, H.; Suzuki, T. *Tetrahedron Lett.* **2004**, *45*, 7011.
26. Kang, S. H.; Kang, S. Y.; Kim, C.; Choi, H.; Jun, H.-S.; Lee, B.; Park, C.; Jeong, J. *Angew. Chem. Int. Ed.* **2003**, *42*, 4779.
27. Scheidt, K. A.; Tasaka, A.; Bannister, T. D.; Wendt, M. D.; Roush, W. R. *Angew. Chem. Int. Ed.* **1999**, *38*, 1652.
28. Liu, B.; Zhou, W. *Tetrahedron Lett.* **2002**, *43*, 4187.
29. Francavilla, C.; Chen, W.; Kinder, F. R. Jr. *Org. Lett.* **2003**, *5*, 1233,
30. Kinder, F. R. (Novartis A-G, Switzerlands Novartis-Erfindungen Verwaltungs G mbH). Process for preparing discodermolide and analogues thereof. PCT Int. Appl. 22, CODEN: PIXXD2 WO 0212220 A2 20020214, 2002.
31. Short, R.; Masamune, S. *J. Am. Chem. Soc.* **1989**, *111*, 1892.
32. The C–Si bond of 1.9 Å is significantly longer than the C–C bond of 1.5 Å. For selected C–Si bond lengths, see (a) Sakurai, H.; Nakadaira, Y.; Tobita, H. *J. Am. Chem. Soc.* **1982**, *104*, 300; (b) Igau, A.; Baceiredo, A.; Grützmacher, H.; Pritzkow, H.; Bertrand, G. *J. Am. Chem. Soc.* **1989**, *111*, 6853.
33. Corey, E. J.; Yu, C.-M.; Kim, S. S. *J. Am. Chem. Soc.* **1989**, *111*, 5495.
34. Pikul, S.; Corey, E. J. *Org. Synth. Coll. Vol. 9*, **1998**, 387.
35. Corey, E. J.; Huang, H.-C. *Tetrahedron Lett.* **1989**, *30*, 5235.
36. Williams, D. R.; Kiryanov, A. A.; Emde, U.; Clark, M. P.; Berliner, M. A.; Reeves, J. T. *Proc. Natl. Acad. Sci. USA*, **2004**, *101*, 12058.
37. Deng, L.-S.; Huang, X.-P.; Zhao, G. *J. Org. Chem.* **2006**, *71*, 4625.
38. Kozlowski, M. C.; Waters, S. P.; Skudlarek, J. W.; Evans, C. A. *Org. Lett.* **2002**, *4*, 4391.
39. Kozlowski, M. C.; Panda, M. *J. Org. Chem.* **2003**, *68*, 2061.

3.2. Silicon Allylation Reaction

Although the addition of allylic trialkylsilanes to carbonyl compounds is analogous to the reaction of allylboranes, it occurs through an acyclic transition state.[1] This is because, in contrast to the boranes, the silicon in allylic trialkylsilanes is a poor Lewis acid and would not be expected to coordinate to the carbonyl oxygen. The transition state changes, however, to a closed one when the silicon becomes sufficiently Lewis acidic, owing to incorporation of electronegative ligand into the silicon atom.[2]

Evidence for a closed transition-state model was gathered on the basis of the diastereoselectivity in reactions of pentacoordinate allylic silicates. Bis(1,2-benzenediolato)allylsilicates **1a** and **1b**, which can be prepared via the reactions of *E*- and *Z*-crotyltrichlorosilanes with dilithium catecholate, react with aromatic aldehydes to give the corresponding homoallylic alcohols **2** in high yields[3] (Scheme 3.2a). Unlike allyltrimethylsilane,[4] the allylation reactions of

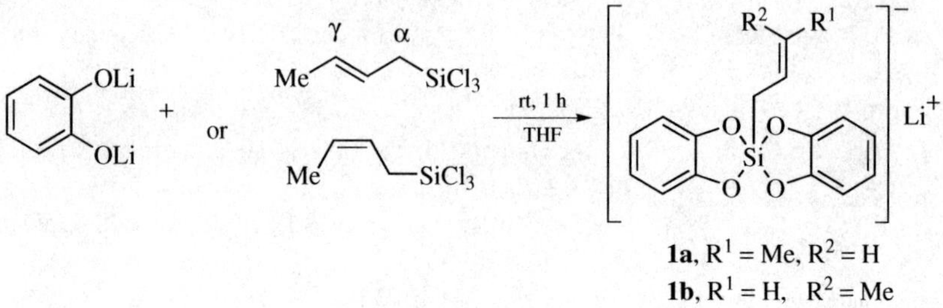

Silicate	anti/syn	Yield, %
1a (E/Z = 88:12)	88:12	82
1b (E/Z = 21:79)	22:78	91

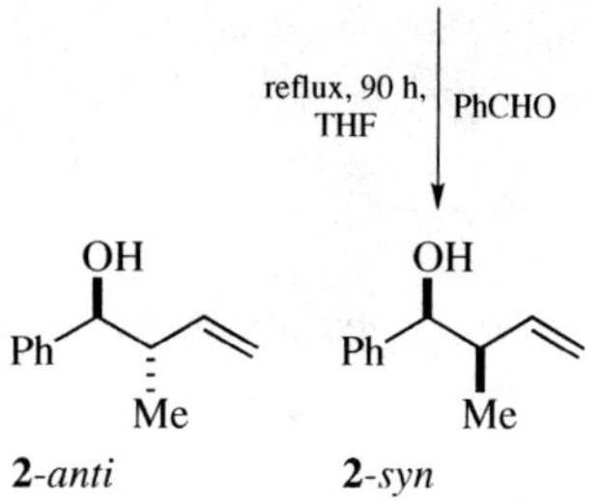

Scheme 3.2a

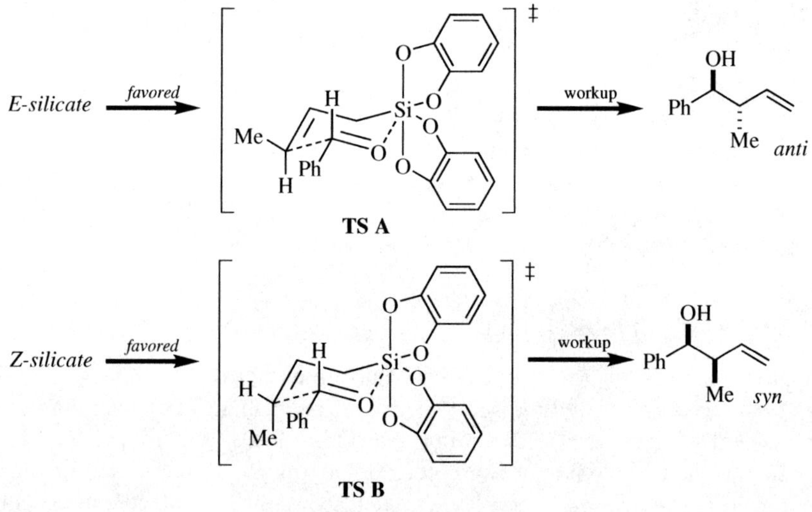

Scheme 3.2b

the allylsilicates **1** do not require external Lewis acids. The reactivity difference is due to the enhanced Lewis acidity of silicon in the silicates **1** over that in allyltrimethylsilane.[5] The pentacoordinate allylsilicates **1** show an extremely high level of diastereoselectivity in the crotylation reaction. For instance, when the crotylsilicate **1a** with an E/Z ratio of 88:12 was used, the corresponding

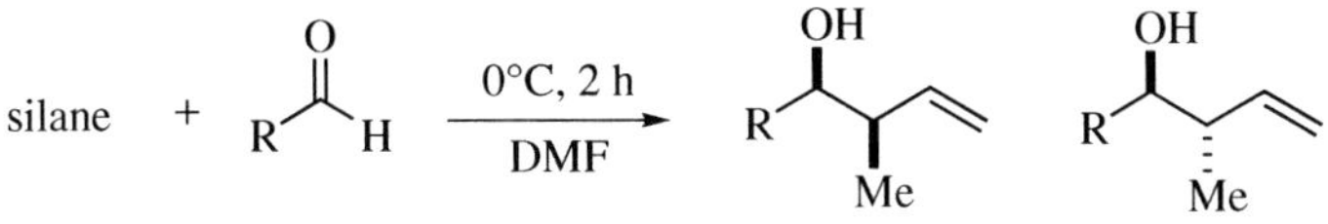

Silane	R	syn/anti	Yield, %
SiCl₃ (E/Z = 97:3)	C₆H₅	3:97	89
	PhCH₂CH₂	3:97	87
	c-C₆H₁₁	4:96	83
SiCl₃ (E/Z = <1:>99)	C₆H₅	>99:1	82
	PhCH₂CH₂	>99:1	90
	c-C₆H₁₁	97:3	85

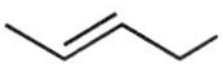

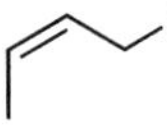

Scheme 3.2c

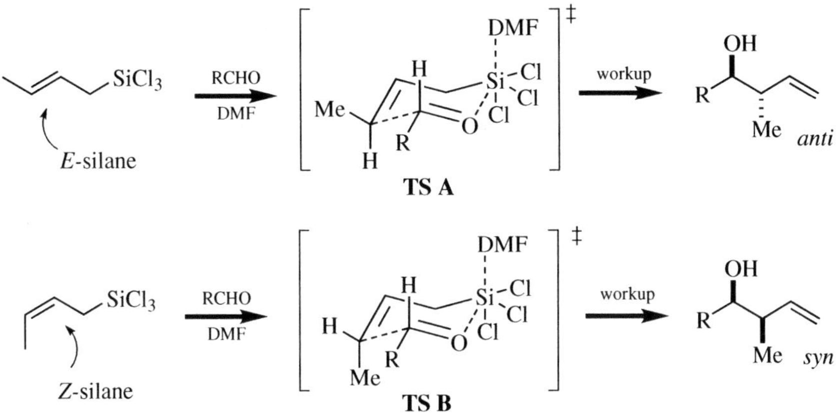

Scheme 3.2d

homoallylic alcohols **2** were obtained with an *anti/syn* ratio of 88:12. The crotylsilicate **1b**, rich in (Z)-isomer (E/Z = 21:79) afforded the alcohol **2** in an *anti/syn* ratio of 22:78. Thus, the reaction is extremely stereoselective, giving *anti*- and *syn*-homoallylic alcohols from the E- and Z-crotylsilicates, respectively.

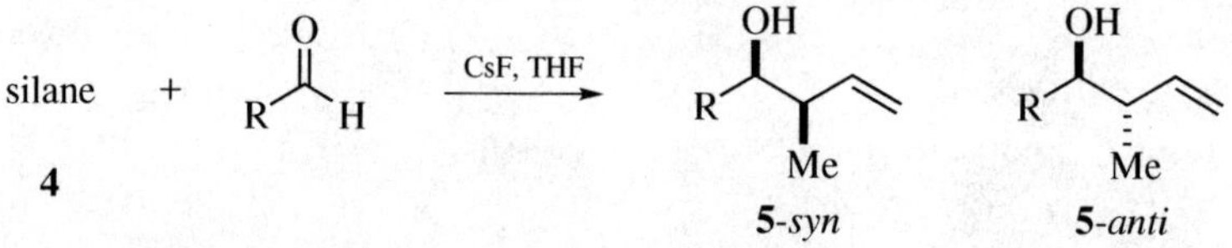

Scheme 3.2e

silane + R−CHO →(CsF, THF) 5-syn (R−CH(OH)−CH(Me)−CH=CH₂) + 5-anti

Silane	R	Temperature (Time)	syn/anti	Yield, %
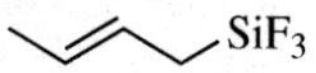 ⟋⟍SiF₃ (4E; E/Z = 99:1)	C₆H₅	0°C (1 h)	1:99	92
	n-C₈H₁₇	rt (4 h)	1:99	96
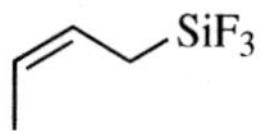 ⟋⟍SiF₃ (4Z; E/Z = 1:99)	C₆H₅	0°C (1 h)	99:1	96
	n-C₈H₁₇	rt (5 h)	92:2	89

Scheme 3.2e

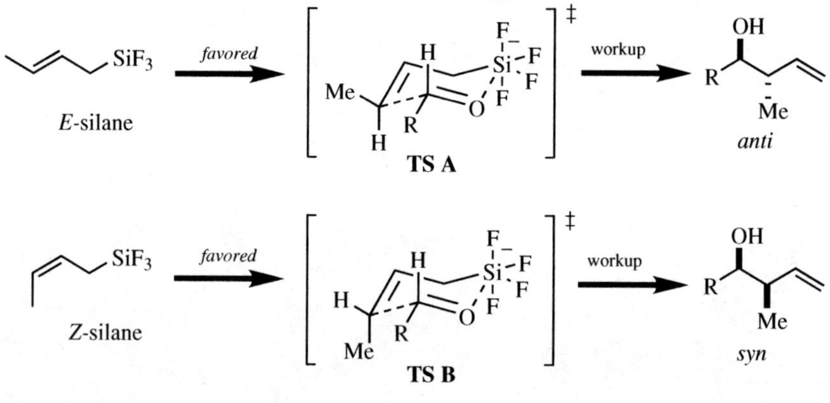

Scheme 3.2f

E-silane →(favored) TS A →(workup) anti product

Z-silane →(favored) TS B →(workup) syn product

Scheme 3.2f

These diastereoselective reactions of pentacoordinate crotyl silicates are reminiscent of those of allylboronates.[6] Consequently, the stereoselectivity of the silicon crotylation is similarly interpreted by the six-membered chairlike transition state (Scheme 3.2b). Transition state **A**, in which the silicon of the (*E*)-crotylsilicate is now hexacoordinated and the phenyl group of benzaldehyde occupies an equatorial position, would give the major *anti*-product. Similarly, the (*Z*)-silicate affords the *syn*-product through transition state **B**. The silicons in transition states **A** and **B** are presumably strongly electron donating to the π-allyl system, thus enhancing the nucleophilicity of the γ-carbon of the allylsilicates. This explains the exclusive formation of the γ-adduct.

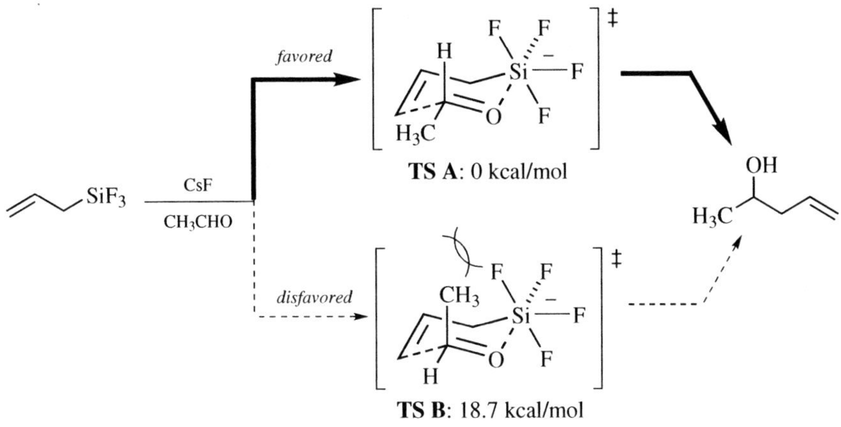

Scheme 3.2g

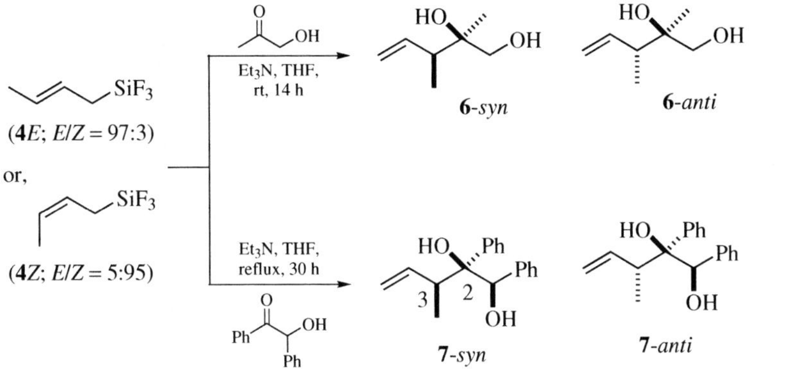

Silane	Ketone[a]	syn/anti[b]	Yield, % (6)	Silane	Ketone[a]	syn/anti[b]	Yield, % (7)
4E	HA	97:3	83	4E	BZ	97:3	71
4Z	HA	5:95	87	4Z	BZ	5:95	75

[a] HA is α-hydroxy acetone and BZ is benzoin.
[b] The ratios of 2,3-*syn* to 2,3-*anti*. The 1,2-*anti* diol was not observed within the product from either **4E** or **4Z**.

Scheme 3.2h

Similar to crotylsilicates, crotyltrichlorosilanes react regioselectively with aldehydes in N,N-dimethylformamide (DMF) without a catalyst to afford the corresponding homoallylic alcohols in high yields (Scheme 3.2c).[7] New carbon–carbon bond formation takes place only at the γ-positions of the crotyltrichlorosilanes. In addition, *syn*-isomers are obtained from Z-crotyltrichlorosilanes, while

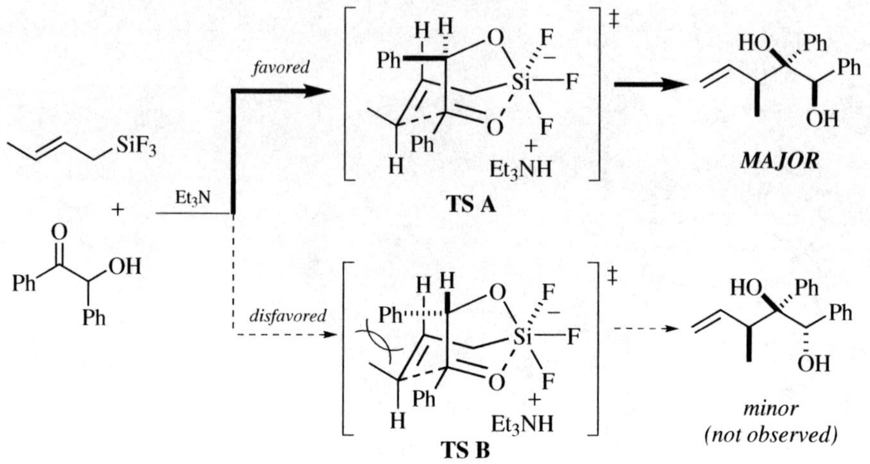

Scheme 3.2i

anti-isomers are produced from *E*-crotyltrichlorosilanes with near-perfect selectivity. Another synthetically useful feature of the reaction is that aromatic and aliphatic aldehydes exhibit the same degree of stereoselectivity.

The intermediate and key species proposed for the reaction in Scheme 3.2c are hypervalent silicates based on the silicon NMR spectra of (*Z*)-crotyltrichlorosilane in DMF. This hypervalent silicate has sufficient Lewis acidity based on the electron-withdrawing chlorine groups as well as nucleophilicity due to electron donation from the hypervalent silicon atom to the allyl systems, which enables the reaction to proceed smoothly. Thus, the high levels of diastereoselectivity can be explained by a six-membered cyclic transition state (Scheme 3.2d).

Another successful method for the highly diastereoselective silicon allylation reaction is the allyltrifluorosilane– cesium fluoride system discovered by Sakurai et al. in 1987[8] (Scheme 3.2e). After a mixture of aldehyde, the allylic trifluorosilane **4**, and cesium fluoride in a ratio of 1:2:2.3 was stirred in THF, the reaction mixture was quenched with a solution of HCl in MeOH to afford the products desired (**5**) in excellent yield and exceptionally high diastereoselectivity. In addition, the reaction is highly regioselective in that the carbon–carbon bond formation occurs exclusively at the γ-carbon of allylic silanes.

The regioselectivity and diastereoselectivity can be interpreted in terms of a six-membered chairlike transition state[8] (Scheme 3.2f). Thus, the nucleophilic attack of a fluoride anion to an allyltrifluorosilane may afford a rather stable pentacoordinate allylsilicate, which then reacts with an aldehyde via a cyclic six-membered transition state. The high level of regioselectivity of the reaction is presumably due to the enhanced nucleophilicity of the γ-carbon of the allylsilicate.

In an effort to explain the high levels of diastereoselectivity observed in the crotylation reactions, Sakurai et al. performed a computational study on the

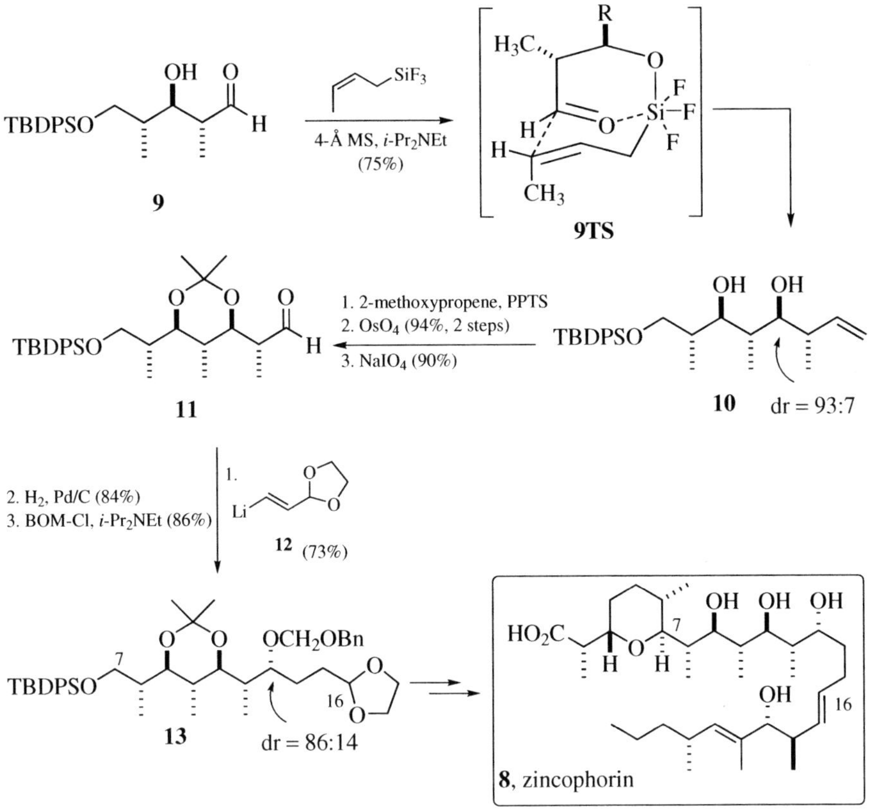

Scheme 3.2j

reaction of acetaldehyde with allyltrifluorosilane[9] (Scheme 3.2g). In the transition states calculated, the geometry around the silicon atom is octahedral and the allyl group of the hexacoordinate silicate occupies the equatorial position of the octahedral geometry. The calculations indicate that transition state **A**, in which the methyl group occupies an equatorial position, is a lot more stable than transition state **B**, by 18.7 kcal/mol. The destabilization in **B** is presumably caused by the nonbonding interactions between the axially oriented methyl group and the Si–F bond. The computational study described above explains the extremely high diastereo selectivity observed with the (E)- and (Z)-crotylsilanes and further confirms the legitimacy of the six-membered chairlike transition state in the allylation reaction with the pentacoordinate allyltrifluorosilane.

In 1989, Sakurai et al. reported that allyltrifluorosilanes react with a variety of α-hydroxy ketones in the presence of stoichiometric amount of triethylamine to yield the corresponding tertiary homoallylic alcohols in an extremely high regio- and diastereoselective manner[10] (Scheme 3.2h). Upon reacting with α-hydroxy acetone, the (E)-crotylsilane **4E** gave **6**-*syn* as a major product in 83% yield with

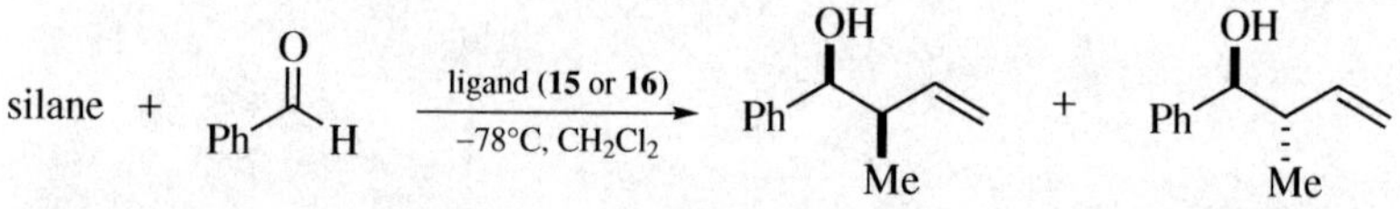

Scheme 3.2k

silane + PhCHO →(ligand (15 or 16), −78°C, CH₂Cl₂)→ 14-syn + 14-anti

Silane	Major Product	ee, %^{a} (syn/anti)	Yield, %^{a}	ee, %^{b} (syn/anti)	Yield, %^{b}
 ⩘SiCl₃	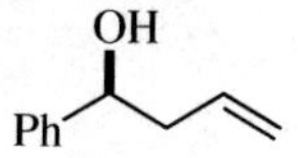 Ph-CH(OH)-CH₂-CH=CH₂	60	81	87	85
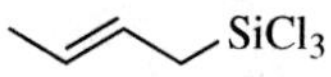 (E)-CH₃CH=CHCH₂SiCl₃ (E/Z = >99:1)	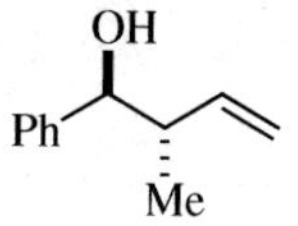 Ph-CH(OH)-CH(Me)-CH=CH₂	66 (2:98)	68	86 (1:99)	82
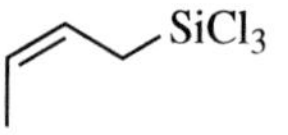 (Z)-CH₃CH=CHCH₂SiCl₃ (E/Z = <1:>99)	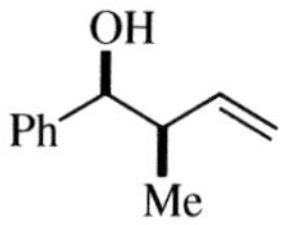 Ph-CH(OH)-CH(Me)-CH=CH₂	60 (98:2)	72	94 (99:1)	89

^{a} One equivalent of **15** was used.
^{b} 5 mol% of **16** was used.

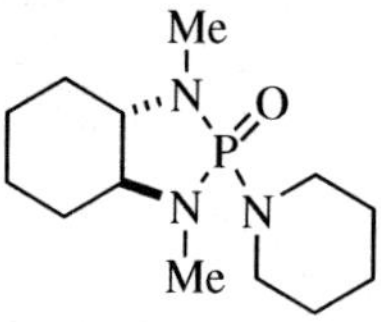

(S,S)-**15** 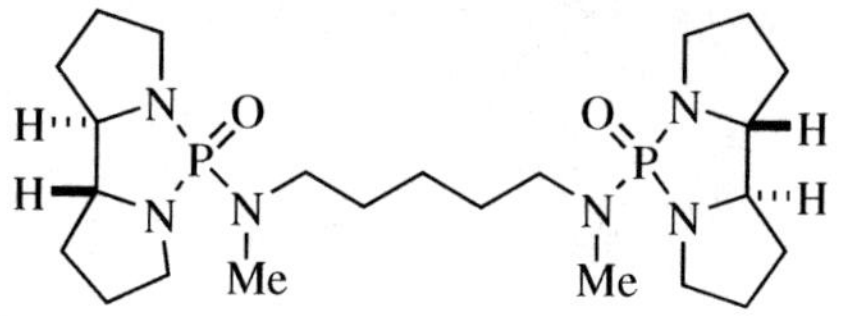   (R,R)-**16**

Scheme 3.2k

a *syn/anti* ratio of 97:3, while the (Z)-isomer **4Z** resulted in the formation of **6-anti** in 87% yield with a *syn/anti* ratio of 5:95. An additional stereoselectivity was discovered when the crotylsilanes reacted with α-substituted-α-hydroxy ketones such as benzoin. The diols **7-syn** and **7-anti** were obtained in good yields and with perfect levels of 1,2-diastereoselectivity from the reactions of (E)- and (Z)-crotylsilanes, respectively.

The stereochemical outcome of the reaction above is explained in terms of a structurally rigid 1,3-bridged six-membered chairlike transition state (Scheme 3.2i). Transition state **B** would be disfavored relative to transition state **A** because the phenyl group of benzoin is engaged in nonbonded interactions with the methyl and the hydrogen of the crotylsilane.

Although β-hydroxy ketones are inert toward allyltrifluorosilanes under the reaction conditions mentioned above,[10] β-hydroxy aldehydes do react with allylic

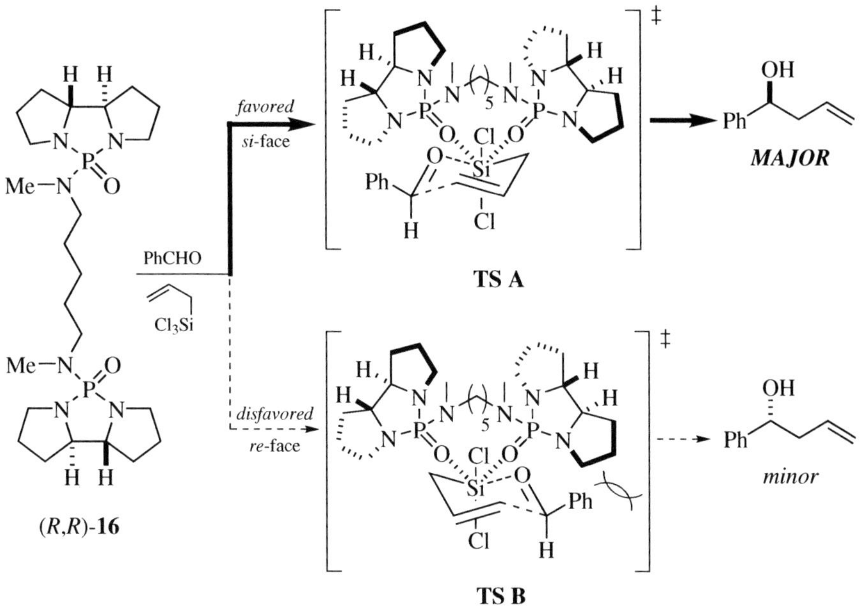

Scheme 3.2l

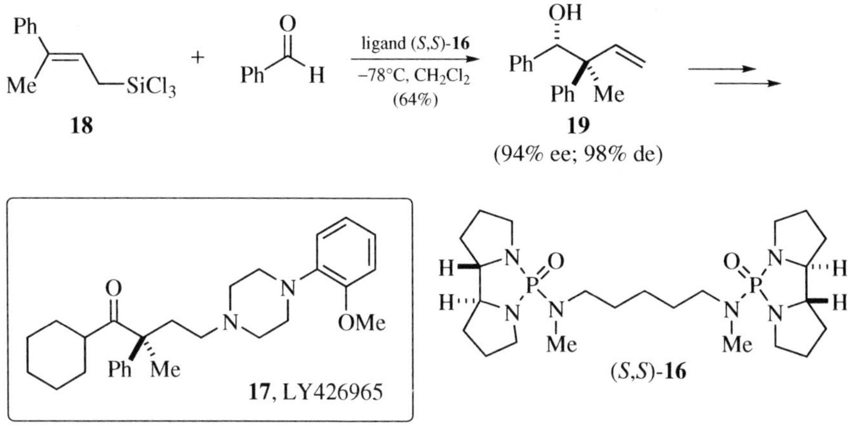

Scheme 3.2m

trifluorosilanes to afford products with high levels of diastereoselectivity. In the synthesis of the C_7–C_{16} segment **13** of the ionophore antibiotic zincophorin **8**, Chemler and Roush performed a reaction of α-methyl-β-hydroxy aldehyde with (Z)-crotyltrifluorosilane to assemble an *anti,anti*-dipropionate stereotriad[11] (Scheme 3.2j). The chiral aldehyde **9** reacted with (Z)-crotyltrifluorosilane to give the desired isomer (**10**) in 93:7 selectivity. The reaction presumably proceeds

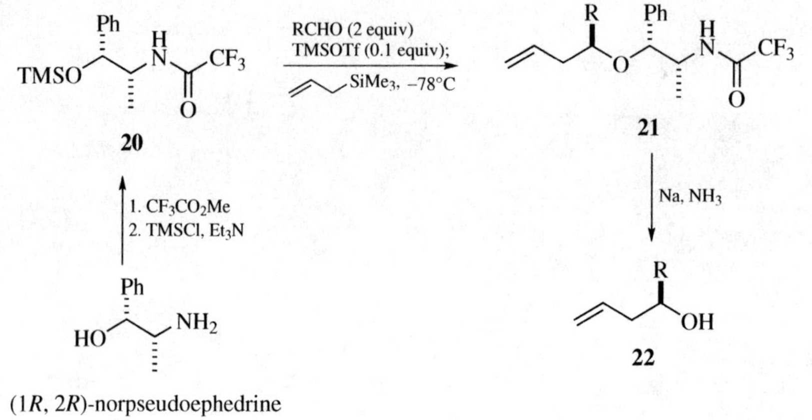

136 METAL ALLYLATION REACTIONS

R	de, % (21)	Yield, % 21	Yield, % 22
CH_3	>99	52	—[a]
$(CH_2)_7CH_3$	>99	65	87
$CH(C_2H_5)_2$	>99	71	88
$c\text{-}C_6H_{11}$	90	49	90
$t\text{-}C_4H_9$	>99	55	82
C_6H_5	56	73	0
$C_6H_4\text{-}p\text{-}OMe$	96	80	75

[a] The yields were not reported.

Scheme 3.2n

through the bicyclic transition state **9TS**, in which the β-hydroxyl group of **9** is coordinated to the silicon center of the (Z)-crotyltrifluorosilane, which requires that the aldehyde alkyl substituent adopt an axial position in the six-membered transition state **9TS**. The crotylsilane would then attack opposite to the aldehyde α-methyl group, resulting in a highly *anti*-selective crotylation.[12] Protection of **10** as an acetonide, dihydroxylation of the terminal olefin, and oxidative cleavage of the resulting diol afforded the aldehyde **11**. The reaction of compound **11** with the vinyllithium species **12** proceeded in 86:14 selectivity to furnish the major diastereomer, which was subjected to hydrogenation and alcohol protection to provide the known C_7–C_{16} segment **13** of zincophorin.[13]

Asymmetric allylation and crotylation reactions using allylic trichlorosilanes and chiral phosphonamides were developed by Denmark and coworkers in 1994 and further refinement of the chiral ligands system was made in 2001[14] (Scheme 3.2k). The influence of the six-membered chairlike transition state is once again evidenced by the excellent correlation of the geometry of the reacting silanes with the diastereomeric composition of the products. Thus, *anti*-isomer is obtained from the *E*-allylic silane, and *syn*-isomer is produced from the *Z*-silane. Based on

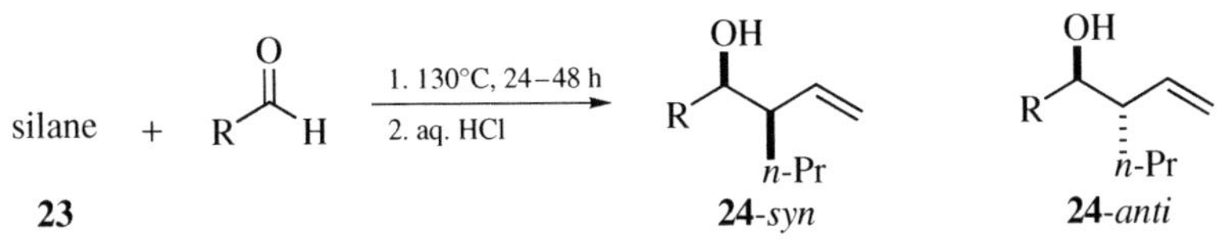

Silane		R	syn/anti	Yield, %
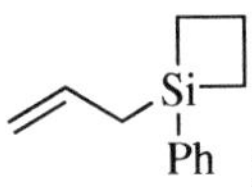 23a		Ph	—	85
		n-C$_6$H$_{13}$	—	58
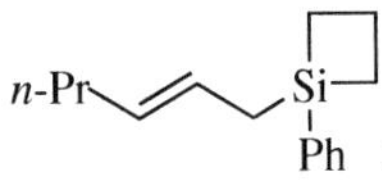 23b		Ph	5:95	68
		n-C$_6$H$_{13}$	10:90	59
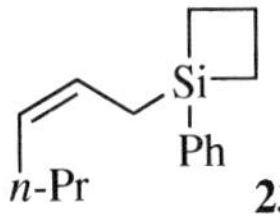 23c		Ph	95:5	66
		n-C$_6$H$_{13}$	80:20	60

Scheme 3.2o

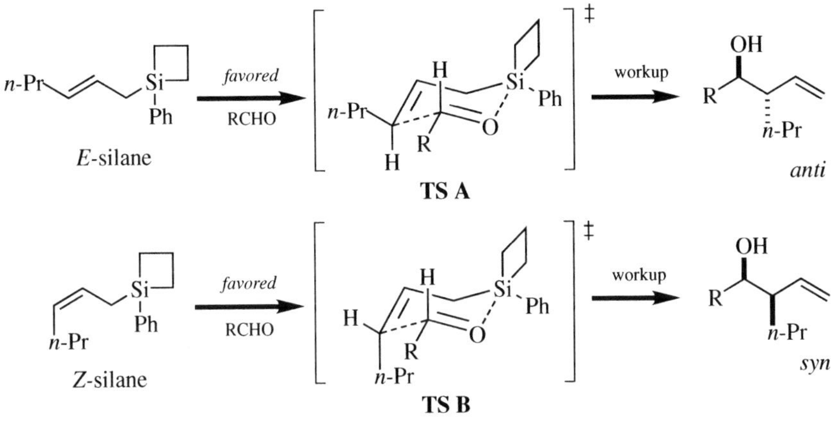

Scheme 3.2p

experimental data,[15a] Denmark et al. proposed a closed six-membered chairlike transition state in rationalizing the enantioselectivity realized with the bisphosphonamide **16**[15b] (Scheme 3.2l). Transition state **B** would be disfavored because the phenyl ring is experiencing a steric collision with a forward-pointing pyrrolidine

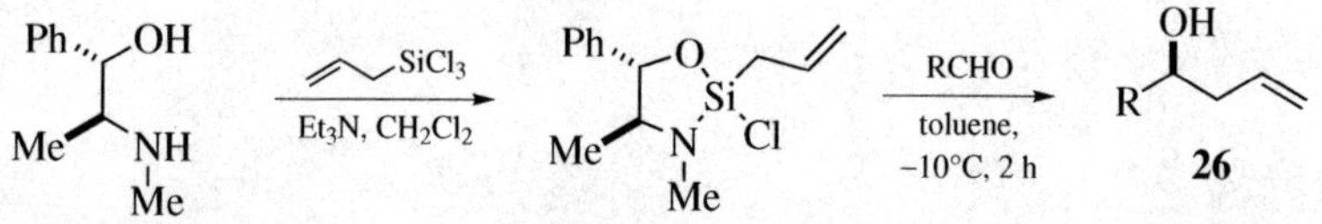

Scheme 3.2q

(1S,2S)-pseudoephedrine → (S,S)-**25** (dr 2:1) → **26**

Reagents: allyl-SiCl₃, Et₃N, CH₂Cl₂; then RCHO, toluene, −10°C, 2 h

R	ee, %	Yield, %
Ph	81	80
PhCH=CH	78	59
PhCH₂CH₂	88	84
c-C₆H₁₁	87	70
t-C₄H₉	96	80
BnOCH₂	88	85

Scheme 3.2r

(R,R)-**27** (from (1R,2R)-1,2-diaminocyclohexane) + RCHO → **28**, CH₂Cl₂, −10°C, 20 h

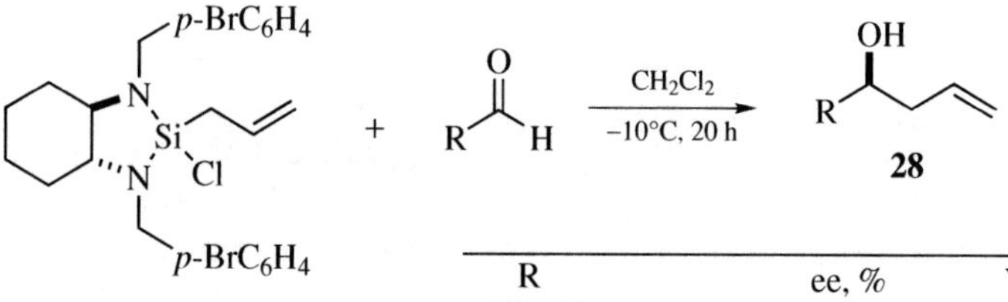

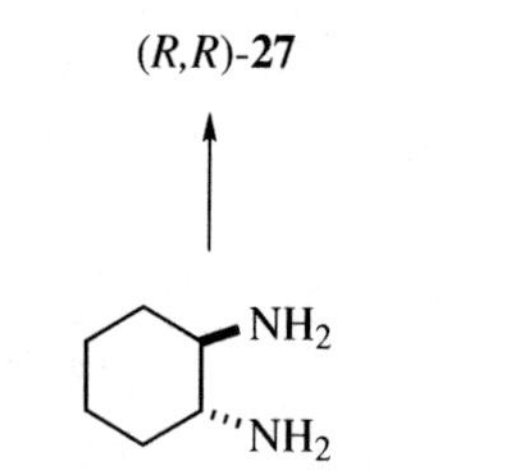

R	ee, %	Yield, %
PhCH₂CH₂	98	90
c-C₆H₁₁	96	93
BnOCH₂	97	67
Ph	98	69
p-MeOC₆H₄	96	62
PhCH=CH	96	75

ring. In the favored transition state, **A**, the allyl group transfer occurs on the si-face of the benzaldehyde to deliver the major enantiomer.

Denmark utilized the asymmetric allylation methodology in the synthesis of the serotonin antagonist LY426965 (**17**), preclinical studies of which suggest its pharmacotherapy use for smoking cessation and depression-related disorders[16] (Scheme 3.2m). The key intermediate alcohol (**19**) was prepared in 94% ee via the asymmetric addition of the silane **18** to benzaldehyde in the presence of the chiral ligand (S,S)-**16**.

In 1992, Tietze et al. discovered an elegant method for the direct and simple preparation of homoallylic ethers with excellent de values (>99%) using the

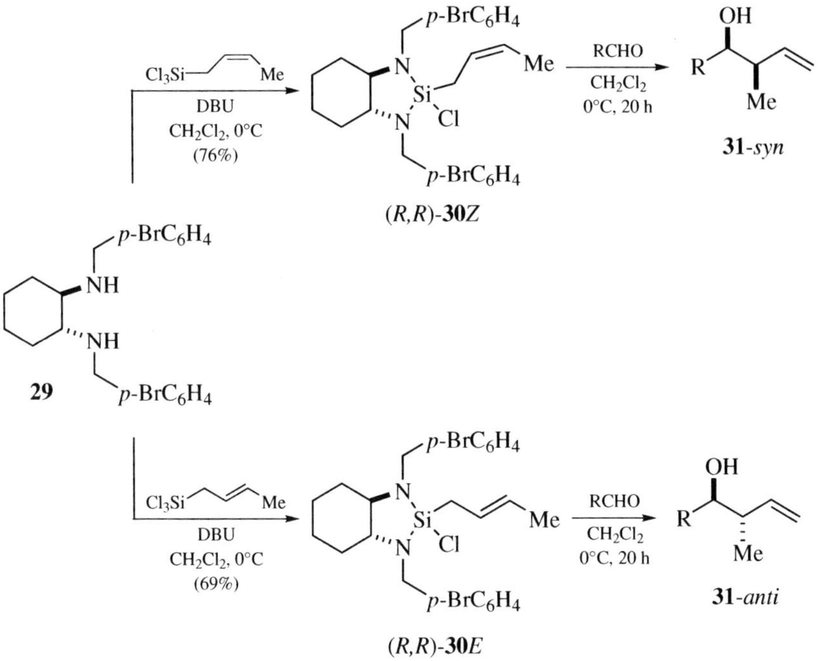

	31-*syn* (w/30Z)		31-*anti* (w/30E)	
R	ee, %	Yield, %	ee, %	Yield, %
PhCH$_2$CH$_2$	97	83	98	81
c-C$_6$H$_{11}$	97	67	97	68
Ph	95	67	93	54
p-CF$_3$C$_6$H$_4$	96	61	96	60
PhCH=CH	95	67	94	52

Scheme 3.2s

trimethylsilyl ether derivative **20** of (1*R*,2*R*)-*N*-trifluoroacetylnorpseudoephedrine[17] (Scheme 3.2n). After aldehydes and the TMS ether **20** were stirred for an hour at −78 °C in the presence of a catalytic amount of TMSOTf, the resulting mixture was treated with 2 equivalents of allyltrimethylsilane at −78 °C for 48 hours. The homoallylic ethers **21** were obtained after workup with nearby perfect diastereoselectivity and in good yields. The homoallylic alcohols **22** were then obtained by reductive cleavage of the chiral auxiliary in **21** with sodium in liquid ammonia. One limitation to Tietze's method is that 1-phenyl-3-buten-1-ol (**22**, R = Ph) cannot be obtained, due to the incompatibility of the reductive cleavage protocol with the corresponding ether (**21**, R = Ph).

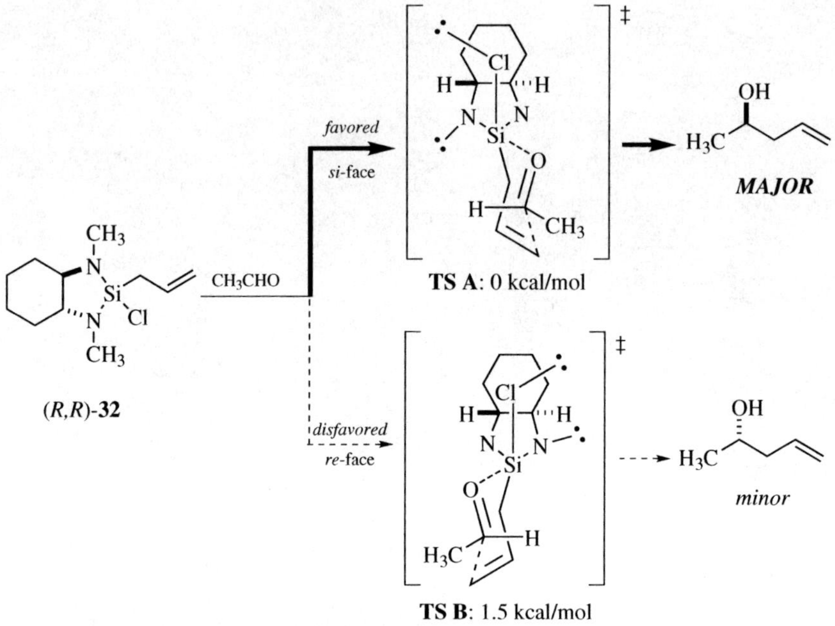

Scheme 3.2t

The incorporation of silicon into a strained ring provides another avenue for allowing allylsilane reagents to participate in six-membered transition states. Although allyldimethylphenylsilane did not add to benzaldehyde even after heating at 160 °C for 24 hours, 1-allyl-1-phenylsilacyclobutane (**23a**) did react with benzaldehyde at 130 °C to provide the homoallylic alcohol **24** in 85% yield[18] (Scheme 3.2o). Not only do the allylic silacyclobutanes undergo allylation reactions with aldehydes, but the corresponding (*E*)- and (*Z*)-crotylsilacyclobutanes (**23b**, **23c**) exhibit a high level of diastereoselectivity in the addition reactions. Thus, (*E*)-1-(2-hexenyl)silacyclobutane (**23b**) reacts with benzaldehyde to provide the corresponding *anti*-homoallylic alcohol **24-*anti*** as a major product in high *anti*/*syn* (95:5) diastereoselectivity. Similarly, (*Z*)-1-(2-hexenyl)silacyclobutane (**23c**) affords a product (**24-*syn***) with an excellent level of diastereoselectivity.

The stereochemical outcome of allylic silacyclobutanes is explained in terms of a six-membered chairlike transition state[18] (Scheme 3.2p). The aldehyde would coordinate to the Lewis acidic silicon[19] to form a pentacoordinated complex, from which the allyl group transfer occurs subsequently to give homoallylic alcohols. In transition states **A** and **B**, the aryl or alkyl group (R) of the aldehydes prefers to occupy an equatorial position to produce *anti*- and *syn*-alcohols from the (*E*)- and (*Z*)-allylic silacyclobutanes, respectively.

Hinted at by Tietze's work and based on the phenomenon that four- or five-membered cyclic silicon compounds exhibit substantial Lewis acidity for

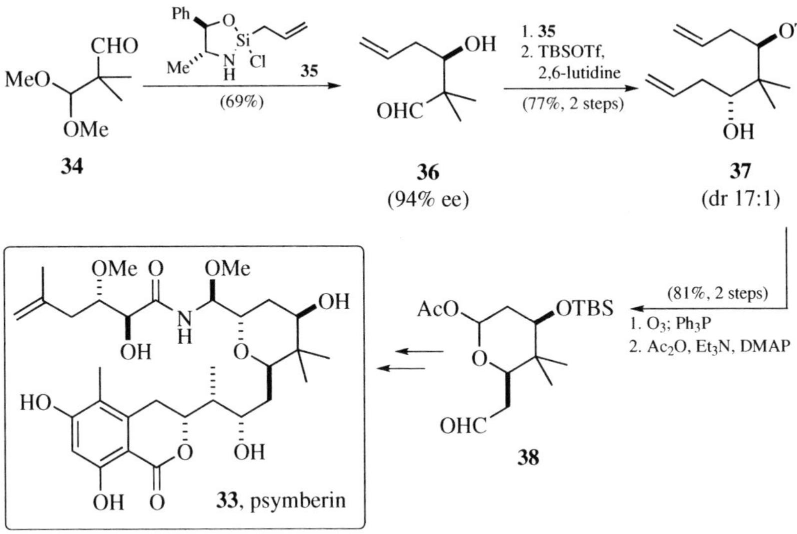

Scheme 3.2u

uncatalyzed allylation reactions, Leighton et al. designed a pseudoephedrine-derived allylsilane reagent (**25**). Treatment of (*1S,2S*)-pseudoephedrine with allyltrichlorosilane afforded **25** as an inseparable 2 : 1 mixture of diastereomers in 88% yield[20] (Scheme 3.2q). Reaction of **25** with benzaldehyde in toluene at −10 °C generated the corresponding homoallylic alcohol with enantioselectivity of 81% ee. Although the reagent **25** is only moderately effective with aromatic and conjugated aldehydes, the levels of enantioselectivity are generally satisfactory for a range of aliphatic aldehydes.

Kubota and Leighton later introduced a diamine-based strained silacycle (**27**) as an enatioselective allylation reagent. Upon reaction with aldehydes, **27** provides chiral homoallylic alcohols in good-to-excellent yields and, more important with uniformly excellent levels of enantioselectivity with both aliphatic and aromatic aldehydes[21] (Scheme 3.2r). Leighton et al. developed two diamine-based reagents, **30*E*** and **30*Z***, for aldehyde crotylation reactions[22] (Scheme 3.2s). Both crotylsilane reagents are easily prepared in bulk and provide the homoallylic alcohols with excellent diastereo- and enantioselectivity. The diamine **29** reacts with *cis*- and *trans*-crotyltrichlorosilanes in the presence of 2 equivalents of DBU to generate reagents (*R,R*)-**30*Z*** and (*R,R*)-**30*E***, respectively. When a variety of aliphatic, aromatic, and α,β-unsaturated aldehydes are treated with the crotylsilane reagent **30*Z***, the corresponding *syn*- diastereomers are obtained in good to excellent diastereoselectivities and greater than 95% ee. Similarly, the reagent **30*E*** reacts with a wide range of aldehydes to afford *anti*-diastereomers with excellent enantioselectivity.

Houk et al. performed a theoretical study on the asymmetric silicon allylation reaction to rationalize the high levels of enantioselectivity realized with

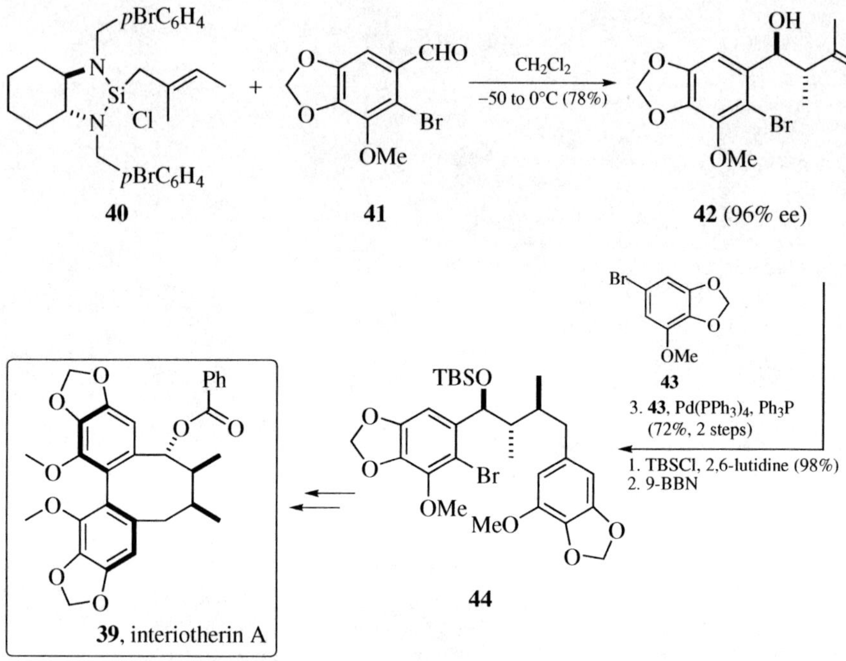

Scheme 3.2v

Leighton's silicon reagent, (*R*,*R*)-**32**[23] (Scheme 3.2t). Calculations indicate that the reaction occurs in a single concerted step through a six-membered chairlike transition state with a pentacoordinate silicon. The two transition states, **A** and **B**, are of the lowest energies that can explain the formation of each enantiomer. Transition State **A** is the lowest-energy transition state because the methyl group of the acetaldehyde occupies an equatorial position in the chairlike transition state and the lone-pair electrons of the nitrogen directly opposite the aldehyde are pointing downward to avoid electronic collision with the lone pair electrons of the chlorine atom. In Transition State **A**, the allyl group transfer occurs to the *si*-face of acetaldehyde to afford the major enantiomer observed experimentally. Transition state **B** which would lead to the formation of a minor enantiomer, is less stable than **A** by 1.5 kcal/mol, due to the unfavorable electron repulsion between the lone-pair electrons of the nitrogen and chlorine atoms.

De Brabander et al. employed Leighton's silane reagent in the synthesis of psymberin **33**, which shows exceptional cell line–specific cytotoxicity[24] (Scheme 3.2u). The monoprotected dialdehyde **34** underwent asymmetric allylation reaction using Leighton's chiral silane reagent **35** with deprotection during workup to give **36** with 94% ee. A second allylation performed on the aldehyde **36** (dr 17 : 1) followed by monosilylation, furnished compound **37**. Ozonolysis of **37** afforded

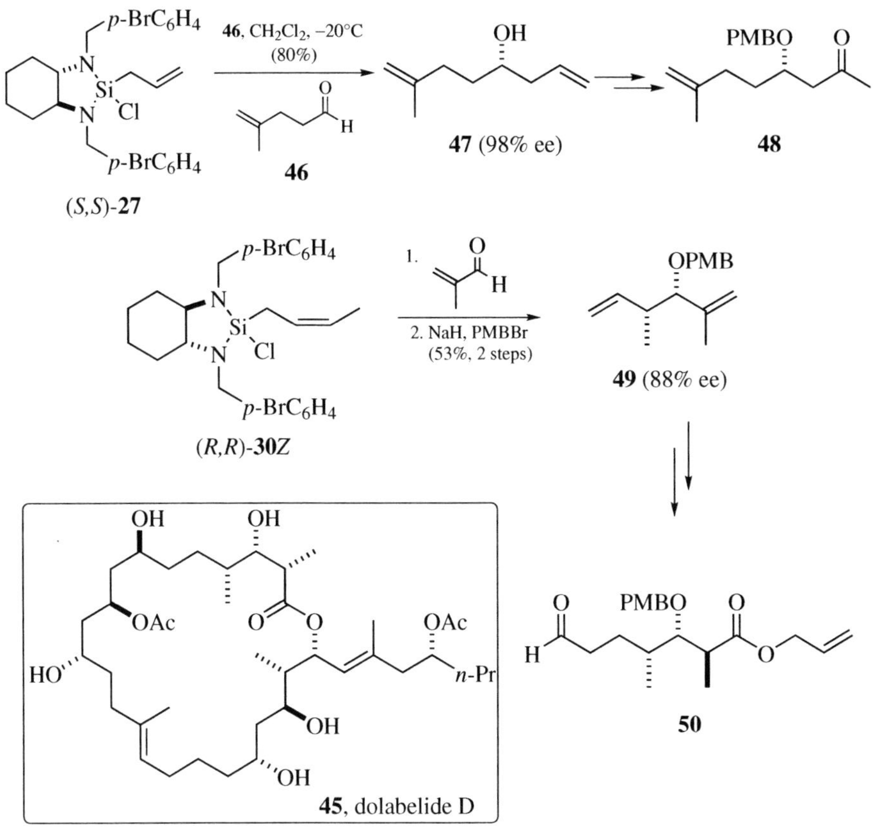

Scheme 3.2w

a lactol, which was then protected as an acetate (**38**), from which psymberin was synthesized.

The total synthesis of dibenzocyclooctadiene lignan natural product interiotherin A (**39**) involved a novel crotylation sequence using the Leighton silane auxiliary[25] (Scheme 3.2v). Thus, the addition of the chiral tiglylsilane **40** to the aldehyde **41** occurred smoothly to afford the *anti*-adduct **42** with complete diastereoselectivity and excellent enantioselection (96% ee). Protection of the benzylic hydroxyl group of **42** as the *tert*-butyldimethylsilyl ether followed by hydroboration of the alkene moiety with 9-BBN furnished the corresponding primary alkylborane, which was then without isolation subjected to a reaction with the aryl bromide **43** under the standard Suzuki–Miyaura coupling conditions to provide 1,4-diarylbutane (**44**).

Leighton et al. used chiral allylsilane and crotylsilane reagents to synthesize two key intermediates, **48** and **50**, in the total synthesis of dolabelide D (**45**), a 24-membered macrolide with cytotoxicity against HeLa-S_3 cells[26] (Scheme 3.2w). Asymmetric allylation of the aldehyde **46** with the chiral reagent (*S*,*S*)-**27** proceeded smoothly to give **47** in 80% yield and 98% ee. The crotylation of methacrolein with (Z)-crotylsilane (R,R)-**30Z** and subsequent protection of the resulting alcohol as a p-methoxybenzyl (PMB) ether provided **49** in 53% yield and 88% ee.

REFERENCES

1. Hayashi, T.; Kabeta, K.; Hamachi, I.; Kumada, M. *Tetrahedron Lett*. **1983**, *24*, 2865.
2. For reviews on silicon allylation reactions, see (a) Hosomi, A. *Acc. Chem. Res*. **1988**, *21*, 200, (b) Fleming, I.; Barbero, A.; Walter, D. *Chem. Rev*. **1997**, *97*, 2063, (c) Miura, K.; Hosomi, A. In *Main Groups Metals in Organic Synthesis*; Yamamoto, H., Oshima, K., Eds.; Wiley-VCH: Weinheim, Denmark, **2004**; Chapt. 10.
3. Kira, M.; Sato, K.; Sakurai, H. *J. Am. Chem. Soc*. **1988**, *110*, 4599.
4. Hosomi, A.; Sakurai, H. *Tetrahedron Lett*. **1976**, *17*, 1295.
5. Fleischer, H. *Eur. J. Inorg. Chem*. **2001**, 393.
6. For reviews on the hypervalent silicon compounds, see (a) Chuit, C.; Corriu, R. J. P.; Reye, C.; Young, J. C. *Chem. Rev*. **1993**, *93*, 1371; (b) Rendler, S.; Oestreich, M. *Synthesis* **2005**, 1727.
7. (a) Kobayashi, S.; Nishio, K. *J. Org. Chem*. **1994**, *59*, 6620; (b) Kobayashi, S.; Nishio, K. *Tetrahedron Lett*. **1993**, *34*, 3453.
8. (a) Kira, M.; Kobayashi, M.; Sakurai, H. *Tetrahedron Lett*. **1987**, *28*, 4081; (b) Kira, M.; Hino, T.; Sakurai, H. *Tetrahedron Lett*. **1989**, *30*, 1099.
9. Kira, M.; Sato, K.; Sakurai, H.; Hada, M.; Izawa, M.; Ushio, J. *Chem. Lett*. **1991**, 387.
10. Sato, K.; Kira, M.; Sakurai, H. *J. Am. Chem. Soc*. **1989**, *111*, 6429.
11. Chemler, S. R.; Roush, W. R. *J. Org. Chem*. **1998**, *63*, 3800.
12. Chemler, S. R.; Roush, W. R. *J. Org. Chem*. **2003**, *68*, 1319.
13. (a) Danishefsky, S. J.; Selnick, H. G.; DeNinno, M. P.; Zelle, R. E. *J. Am. Chem. Soc*. **1987**, *109*, 1572; (b) Danishefsky, S. J.; Selnick, H. G.; Zelle, R. E.; DeNinno, M. P. *J. Am. Chem. Soc*. **1988**, *110*, 4368.
14. (a) Denmark, S. E.; Coe, D. M.; Pratt, N. E.; Griedel, B. D. *J. Org. Chem*. **1994**, *59*, 6161; (b) Denmark, S. E.; Fu, J. *J. Am. Chem. Soc*. **2001**, *123*, 9488.
15. (a) Denmark, S. E.; Fu, J.; Coe, D. M.; Su, X.; Pratt, N. E.; Griedel, B. D. *J. Org. Chem*. **2006**, *71*, 1513; (b) Denmark, S. E.; Fu, J.; Lawler, M. J. *J. Org. Chem*. **2006**, *71*, 1523.
16. Denmark, S. E.; Fu, J. *Org. Lett*. **2002**, *4*, 1951.
17. (a) Tietze, L. F.; Dölle, A.; Schiemann, K. *Angew. Chem. Int. Ed*. **1992**, *31*, 1372; (b) Tietze, L. F.; Wulff, C.; Wegner, C.; Schuffenhauer, A.; Schiemann, K. *J. Am. Chem. Soc*. **1998**, *120*, 4276.
18. Matsumoto, K.; Oshima, K.; Utimoto, K. *J. Org. Chem*. **1994**, *59*, 7152.
19. (a) Myers, A. G.; Kephart, S. E.; Chen, H. *J. Am. Chem. Soc*. **1992**, *114*, 7922. (b) Denmark, S. E.; Griedel, B. D.; Coe, D. M. *J. Org. Chem*. **1993**, *58*, 988.
20. Kinnaird, J. W. A.; Ng, P. Y.; Kubota, K.; Wang, X.; Leighton, J. L. *J. Am. Chem. Soc*. **2002**, *124*, 7920.
21. Kubota, K.; Leighton, J. L. *Angew. Chem. Int. Ed*. **2003**, *42*, 946.

22. Hackman, B. M.; Lombardi, P. J.; Leighton, J. L. *Org. Lett.* **2004**, *6*, 4375.
23. Zhang, X.; Houk, K. N.; Leighton, J. L. *Angew. Chem. Int. Ed.* **2005**, *44*, 938.
24. Jiang, X.; Garcia-Fortanet, J.; De Brabander, J. K. *J. Am. Chem. Soc.* **2005**, *127*, 11254.
25. Coleman, R. S.; Gurrala, S. R.; Mitra, S.; Raao, A. *J. Org. Chem.* **2005**, *70*, 8932.
26. Park, P. K.; O' Malley, S. J.; Schmidt, D. R.; Leighton, J. L. *J. Am. Chem. Soc.* **2006**, *128*, 2796.

4 Stereoselective Reductions

GENERAL CONSIDERATIONS

Since the discovery of sodium borohydride ($NaBH_4$) in 1942 and lithium aluminum hydride ($LiAlH_4$) in 1945, the reduction of carbonyl compounds has become one of the most versatile reactions in organic synthesis.[1] Many variants of these two metal hydrides have been created to provide synthetic organic chemists with a means to access organic compounds with a variety of functional groups.[2] Lithium aluminum hydride is a powerful reagent for reducing ketones to alcohols.[3] In 1976, Ashby and Boone conducted a series of kinetic experiments to better understand the mechanism of lithium aluminum hydride reduction of ketones. They observed that the rate of reduction depends on the type of countercation associated with aluminum hydride[4] (Scheme 4.I). When the mesityl phenyl ketone **1** was treated with lithium and sodium aluminum hydrides in THF at 25°C, $LiAlH_4$ reacted 10 times faster than $NaAlH_4$ to produce the alcohol **2**.

Based on the accompanying kinetic data and an observation that lithium cation is essential in the lithium aluminum hydride reduction,[5] Ashby and Boone proposed that the reduction would occur via a six-membered transition state in which the lithium cation is involved[4] (Scheme 4.II). Because the aluminum in the boat transition state **TS**-*boat* is proximal to the carbonyl oxygen, the boat transition state might be of lower energy than the chairlike transition state **TS**-*chair*. Furthermore, the boatlike transition state would be a favored states as it results in direct formation of the lithium alkoxyaluminum hydride intermediate.

Ashby and Boone's proposed mechanism was not verified until 2001, when Luibrand et al. carried out a theoretical study on the lithium aluminum hydride reduction of formaldehyde[6] (Scheme 4.III). Two types of complexes are possible between formaldehyde and $LiAlH_4$, depending on the geometry of $LiAlH_4$; complex **A** is from tridentate η^3-$LiAlH_4$, and complex **B** is from bidentate η^2-$LiAlH_4$.[7] Calculations indicate that the intermediate product **OLiAl** would be formed via six-membered transition state **A**, derived from complex **A**, which is more stable than transition state **B** by 2.0 kcal/mol.

Six-Membered Transition States in Organic Synthesis, By Jaemoon Yang
Copyright © 2008 John Wiley & Sons, Inc.

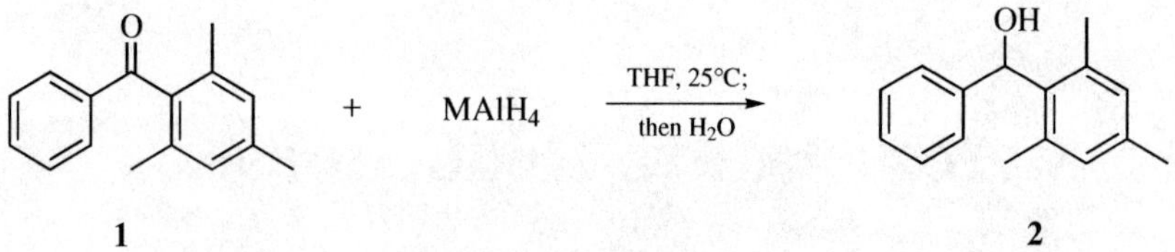

148 STEREOSELECTIVE REDUCTIONS

M	Rel. Rate
Li	10
Na	1

Scheme 4.I

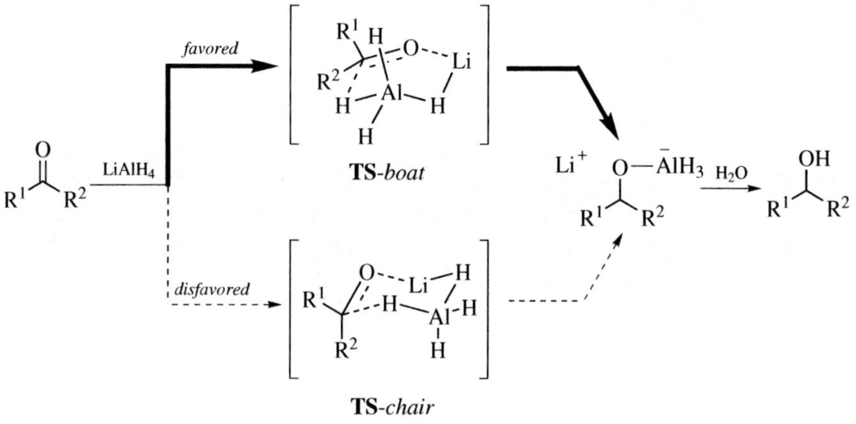

Scheme 4.II

The enantioselective reduction of unsymmetrical ketones to produce optically active secondary alcohols has been one of the most vibrant topics in organic synthesis.[8] Perhaps Tatchell et al. were first (in 1964) to employ lithium aluminum hydride to achieve the asymmetric reduction of ketones[9] (Scheme 4.IV). When pinacolone and acetophenone were treated with the chiral lithium alkoxyaluminum hydride reagent **3**, generated from 1.2 equivalents of 1,2-*O*-cyclohexylidene-D-glucofuranose and 1 equivalent of LiAlH$_4$, the alcohol **4** was obtained in 5 and 14% ee, respectively. Tatchell improved the enantioselectivity in the reduction of acetophenone to 70% ee with an ethanol-modified lithium aluminum hydride–sugar complex.[10]

Since the seminal work by Tatchell's group, many efforts have been directed to developing a highly enatioselective reduction of ketones.[11] One successful

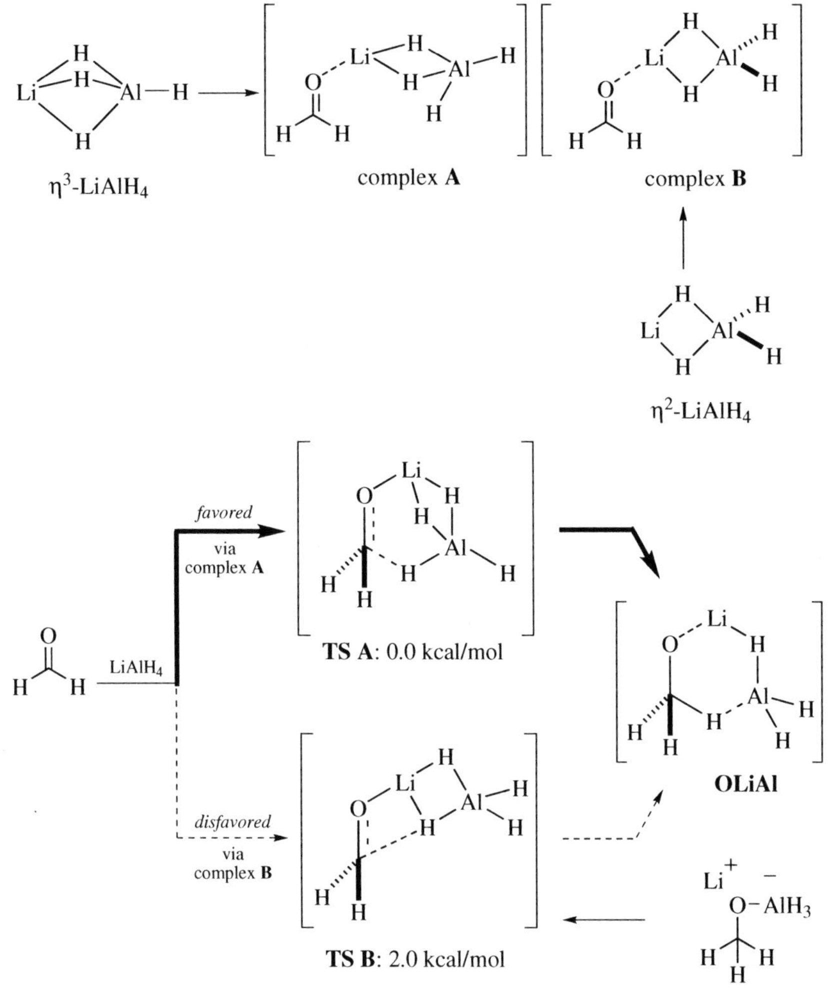

Scheme 4.III

system is the one reported by Jacquet and Vigneron in 1974.[12] The chiral complex **5**, generated by reacting lithium aluminum hydride with 1 equivalent of (−)-*N*-methylephedrine and 2 equivalents of 3,5-dimethylphenol, has been used in the asymmetric reduction of ketones (Scheme 4.V). The chiral alcohol **6** was obtained with high levels of enantioselectivity for a variety of phenyl alkyl ketones. Further improvements have been made in the design of chiral reducing reagents incorporating lithium aluminum hydride. A highly enantioselective chiral reagent is the one developed by Noyori, discussed in Section 4.3.

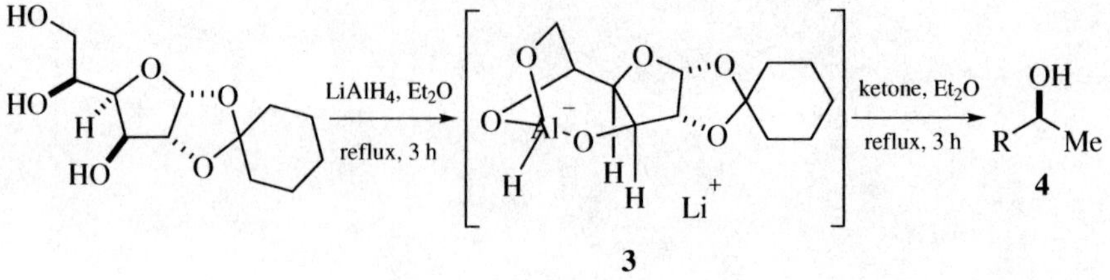

Scheme 4.IV

Ketone	ee, %	Yield, %
t-BuCOMe	5.2	52
PhCOMe	14.3	66

Scheme 4.V

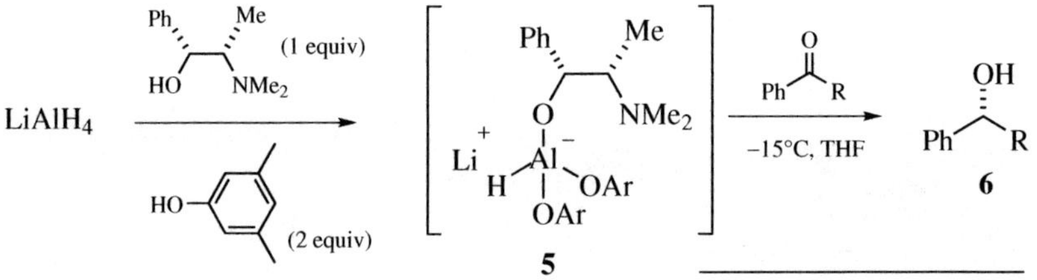

R	ee, %
CH_3	83
C_2H_5	85
$n\text{-}C_3H_7$	89
$n\text{-}C_4H_9$	78

REFERENCES

1. (a) http://nobelprize.org/nobel_prizes/chemistry/laureates/1979/brown-lecture.pdf; (b) Brown, H. C.; Krishnamurthy, S. *Tetrahedron* **1979**, *35*, 567.
2. Seyden-Penne, J. *Reductions by the Alumino- and Borohydrides in Organic Synthesis*, 2nd ed.; Wiley-VCH: New York, **1997**.
3. House, H. O. *Modern Synthetic Reactions*, 2nd ed.; Benjamin-Cummings: Menlo Park, CA, **1972**; Chap. 2.
4. Ashby, E. C.; Boone, J. R. *J. Am. Chem. Soc.* **1976**, *98*, 5524.
5. Pierre, J. L.; Handel, M.; Perrand, R. *J. Tetrahedron* **1975**, *31*, 2795.
6. Luibrand, R. T.; Taigounov, I. R.; Taigounov, A. A. *J. Org. Chem.* **2001**, *66*, 7254.
7. Demachy, I.; Volatron, F. *Inorg. Chem.* **1994**, *33*, 3965.

8. (a) Morrison, J. D.; Mosher, H. S. *Asymmetric Organic Reactions*; Prentice Hall: London, **1971**; pp. 177–202; (b) Mosher, H. S.; Morrison, J. D. *Science* **1983**, *221*, 1013; (c) Yoon, N. M. *Pure Appl. Chem.* **1996**, *68*, 843; (d) Cho, B. T. *Aldrichimica Acta* **2002**, *35*, 3.
9. Landor, S. R.; Miller, B. J.; Tatchell, A. R. *Proc. Chem. Soc.* **1964**, 227.
10. Landor, S. R.; Miller, B. J.; Tatchell, A. R. *J. Chem. Soc.* (C) **1967**, 197.
11. (a) Singh, V. K. *Synthesis* **1992**, 605; (b) Kim, J.; Suri, J. T.; Corde, D. B.; Singaram, B. *Org. Proc. Res. Dev.* **2006**, *10*, 949.
12. Jacquet, I.; Vigneron, J. P. *Tetrahedron Lett.* **1974**, *15*, 2065.

REACTIONS

4.1. Diastereoselective *Syn*-Reduction of β-Hydroxy Ketones

In 1984, Narasaka and Pai demonstrated that high levels of 1,3-asymmetric induction can be realized in the reduction of acyclic β-hydroxy ketones via boron chelates.[1] Treatment of the β-hydroxy ketones **1** with tributylborane in the presence of a catalytic amount of air results in formation of the chelated dibutylboric ester complex **1BBu$_2$**. Upon reaction with sodium borohydride at −78°C for 2 to 6 hours, *syn*-1,3-diols **2-*syn*** are obtained after oxidative workup with good to high levels of diastereoselectivity (Scheme 4.1a). When the reaction is carried out at −100 °C, excellent diastereoselectivity is obtained in the reduction to afford *meso*-undecane-5,7-diol. The high level of 1,3-asymmetric induction can be explained by considering the six-membered transition states in Scheme 4.1b.[2]

The hydride nucleophile prefers to attack the carbonyl group of the borane complex **1BBu$_2$** from the top face following the Bürgi–Dunitz trajectory.[3] The

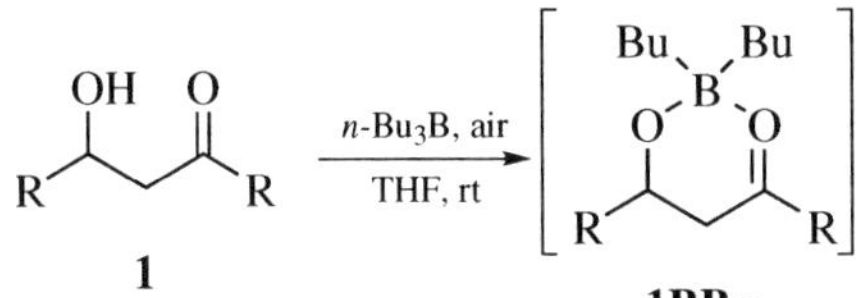

R	syn/anti	Yield, %
Ph	98:2	86
n-C$_4$H$_9$	88:12	91
	(96:4)^a	90^a
c-C$_6$H$_{11}$	73:27	95

^{a}Reaction at −100°C.

Scheme 4.1a

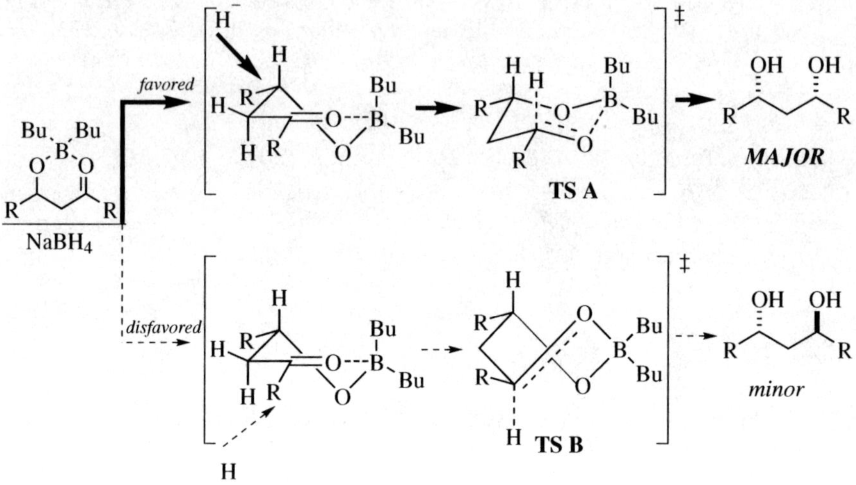

Scheme 4.1b

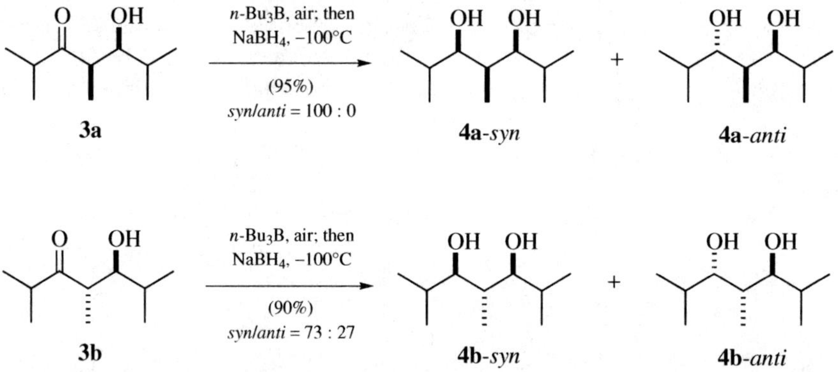

Scheme 4.1c

resulting six-membered chairlike transition state **A**, in which both bulky R groups occupy an equatorial position, is more stable than transition state **B**.[4] The approach of the hydride from the bottom face would be sterically disfavored due to the presence of the axial hydrogen of the α-carbon.[1] Furthermore, the bottom attack would lead to a twist-boatlike transition state **B**, which is of higher energy than the chairlike transition state **A**, due to interactions between the R group and β-hydrogens.[5] Therefore, the 1,3-*syn*-diols are formed as a major diastereomer via transition state **A**.

To determine the effect of an α-substituent on the diastereoselectivity of the BBu$_3$/NaBH$_4$ system, Narasaka and Pai carried out the reduction of α-methyl-β-hydroxy ketones[1] (Scheme 4.1c). Whereas the α,β-*syn*-β-hydroxy

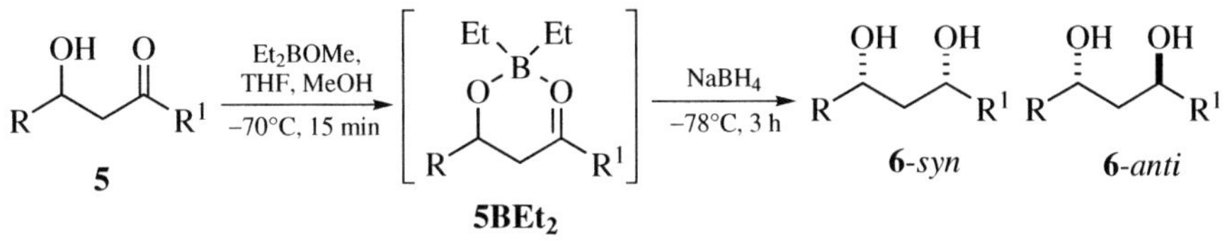

Scheme 4.1d

R	R^1	syn/anti	Yield, %
Ph	Ph	99:1	95
Ph	CH$_2$CO$_2$Et	98:2	85
n-C$_4$H$_9$	n-C$_4$H$_9$	99:1	99

ketone **3a** underwent reduction in excellent yields and with very high stereoselectivity, the diastereomeric α,β-*anti*-β-hydroxy ketone **3b** exhibited lower *syn*-selectivity. The interference of an α-substituent on the *syn*-diastereoselectivity is rationalized by considering the conformation of dibutylborinate complexes of **3a** and **3b**.[1]

Three years after Narasaka and Pai's disclosure, Prasad et al. developed a modified procedure to improve *syn*-diastereoselectivity in the reduction of certain β-hydroxy ketones[6] (Scheme 4.1d). When methoxydiethylborane, in lieu of tributylborane, reacts with β-hydroxy ketones at −70 C in anhydrous methanol, the complex **5BEt$_2$** is formed. Subsequent treatment of the complex with sodium borohydride and quenching the reaction mixture with acetic acid affords *syn*-diols in excellent levels of diastereoselectivity regardless of the structure of β-hydroxy ketones. Another practical advantage of Prasad et al.'s modification may be an enhanced safety feature, as methoxydiethylborane is generally less hazardous to handle than triethylborane.[6]

An excellent application of the Narasaka reduction is a diastereoselective synthesis by Merck scientists of **7**, a structurally novel analog of the natural product compactin (**8**)[7], which is a potent inhibitor of 3-hydroxy-3-methylglutaryl coenzyme A (HMG-CoA) reductase[8] (Scheme 4.1e). The key step in the construction of the β-hydroxy-δ-lactone moiety in **7** is the highly diastereoselective reduction of the β-hydroxy ketone **9** using a triethyl borane/sodium borohydride system. The *syn*-diol **10** was obtained in high yield and with a remarkably high level of diastereoselectivity.

In the stereoselective synthesis of epothilone A (**11**), Carreira used a *syn*-reduction methodology in the synthesis of the key intermediate (**14**)[9] (Scheme 4.1f). Reduction of the isoxazoline **12** with samarium iodide at 0 C in THF gave the ketone **13**. Narasaka reduction of the β-hydroxy ketone **13** using triethylborane/sodium borohydride afforded the *syn*-diol **14** in high yield and with high diastereoselectivity.

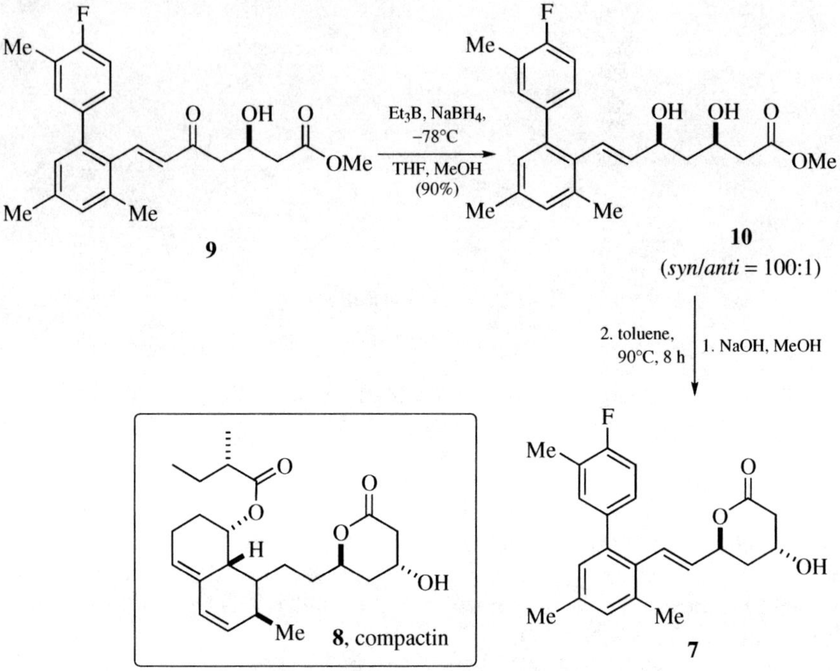

Scheme 4.1e

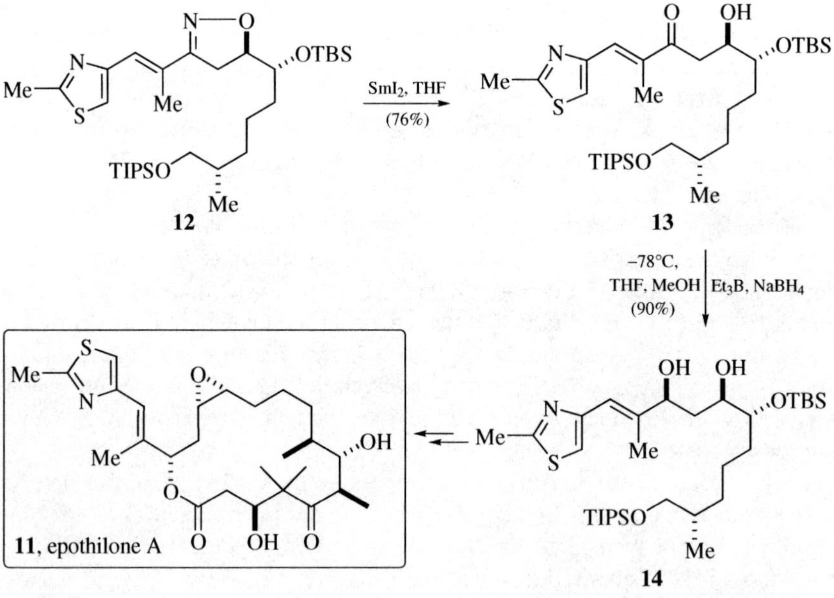

Scheme 4.1f

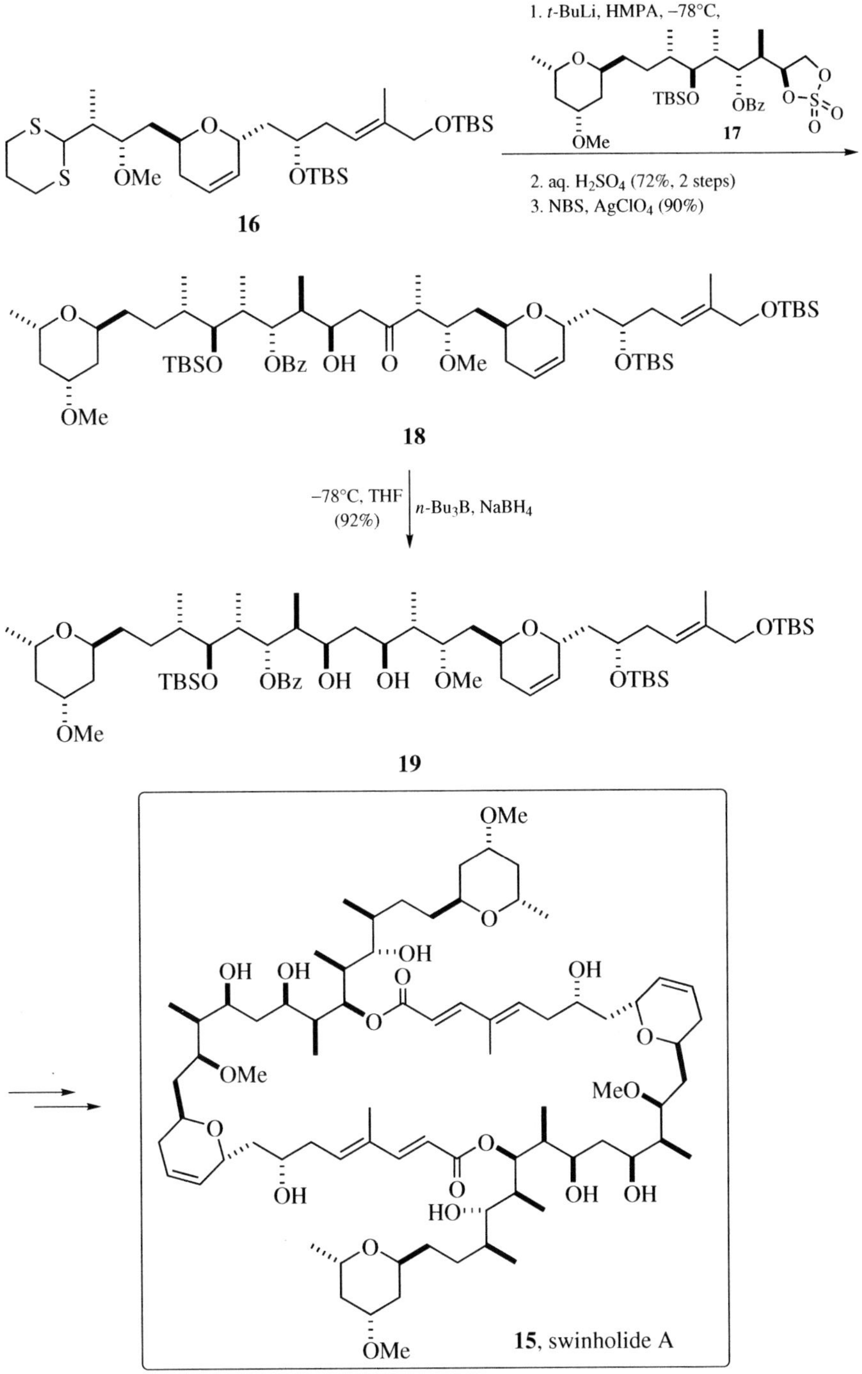
1. t-BuLi, HMPA, −78°C,
TBSO
OBz
OMe
17
2. aq. H2SO4 (72%, 2 steps)
3. NBS, AgClO4 (90%)
S
S
OMe
O
OTBS
OTBS
16
TBSO
OBz OH O
OMe
OTBS
OTBS
OMe
18
−78°C, THF
(92%)
n-Bu3B, NaBH4
TBSO
OBz OH OH OMe
OTBS
OTBS
OMe
19
OMe
OH
OH OH
OH
OMe
MeO
OH
OH OH
HO
OMe
15, swinholide A

Scheme 4.1g

Nicolaou et al. exercised the stereoselective *syn*-reduction of β-hydroxy ketones in the total synthesis of swinholide A (**15**), a marine natural product that displays a range of biological properties, including antifungal activity and potent cytotoxicity against a number of tumor cell lines[10] (Scheme 4.1g). Treatment of **16** with *t*-BuLi in the presence of HMPA generated the lithio derivative of dithiane, which underwent a coupling reaction with the cyclic sulfate **17**. After aqueous acid treatment and removal of the dithiane moiety with *N*-bromosuccinimide (NBS) and AgClO$_4$, the β-hydroxy ketone **18** was obtained, which upon reduction with the NaBH$_4$/*n*-Bu$_3$B system resulted in formation of the 1,3-*syn*-diol **19** in 92% yield.

Fleming and Ghosh employed the β-hydroxy ketone *syn*-selective reduction method to synthesize the intermediate **23** during the synthesis of the nonactin **20**[11] (Scheme 4.1h). Treatment of the starting acid **21** with diazomethane followed by deprotection of the ketal in the presence of pyridinum *p*-toluenesulfonate (PPTS) furnished the β-hydroxy ketone **22**, which was then reduced successfully by employing the reagents NaBH$_4$/Bu$_2$BOMe to give the *syn*-1,3-diol as a 90:10 mixture, of which the major diastereomer was separated from the minor isomer and protected as the acetonide to afford **23** in good yield.

The *syn*-reduction methodology was utilized in the total synthesis of anachelin H (**24**), which was isolated from the freshwater cyanobacterium *Anabaena cylindrical* and postulated to serve as a bacterial growth factor facilitating iron uptake[12] (Scheme 4.1i). The reduction of the β-hydroxy ketone **25** was carried out by precomplexing the substrate with Et$_2$BOMe, formed from Et$_3$B, pivalic

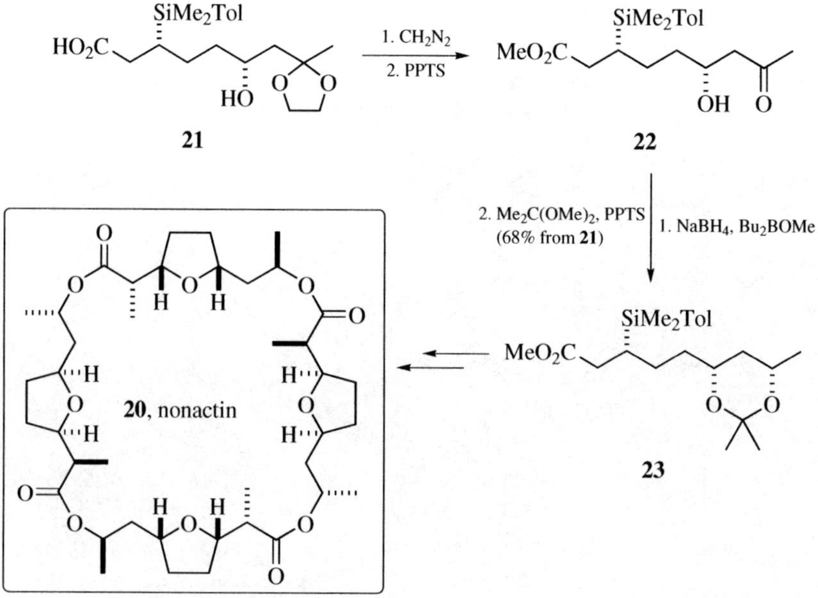

Scheme 4.1h

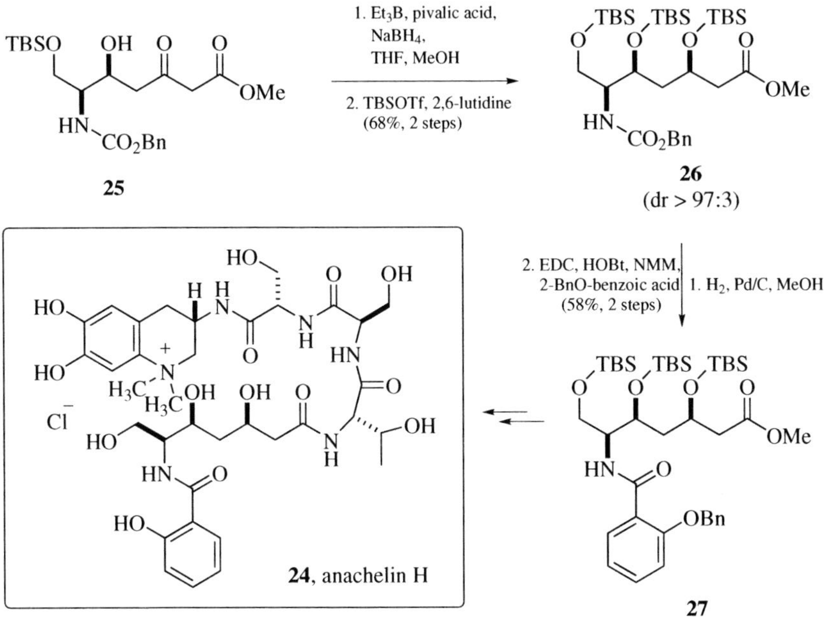

Scheme 4.1i

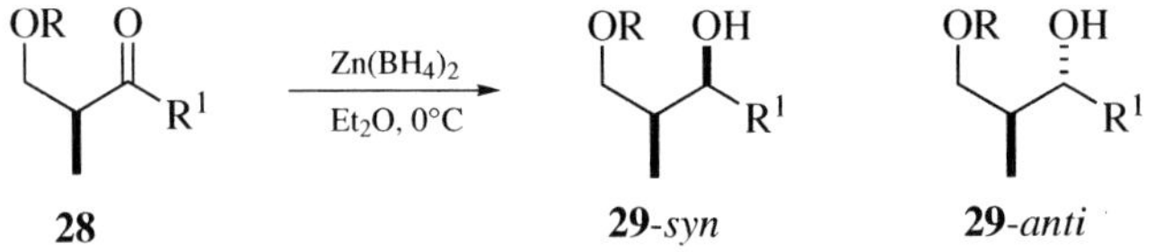

R	R¹	syn/anti	Yield, %
H	Ph	25:1	95
H	CH$_2$=C(Me)	25:1	91
H	PhCH$_2$CH$_2$	1.3:1	97
Me	Ph	33:1	99

Scheme 4.1j

acid, and MeOH, then reducing with NaBH$_4$. The resulting *syn*-diol compound was isolated in good yield and greater than 97:3 diastereoselectivity after subsequent TBS protection. Cleavage of the benzyloxycarbonyl group of **26** followed by condensation with *O*-benzyl salicylic acid afforded the intermediate **27**.

In 1984, Nakata and co-workers reported that zinc borohydride reduces the α-methyl-β-hydroxy or α-methyl-β-methoxy ketone **28** stereoselectively to afford a *syn*-isomer as the major product[13] (Scheme 4.1j). The levels of diastereoselectivity are exceptionally high when the ketones are conjugated. Based on the x-ray crystal structure of $Cp_2Nb(CO)H \cdot Zn(BH_4)_2$,[14] Oishi and Nakata proposed that the zinc cation is coordinated to oxygens of both the alcohol and the carbonyl to form a structurally rigid six-membered chelated complex (**28ZnB**)[15] (Scheme 4.1k). The hydride transfer occurs intramolecularly to form the zinc complex **28ZnO** of the 1,3-diol product. Of the two possible transition states, **A** and **B**, transition state **A** would be the favored one in which the hydride transfer occurs opposite the α-methyl substituent. Transition state **B** would be less favorable than **A**, as the incoming hydride nucleophile experiences a steric interaction with the α-methyl group.[16]

Zinc borohydride has found many synthetic applications in the context of a "chelation-controlled" reduction.[17] In the synthesis of the antibiotic tirandamycin **30**, DeShong et al. prepared a key intermediate (**32**) via stereoselective reduction of a β-silyloxy ketone[18] (Scheme 4.1l). Reduction of **31** with $Zn(BH_4)_2$ gave the mono-TBS-protected 1,2-*syn*-2,3-*anti*-diol **32** stereoselectively. Oxidation of

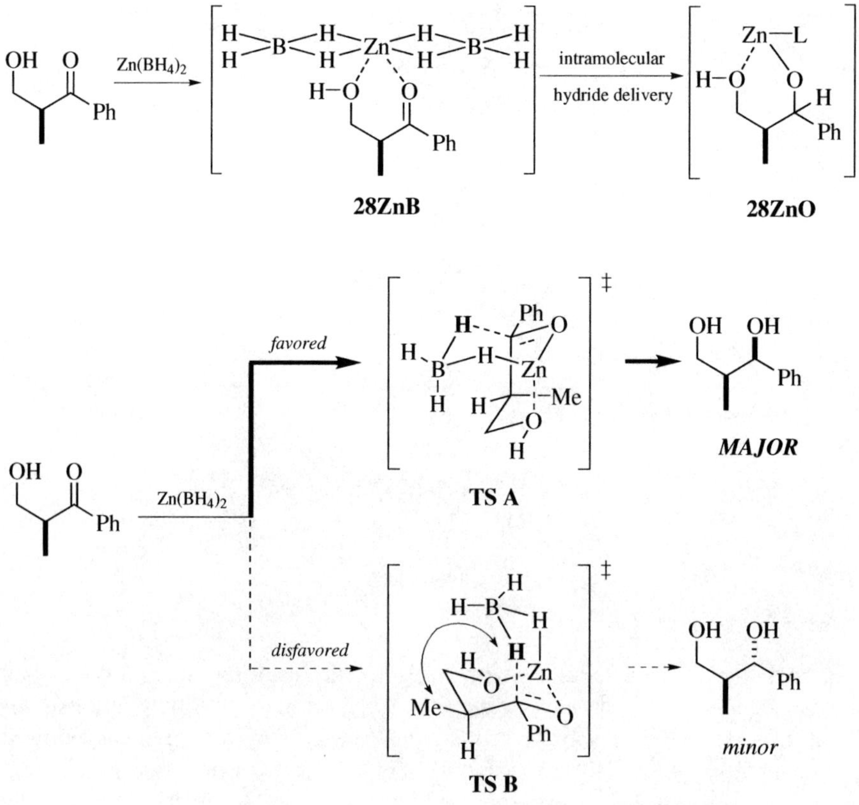

Scheme 4.1k

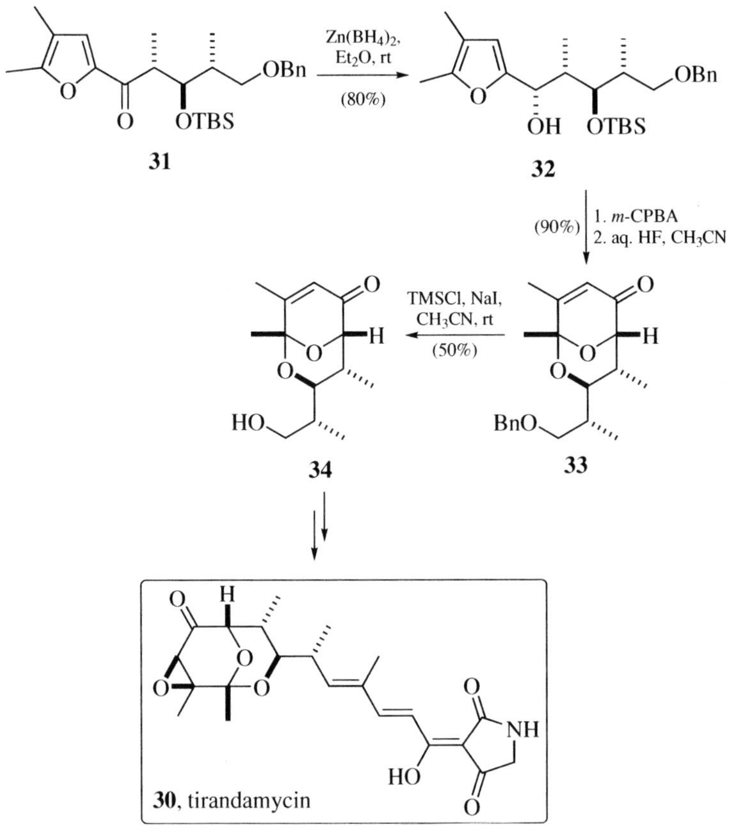

Scheme 4.1l

32 with *m*-chloroperbenzoic acid followed by deprotection afforded the bicyclic enone **33**. Removal of the benzyl ether with trimethylsilyl iodide generated in situ from TMSCl and NaI then provided the alcohol **34**.

In 2005, Carreira et al. reported the total synthesis of erythonolide A (**35**), one of the most popular target molecules in organic synthesis[19] (Scheme 4.1m). The *syn*-reduction of the hydroxy ketone **36** with $Zn(BH_4)_2$ in CH_2Cl_2 at $-30\,°C$ produced the *syn*-diol **37** as a single diastereomer. Protection of the diol with benzaldehyde dimethylacetal in the presence of camphorsulfonic acid (CSA) gave **38** in good yield.

(+)-Conagenin (**39**) is a promising anticancer natural product isolated in 1991 from the culture broths of *Streptomyces roseosporus*[20] (Scheme 4.1n). The chelation-controlled $Zn(BH_4)_2$ reduction of the β-hydroxy ketone **40** produced the 1,3-*syn*-diol **41** in 30 : 1 diastereoselectivity. Protection of the hydroxyl groups gave the bis-acetate **42**, which was then subjected to oxidative cleavage to afford the acid **43**.

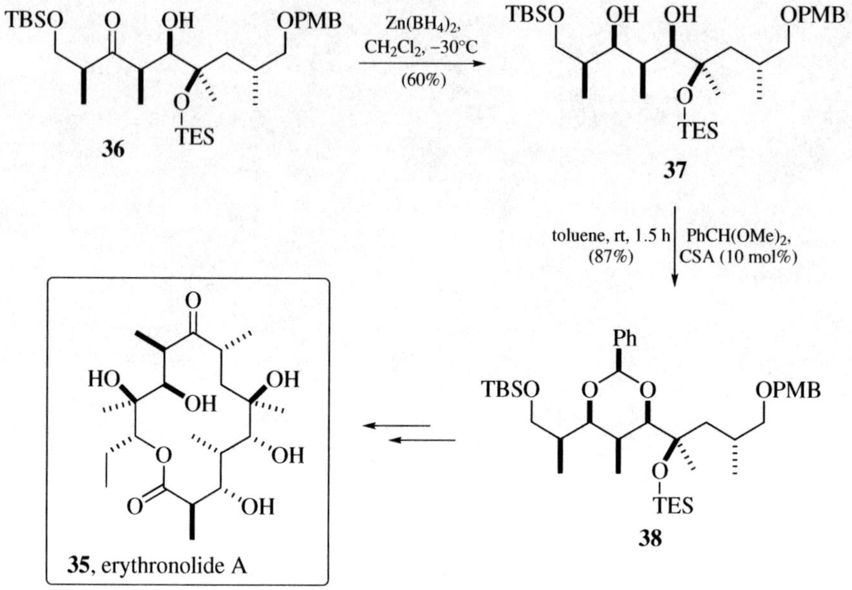

Scheme 4.1m

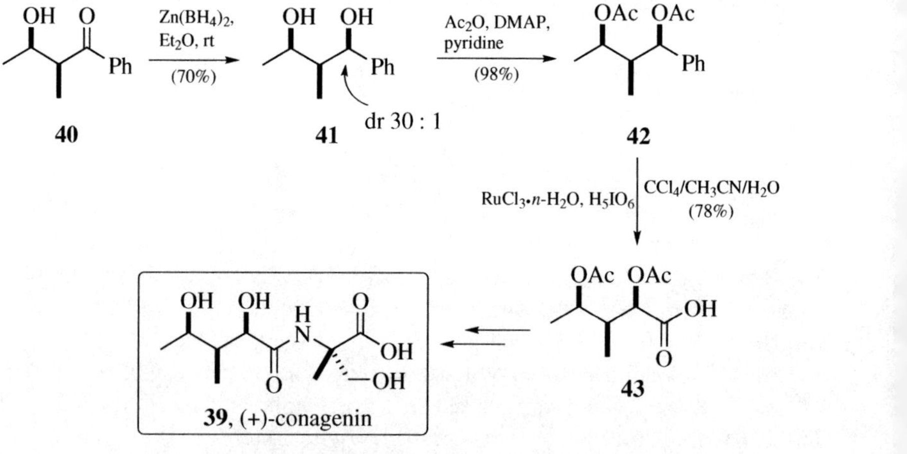

Scheme 4.1n

REFERENCES

1. Narasaka, K.; Pai, F.-C. *Tetrahedron* **1984**, *40*, 2233.
2. For examples of chelation-controlled reactions, see (a) Leitereg, T. J.; Cram, D. J. *J. Am. Chem. Soc.* **1968**, *90*, 4019; (b) Still, W. C.; McDonald, J. H., III. *Tetrahedron Lett.* **1980**, *21*, 1031; (c) Still, W. C.; Schneider, J. A. *Tetrahedron Lett.* **1980**, *21*, 1035.

3. (a) Bürgi, H. B.; Dunitz, J. D.; Shefter, E. J. *J. Am. Chem. Soc.* **1973**, *95*, 5065; (b) Bürgi, H. B.; Dunitz, J. D.; Lehn, J. M.; Wipff, G. *Tetrahedron* **1974**, *30*, 1563; (c) Bürgi, H. B.; Lehn, J. M.; Wipff, G. *J. Am. Chem. Soc.* **1974**, *96*, 1956.

4. For an example of a preferential axial attack of nucleophile on the oxonium ion of six-membered ring, see Lewis, M. D.; Cha, J. K.; Kishi, Y. *J. Am. Chem. Soc.* **1982**, *104*, 4976.

5. (a) Velluz, L.; Valls, J.; Nomine, G. *Angew. Chem. Int. Ed.* **1965**, *4*, 181; (b) Hammond, G. S. *J. Am. Chem. Soc.* **1955**, *77*, 334; (c) Corey, E. J.; Sneen, R. A. *J. Am. Chem. Soc.* **1956**, *78*, 6269; (d) House, H. O.; Umen, M. J. *J. Org. Chem.* **1973**, *38*, 1000.

6. Chen, K.-M.; Hardtmann, G. E.; Prasad, K.; Repic, O.; Shapiro, M. J. *Tetrahedron Lett.* **1987**, *28*, 155.

7. For selected total syntheses of compactin, see (a) Robichaud, J.; Tremblay, F. *Org. Lett.* **2006**, *8*, 597; (b) Grieco, P. A.; Zelle, R. E.; Lis, R.; Finn, J. *J. Am. Chem. Soc.* **1983**, *105*, 1403; (c) Kozikowski, A. P.; Li, C.-S. *J. Org. Chem.* **1987**, *52*, 3541; (d) Danishefsky, S. J.; Simoneau, B. *J. Am. Chem. Soc.* **1989**, *111*, 2599.

8. Sletzinger, M.; Verhoeven, T. R.; Volante, R. P.; McNamara, J. M. *Tetrahedron Lett.* **1985**, *26*, 2951.

9. Bode, J. W.; Carreira, E. M. *J. Org. Chem.* **2001**, *66*, 6410.

10. Nicolaou, K. C.; Ajito, K.; Patron, A. P.; Khatuya, H.; Richter, P. K.; Bertinato, P. *J. Am. Chem. Soc.* **1996**, *118*, 3059.

11. Fleming, I.; Ghosh, S. K. *J. Chem. Soc. Perkin Trans. 1*, **1998**, 2733.

12. Gademann, K.; Bethuel, Y. *Org. Lett.* **2004**, *6*, 4707.

13. Nakata, T.; Tani, Y.; Hatozaki, M.; Oishi, T. *Chem. Pharm. Bull.* **1984**, *32*, 1411.

14. For the x-ray crystal structure of MeZn(BH_4), see Aldridge, S.; Blake, A. J.; Downs, A. J.; Parsons, S.; Pulham, C. R. *J. Chem. Soc. Dalton Trans.* **1996**, 853.

15. Oishi, T.; Nakata, T. *Acc. Chem. Res.* **1984**, *17*, 338.

16. (a) Cimarelli, C.; Palmieri, G. *Tetrahedron: Asymmetry* **2000**, *11*, 2555; (b) Notz, W.; Hartel, C.; Waldscheck, B.; Schmidt, R. R. *J. Org. Chem.* **2001**, *66*, 4250; (c) Ravikumar, K. S.; Sinha, S.; Chandrasekaran, S. *J. Org. Chem.* **1999**, *64*, 5841.

17. Narasimhan, S.; Balakumar, R. *Aldrichimica Acta* **1998**, *31*, 19.

18. DeShong, P.; Ramesh, S.; Perez, J. J. *J. Org. Chem.* **1983**, *48*, 2117.

19. Muri, D.; Lohse-Fraefel, N.; Carreira, E. M. *Angew. Chem. Int. Ed.* **2005**, *44*, 4036.

20. Matsukawa, Y.; Isobe, M.; Kotsuki, H.; Ichikawa, Y. *J. Org. Chem.* **2005**, *70*, 5339.

4.2. Diastereoselective *Anti*-Reduction of β-Hydroxy Ketones

In 1986, Evans and colleagues introduced tetramethylammonium triacetoxyborohydride, a mild reducing reagent for the highly diastereoselective synthesis of 1,3-*anti*-diols from acyclic β-hydroxy ketones[1] (Scheme 4.2a). The reaction, commonly called the *Evans–Chapman–Carreira reduction*, is typically carried out in a 1 : 1 mixture of anhydrous acetic acid and acetonitrile, as the reaction needs to be run at low temperature in the presence of acetic acid. An important observation is that the *anti*-diastereoselectivity is realized regardless of the stereochemistry of the α-alkyl substituent: Both *anti*- and *syn*-α-methyl-β-hydroxy

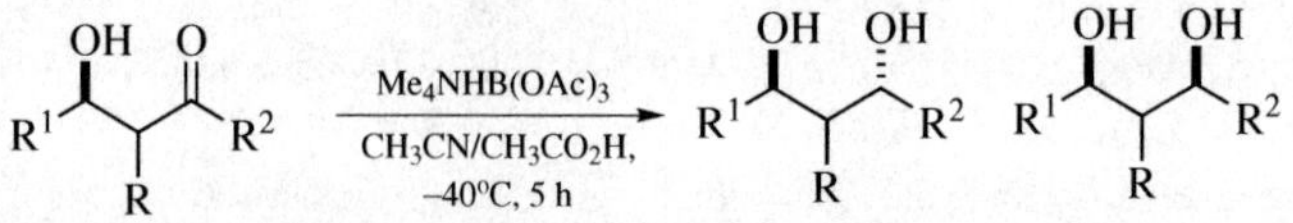

Scheme 4.2a

Reactant	Product	anti/syn	Yield, %
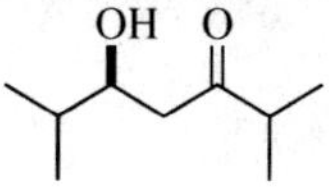 (OH O, iPr-CH(OH)-CH₂-C(O)-iPr)	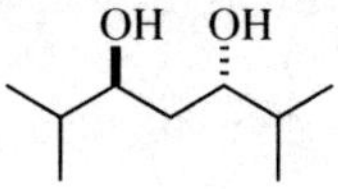 (OH OH diol)	96:4	86
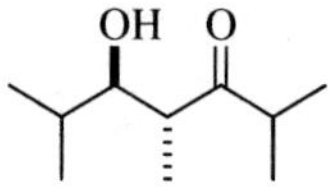 (OH O, with β-Me)	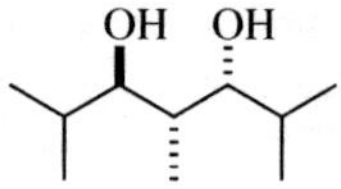 (OH OH diol)	98:2[a]	92
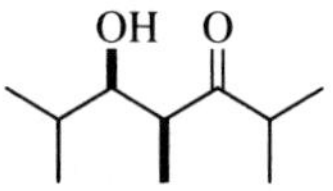 (OH O, with α-Me)	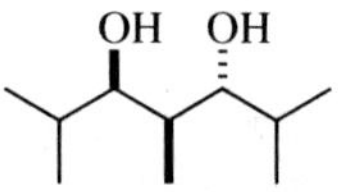 (OH OH diol)	98:2[a]	84
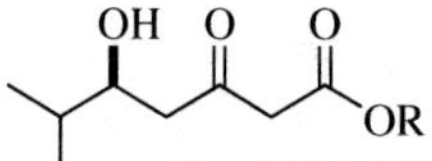 (OH O O, keto ester, [R = (CH₂)₃Ph])	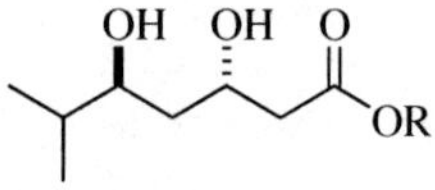 (OH OH O)	95:5	92
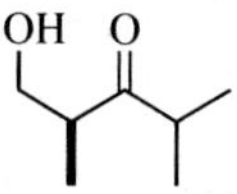 (OH O, α-methyl, no β-substituent)	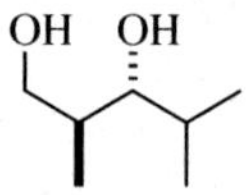 (OH OH)	83:17	—

[a] Reaction at –20°C for 18 h.

ketones undergo reductions with remarkably high levels of diastereoselectivity, favoring the *anti*-diol diastereoisomers. When the substrate lacks a β-substituent, a diminished selectivity is observed, hinting that the β-substituent in the starting material is a key structural element in attaining high stereoselectivity.

The diastereoselectivity of this reaction reflects competition between two chair-like transition states **A** and **B**, each of which involves intramolecular hydride delivery as well as activation by acid catalysis. The 1,3-diaxial interaction between R^2 and acetoxy groups would destabilize transition state **B** to a greater extent than the analogous interaction between hydroxy and acetoxy groups in the favored transition state, $\mathbf{A}^1$ (Scheme 4.2b).

The Evans–Chapman–Carreira reaction was used in the total synthesis of macrolactin A(**3**), one of the polyene macrolide antibiotics, which shows inhibition of HIV replication in T-lymphoblast cells in preliminary studies[2] (Scheme 4.2c).

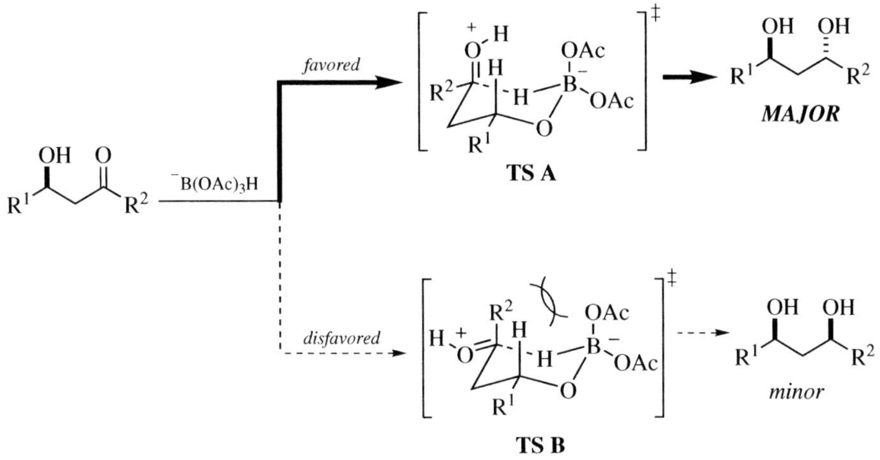

Scheme 4.2b

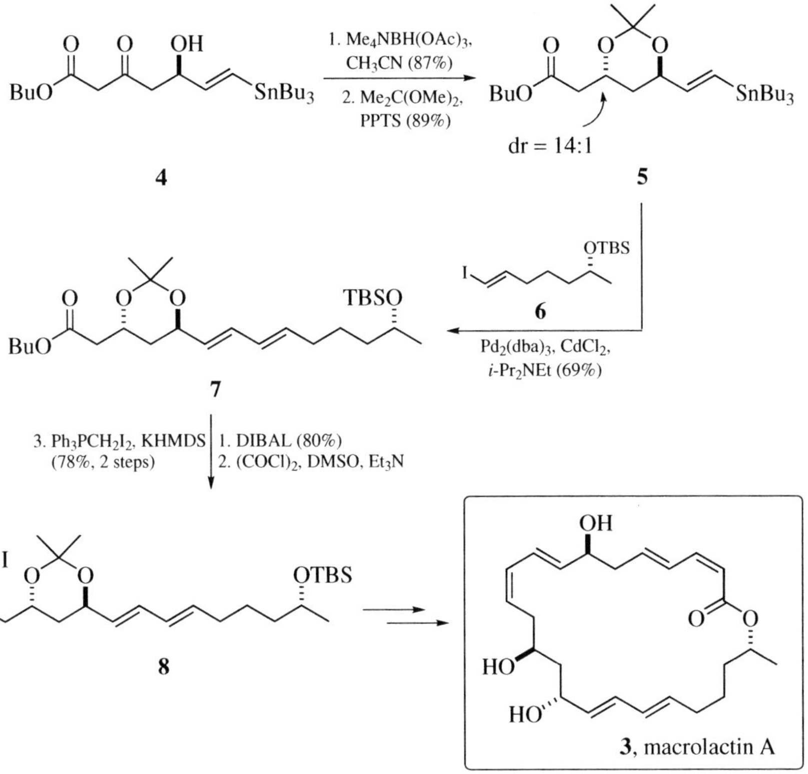

Scheme 4.2c

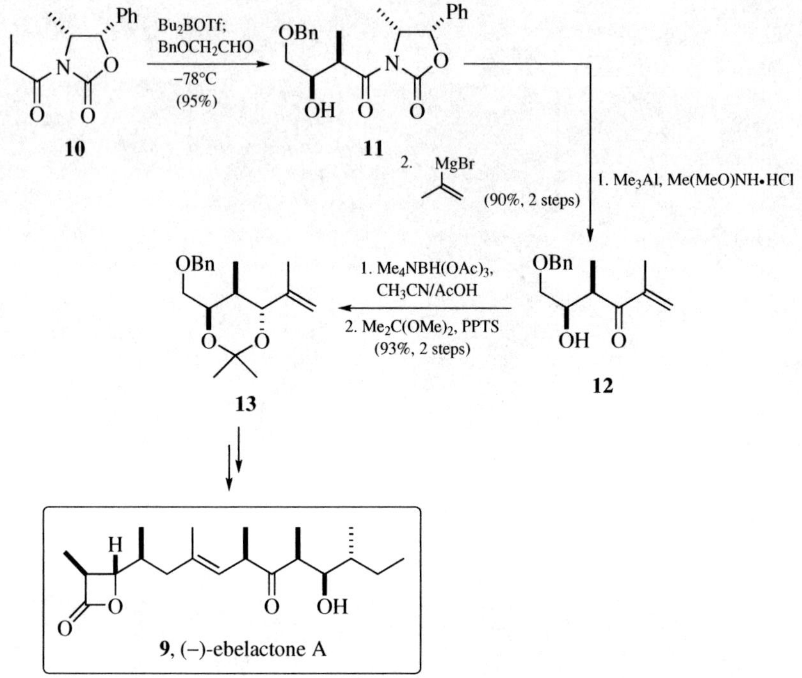

Scheme 4.2d

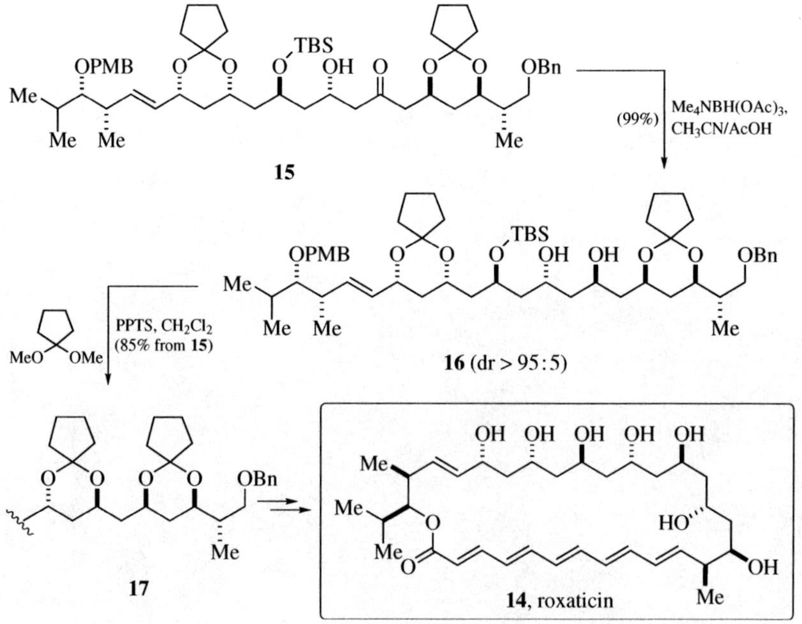

Scheme 4.2e

The selective reduction of the δ-hydroxy-β-keto ester **4** with Me₄NBH(OAc)₃ afforded the corresponding 1,3-*anti*-diol in 87% yield and 14:1 diastereoselectivity. The 1,3-*anti*-diol was protected as the acetonide **5**, followed by a Pd-catalyzed coupling reaction with the vinyl iodide **6** to provide the diene **7** in 69% yield. Reduction of the ester, Swern oxidation, and finally, Wittig olefination afforded the (Z)-vinyl iodide **8**.

Mandal completed the total synthesis of (−)-ebelactone A (**9**), which is an inhibitor of esterases, lipases, and *N*-formylmethionine aminopeptidases located on the cellular membrane of various cell strains[3] (Scheme 4.2d). The synthesis began with Evans's *syn*-aldol reaction between *N*-propionyloxazolidinone (**10**) and benzyloxyacetaldehyde to afford the *syn*-aldol adduct **11** in 95%

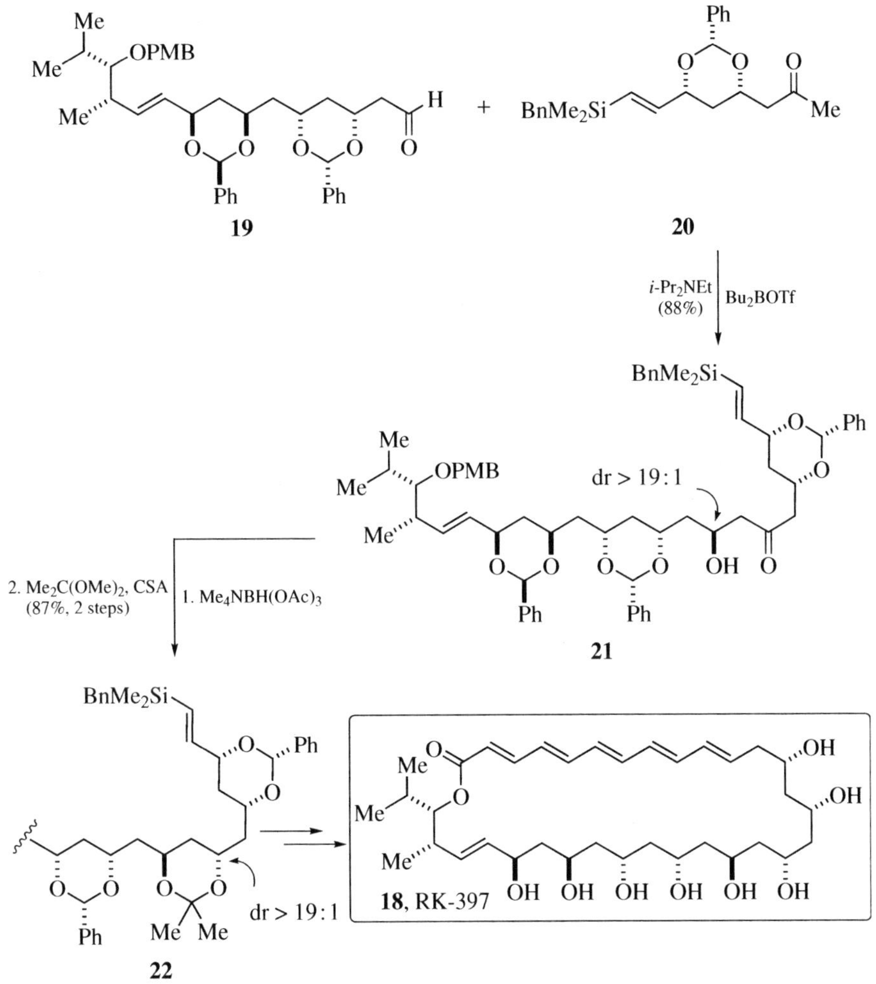

Scheme 4.2f

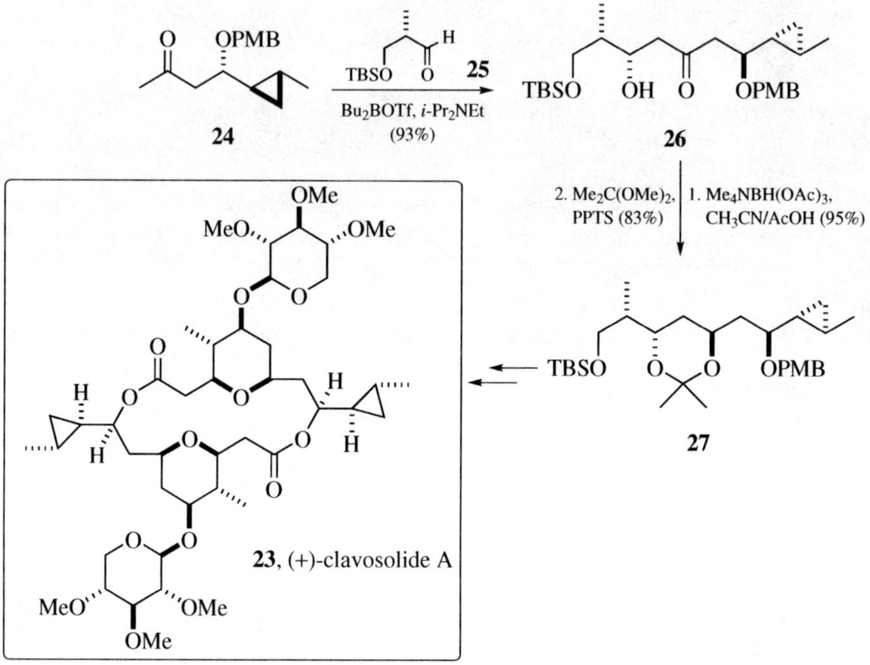

Scheme 4.2g

yield as a single diastereomer. Conversion of **11** to the Weinreb amide followed by addition of 2-propenylmagnesium bromide yielded the enone **12**. The Evans–Chapman–Carreira 1,3-*anti*-reduction of the β-hydroxy ketone **12** with Me$_4$NBH(OAc)$_3$ resulted in the 1,3-*anti*-diol, which was then converted to the acetonide **13** in 93% yield over two steps.

The 1,3-*anti*-reduction of β-hydroxy ketones was utilized in the total synthesis of (+)-roxaticin (**14**), a pentaene macrolide isolated from streptomycete X-14994[4] (Scheme 4.2e). The β-hydroxy ketone **15** underwent 1,3-*anti*-reduction to afford the diol **16** in 99% yield and greater than 95:5 diastereoselectivity. The resulting diol was then protected as a cyclopentylidene ketal (**17**) by using cyclopentylidene dimethyl ketal and pyridinium *p*-toluenesulfonate (PPTS).

In the synthesis of RK-397 (**18**), Denmark and Fujimori prepared an *anti*-diol using the Evans–Chapman–Carreira protocol[5] (Scheme 4.2f). The β-hydroxy ketone **21**, obtained by a diastereoselective boron aldol reaction between **19** and **20**, was reduced with tetramethylammonium triacetoxyborohydride to afford the *anti*-diol derivative **22** in greater than 19:1 diastereoselectivity.

The 1,3-*anti*-selective reduction was utilized in the total synthesis of the structurally unique compound (+)-clavosolide A (**23**)[6] (Scheme 4.2g). The 1,5-*anti*-aldol reaction of a dibutylboron enolate of **24** with the aldehyde **25** proceeded smoothly to afford the β-hydroxy ketone **26** in 93% yield and >96:4 diastereoselectivity. Compound **26** was subsequently treated with

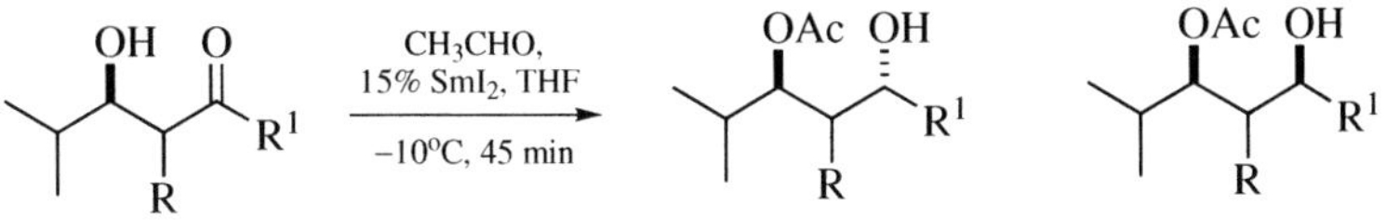

Reactant	Product	anti/syn	Yield, %
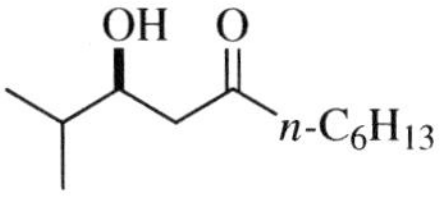 OH O, n-C₆H₁₃ (isopropyl β-hydroxy ketone)	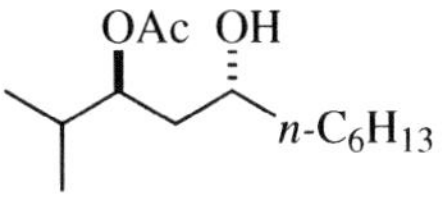 OAc OH, n-C₆H₁₃	> 99:1	96
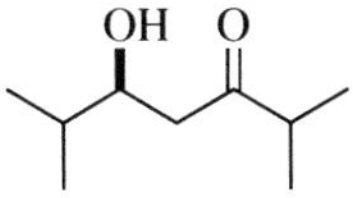 OH O (isopropyl β-hydroxy isopropyl ketone)	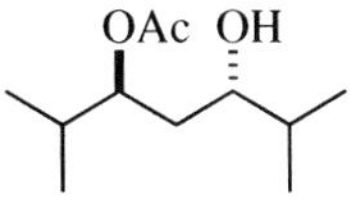 OAc OH	> 99:1	85
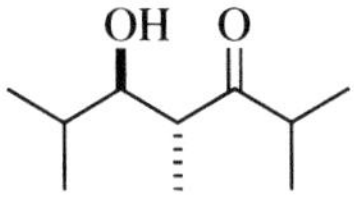 OH O (with α-methyl, syn)	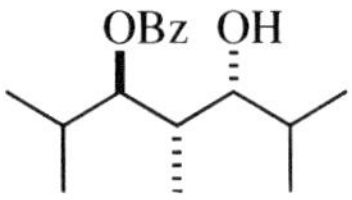 OBz OH	> 99:1[a]	95[a]
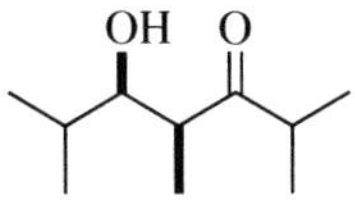 OH O (with α-methyl, anti)	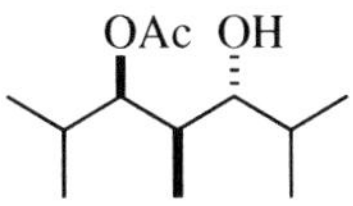 OAc OH	> 99:1	85

[a] Benzaldehyde was used.

Scheme 4.2h

Me₄NBH(OAc)₃ in CH₃CN–AcOH, followed by protection of the resulting 1,3-*anti*-diol with 2,2-dimethoxypropane to provide the acetonide **27**.

In 1990, Evans and Hoveyda disclosed another novel method for the synthesis of *anti*-1,3-diol using samarium iodide[7] (Scheme 4.2h). When the β-hydroxy ketone **28** was treated with 4 to 8 equivalents of aldehyde and a catalytic amount (15 mol %) of freshly prepared SmI₂ in THF at −10 °C, the corresponding 1,3-*anti*-diol monoester **29**-*anti* was formed in high yields and with superb levels of diastereoselectivity. The *anti*-diastereoselectivity is not affected by the presence of a chiral substituent at the C₂ position, as the reduction of both *syn*- and *anti*-α-methyl-β-hydroxy ketones follows the same stereochemical course with equally high asymmetric induction. One advantage of the current samarium iodide–catalyzed Evans–Tishchenko reduction over the triacetoxyborohydride reduction is that the two alcohol functional groups are in a different protection format in the product, enabling further synthetic operations on the 1,3-*anti*-diol monoester.

To explain the high levels of *anti*-selectivity, Evans and Hoveyda proposed that the reduction occurs through a hydride-bridged six-membered transition

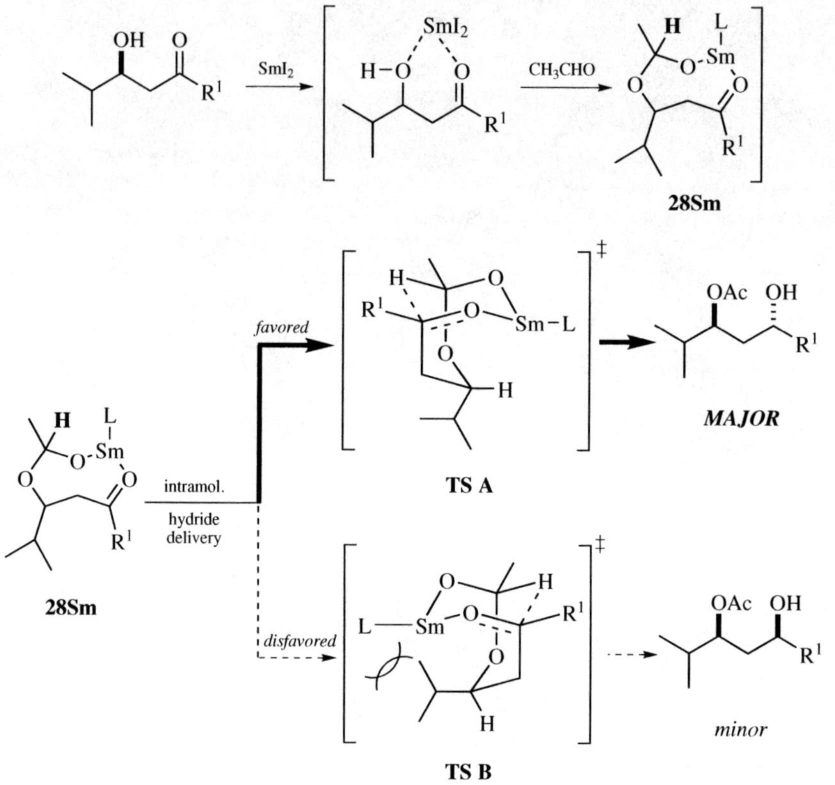

Scheme 4.2i

state[7] (Scheme 4.2i). The samarium-catalyzed reduction may involve coordination of the hydroxy ketone to the catalyst and subsequent formation of an eight-membered samarium-containing hemiacetal (**28Sm**).[8] The stereochemistry-determining stage would be the intramolecular hydride delivery step. Transition state **A**, leading to the formation of the major isomer, would be favored because the isopropyl group adopts an equatorial position, whereas the isopropyl substituent in transition state **B** occupies an energetically unfavorable axial position.[9]

Schreiber et al. used the Evans–Tishchenko reduction in the total synthesis of (−)-rapamycin (**30**), which complexes with an intracellular receptor FKBP12 to interfere potently with distinct signaling components of the cell cycle[10] (Scheme 4.2j). The sulfone **31** underwent olefination reaction, followed by regioselective dihydroxylation and periodate cleavage, to furnish the β-hydroxy ketone **32** in good overall yield. A 1:4 mixture of the ketone **32** and (*S*)-*N*-Boc-pipecolinal (**33**) was treated with 30 mol % of PhCHO-SmI$_2$, and the Evans–Tishchenko reduction product **34** was obtained in 95% yield as a mixture of > 20:1 *anti/syn* 1,3-diol monoesters.

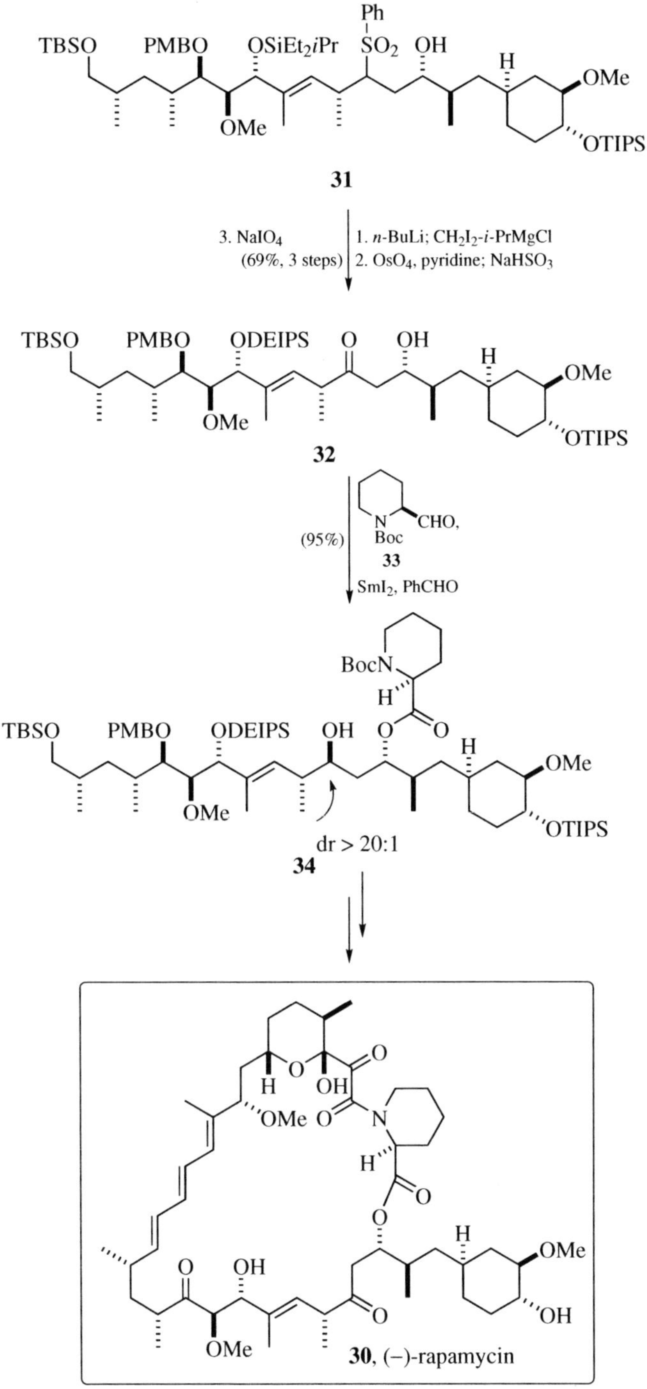
Ph
TBSO
PMBO
OSiEt2iPr
SO2
OH
H
OMe
OMe
OTIPS
31
3. NaIO4
(69%, 3 steps)
1. n-BuLi; CH2I2-i-PrMgCl
2. OsO4, pyridine; NaHSO3
TBSO
PMBO
ODEIPS
O
OH
H
OMe
OMe
OTIPS
32
CHO,
Boc
33
(95%)
SmI2, PhCHO
BocN
H
TBSO
PMBO
ODEIPS
OH
O
O
H
OMe
OMe
OTIPS
dr > 20:1
34
O
H
OH
O
OMe
O
N
H
O
O
H
OMe
O
OH
O
OH
OMe
30, (−)-rapamycin

Scheme 4.2j

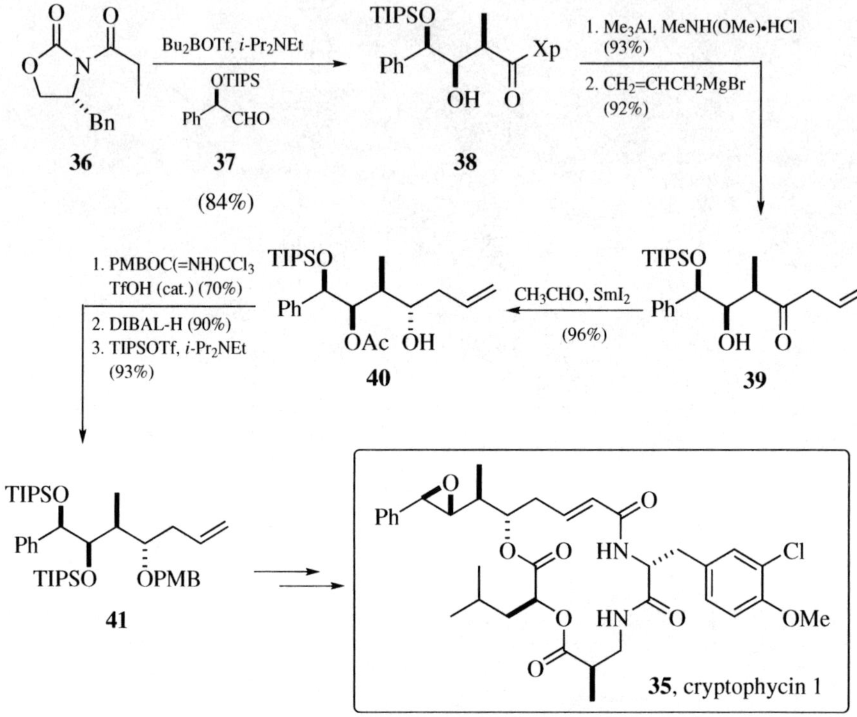

Scheme 4.2k

Gardinier and Leahy employed a samarium-catalyzed *anti*-reduction protocol to access one of the key intermediates en route to the synthesis of cryptophycin 1 (**35**), which exhibits extraordinary activity against a variety of tumor cell lines[11] (Scheme 4.2k). Thus, the *syn*-aldol reaction between *N*-propionyloxazolidinone (**36**) and the chiral aldehyde **37** proceeded smoothly to afford the adduct desired (**38**) as a single product in high yield. Formation of Weinreb amide and subsequent addition of allylmagnesium bromide provide the β-hydroxy ketone **39** which was then subjected to Evans–Tishchenko reduction conditions to produce the monoacetate **40** cleanly in 96% isolated yield. Protection of the alcohol as a *p*-methoxybenzyl (PMB) ether, DIBAL reduction of the acetate, and treatment of the resulting alcohol with triisopropylsilyl triflate furnished compound **41**.

The samarium-catalyzed reduction was utilized in the asymmetric synthesis of the marine macrolide bryostatin 2 (**42**) to furnish an intermediate (**46**)[12] (Scheme 4.2l). The ketone **43** underwent an aldol reaction with the ketoaldehyde **44** via the isopinylboryl enolate to give the aldol adduct **45** in good yield and 93:7 diastereoselectivity. Subsequent samarium-catalyzed Evans–Tishchenko reduction of the β-hydroxy ketone **45** provided the *p*-nitrobenzoate **46** with excellent stereoselectivity. Silylation and saponification readily converted compound **46** into the alcohol **47** in 88% yield over two steps.

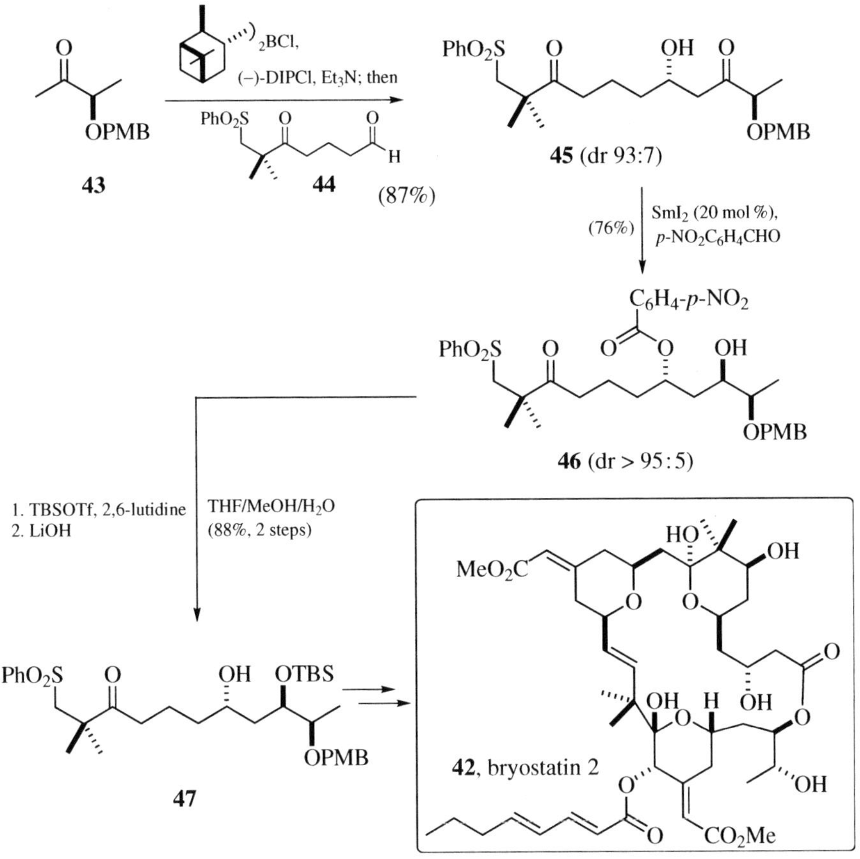

Scheme 4.2l

Dermostatin A (**48**) is a 36-membered macrolide that shows potent antifungal activity against a large number of human pathogens and has been used clinically as a treatment for deep vein mycoses. During the total synthesis of dermostatin A, Sinz and Rychnovsky utilized the Evans–Tishchenko reduction methodology to furnish a 1,3-*anti*-diol intermediate (**52**)[13] (Scheme 4.2m). The enol silane **49** and aldehyde **50** were subjected to Mukaiyama aldol coupling to yield the 1,3-*anti* adduct desired (**51**) with a modest 3.3:1 diastereomeric ratio. The Evans–Tishchenko *anti*-reduction of the β-hydroxy ketone **51** with isobutyraldehyde and subsequent reductive cleavage of the monoester provided the 1,3-*anti*-diol **52** with excellent diastereoselectivity. Desilylation under acidic condition followed by protection of the tetraol afforded the bis-acetonide **53**.

In the synthesis of the marine macrolide leucascandrolide A (**54**), Kozmin used the samarium-catalyzed diastereoselective ketone reduction method in a highly stereocontrolled synthesis of the C_1–C_{15} fragment **58**[14] (Scheme 4.2n). Generation of the dicyclohexylboron enolate of the ketone **55** followed by addition

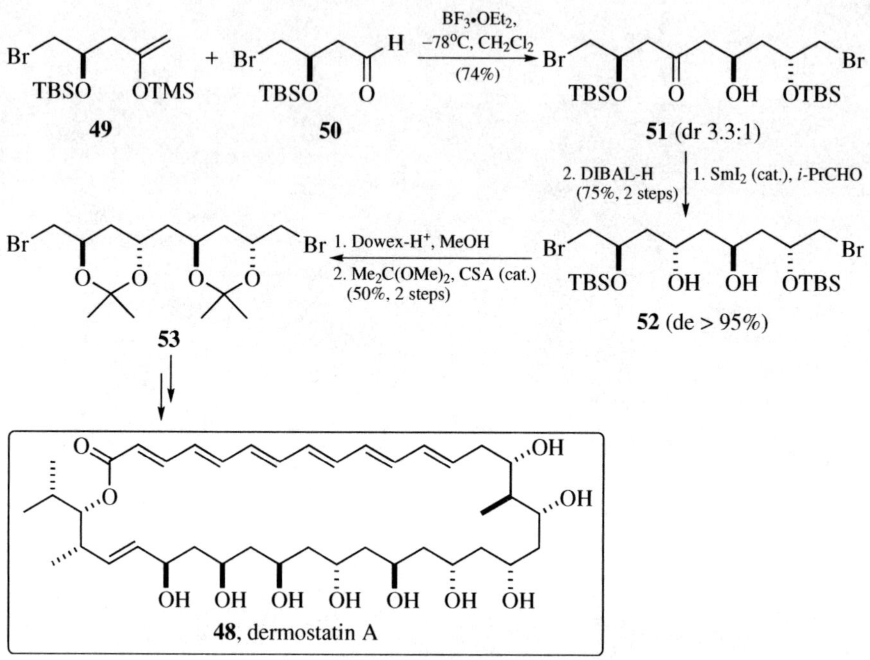

Scheme 4.2m

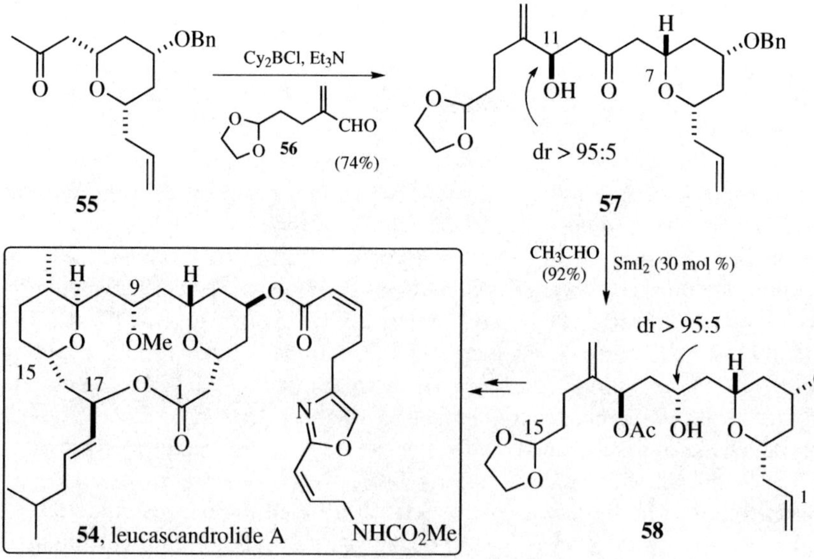

Scheme 4.2n

of the aldehyde **56** provided the aldol adduct desired (**57**) as a single diastereomer. Diastereoselective β-hydroxy ketone reduction of compound **57** using an Evans–Tishchenko protocol cleanly afforded the monoacetate **58** in greater than 95:5 diastereoselectivity.[15]

REFERENCES

1. (a) Evans, D. A.; Chapman, K. T. *Tetrahedron Lett.* **1986**, *27*, 5939; (b) Evans, D. A.; Chapman, K. T.; Carreira, E. M. *J. Am. Chem. Soc.* **1988**, *110*, 3560.
2. Kim, Y.; Singer, R. A.; Carreira, E. M. *Angew. Chem. Int. Ed.* **1998**, *37*, 1261.
3. Mandal, A. K. *Org. Lett.* **2002**, *4*, 2043.
4. Evans, D. A.; Connell, B. T. *J. Am. Chem. Soc.* **2003**, *125*, 10899.
5. Denmark, S. E.; Fujimori, S. *J. Am. Chem. Soc.* **2005**, *127*, 8971.
6. Son, J. B.; Kim, S. N.; Kim, N. Y.; Lee, D. H. *Org. Lett.* **2006**, *8*, 661.
7. Evans, D. A.; Hoveyda, A. H. *J. Am. Chem. Soc.* **1990**, *112*, 6447.
8. (a) Molander, G. A.; Etter, J. B. *J. Am. Chem. Soc.* **1987**, *109*, 6556; (b) Keck, G. E.; Wager, C. A.; Sell, T.; Wager, T. T. *J. Org. Chem.* **1999**, *64*, 2172.
9. Abu-Hasanayn, F.; Streitwieser, A. *J. Org. Chem.* **1998**, *63*, 2954.
10. Romo, D.; Meyer, S. D.; Johnson, D. D.; Schreiber, S. L. *J. Am. Chem. Soc.* **1993**, *115*, 7906.
11. Gardinier, K. M.; Leahy, J. W. *J. Org. Chem.* **1997**, *62*, 7098.
12. Evans, D. A.; Carter, P. H.; Carreira, E. M.; Prunet, J. A.; Charette, A. B.; Lautens, M. *Angew. Chem. Int. Ed.* **1998**, *37*, 2354.
13. Sinz, C. J.; Rychnovsky, S. D. *Angew. Chem. Int. Ed.* **2001**, *40*, 3224.
14. Kozmin, S. A. *Org. Lett.* **2001**, *3*, 755.
15. Wang, Y.; Janjic, J.; Kozmin, S. A. *J. Am. Chem. Soc.* **2002**, *124*, 13670.

4.3. Asymmetric Reduction

In 1979, Noyori and co-workers invented a new type of chiral aluminum hydride reagent (**1**), which is prepared in situ from LiAlH$_4$, (*S*)-1,1'-bi-2-naphthol (BINOL), and ethanol. The reagent, called *binaphthol-modified lithium aluminum hydride* (BINAL-H), affects asymmetric reduction of a variety of phenyl alkyl ketones to produce the alcohols **2** with very high to perfect levels of enantioselectivity when the alkyl groups are methyl or primary[1] (Scheme 4.3a).

The binaphthol-modified lithium aluminum hydride reagents (BINAL-Hs) are also effective in enantioselective reduction of a variety of alkynyl and alkenyl ketones[2] (Scheme 4.3b). When the reaction is carried out with 3 equivalents of (*S*)-BINAL-H at −100 to −78 °C, the corresponding propargylic alcohol **3** and allylic alcohol **4** are obtained in high chemical yields with good to excellent levels of enantioselectivity. As is the case with aryl alkyl ketones, the alcohols with (*S*)-configuration are obtained when (*S*)-BINAL-H is employed.

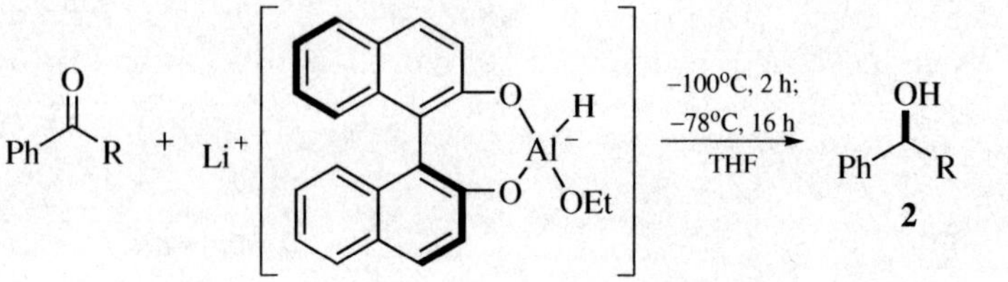

Scheme 4.3a

Ph-CO-R + Li⁺[(S)-BINAL-H with OEt] →(−100°C, 2 h; −78°C, 16 h; THF)→ Ph-CH(OH)-R (**2**)

1, (S)-BINAL-H

R	ee, %	Yield, %
CH$_3$	95	61
C$_2$H$_5$	98	62
n-C$_3$H$_7$	100	92
n-C$_4$H$_9$	100	64
i-C$_3$H$_7$	71	68
t-C$_4$H$_9$	44	80

Scheme 4.3a

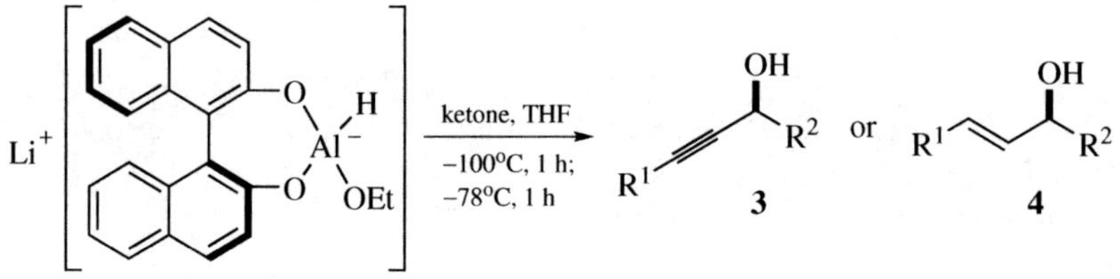

Scheme 4.3b

Li⁺[(S)-BINAL-H, OEt] + ketone, THF, −100°C, 1 h; −78°C, 1 h →

R^1–C≡C–CH(OH)–R^2 (**3**) or R^1–CH=CH–CH(OH)–R^2 (**4**)

1, (S)-BINAL-H

Ketone	Alcohol	ee, %	Yield, %
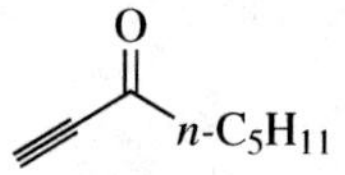 HC≡C–CO–n-C$_5$H$_{11}$	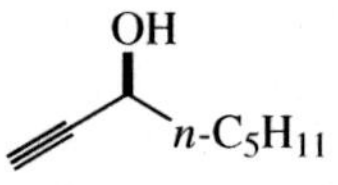 HC≡C–CH(OH)–n-C$_5$H$_{11}$	84	71
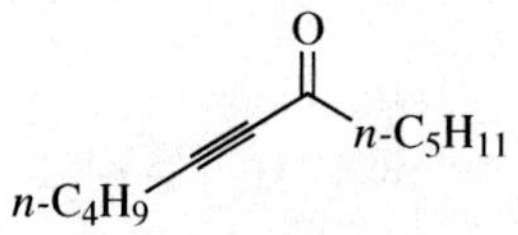 n-C$_4$H$_9$–C≡C–CO–n-C$_5$H$_{11}$	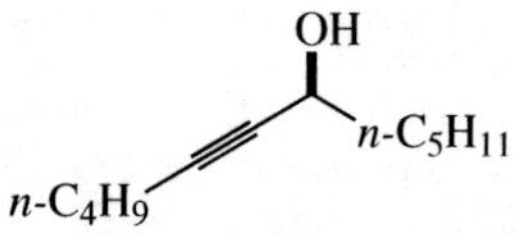 n-C$_4$H$_9$–C≡C–CH(OH)–n-C$_5$H$_{11}$	90[a]	90[a]
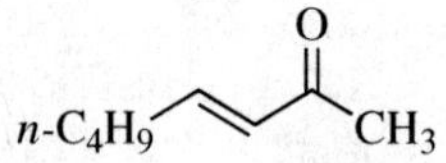 n-C$_4$H$_9$–CH=CH–CO–CH$_3$	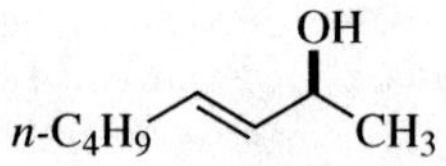 n-C$_4$H$_9$–CH=CH–CH(OH)–CH$_3$	79	47
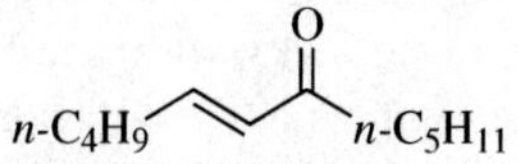 n-C$_4$H$_9$–CH=CH–CO–n-C$_5$H$_{11}$	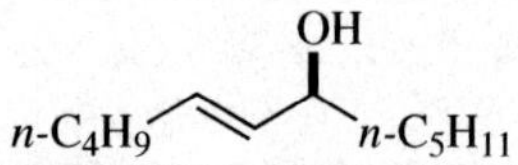 n-C$_4$H$_9$–CH=CH–CH(OH)–n-C$_5$H$_{11}$	91	91

[a] BINAL-H from methanol was used.

Scheme 4.3b

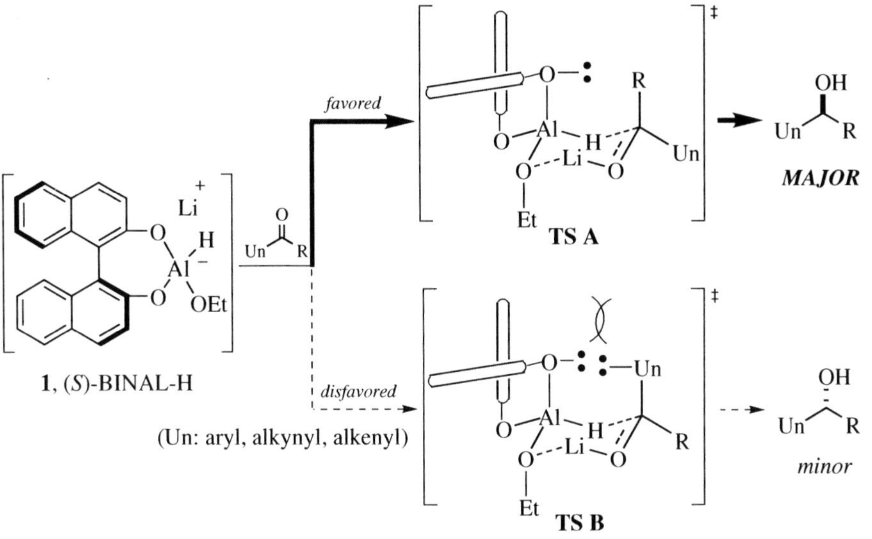

Scheme 4.3c

Noyori et al. proposed that the reaction would be initiated by complexation of the Lewis acidic lithium cation to the ketone oxygen atom; then hydride transfer occurs from aluminum to the carbonyl carbon by way of a six-membered chair-like transition state[3] (Scheme 4.3c). Between the two competing six-membered chairlike transition states **A** and **B**, transition state **B** is disfavored, due to the substantial n/π-type electronic repulsion between the axially oriented binaphthoxyl oxygen and the unsaturated phenyl or alkenyl moiety. Although there is a 1,3-diaxial steric interaction between the Al–O and C–R bonds in transition state **A**, the absence of electronic repulsion would make transition state **A** preferred to **B**, giving the (S)-enantiomer as the major product. However, there is a delicate balance between electronic and steric factors. Hence, the decrease in enantioselectivity seen in the reduction of *tert*-butyl phenyl ketone reflects greater steric interactions in transition state **A** between the Al–O bond and the *tert*-butyl group.

The BINAL-H reduction protocol was utilized in a highly enantioselective prostaglandin synthesis[4] (Scheme 4.3d). Asymmetric reduction of the iodovinyl ketone **6** with (S)-BINAL-H proceeded well, with superb enantioselectivity and in high yield.[5] Protection of the resulting alcohol gave the TBS ether **7**. Conjugate addition of the cuprate reagent generated from **7** into cyclopentenone **8** followed by alkylation with **9** afforded the cyclopentanone **10**. Desilylation followed by enzymatic hydrolysis then produced the natural prostaglandin E_2 (PGE2) **5**.

Marko et al. employed an enantioselective Noyori BINAL-H reduction in the synthesis of methyl monate C (**11**), the methyl ester derivative of the potent antibiotic pseudomonic acid C[6] (Scheme 4.3e). The α,β-unsaturated ketone **12** underwent the Noyori reduction with the (S)-BINAL-H reagent to give the product desired (**13**) in 70% yield and 95% ee. The chiral alcohol was then condensed

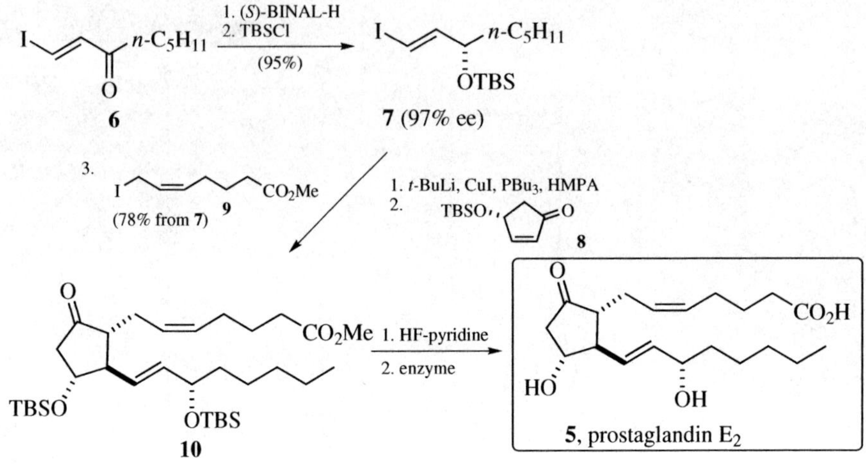

176 STEREOSELECTIVE REDUCTIONS

Scheme 4.3d

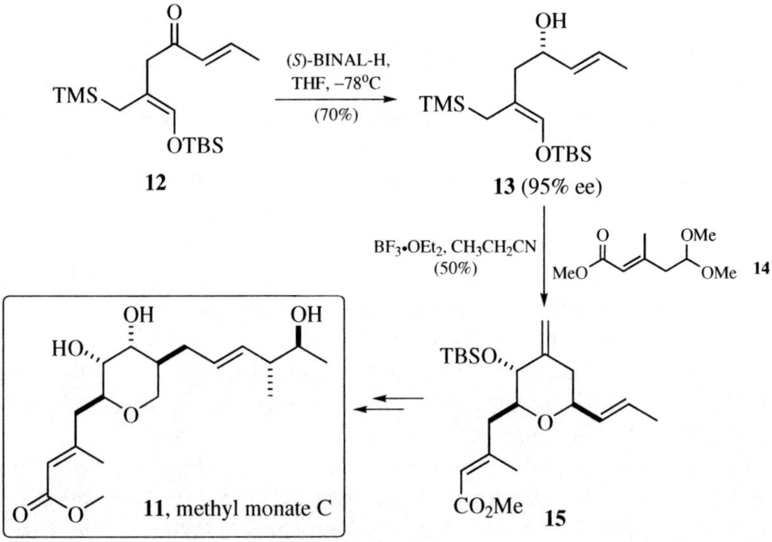

Scheme 4.3e

with compound **14** in the presence of $BF_3 \cdot OEt_2$ to provide the tetrahydropyran **15** as a single diastereomer in 50% yield.

Noyori et al. demonstrated the effectiveness of the BINAL-H reduction method by synthesizing the Japanese beetle pheromone (R)-**16**[7] (Scheme 4.3f). The alkynyl ketone **17** was treated with 3 equivalents of (R)-BINAL-H at $-100\,°C$ for 1 hour and then held at $-78\,°C$ for 2 hours. The propargylic alcohol **18** was obtained in good yield and with moderate enantioselectivity of 84% ee. Exposure

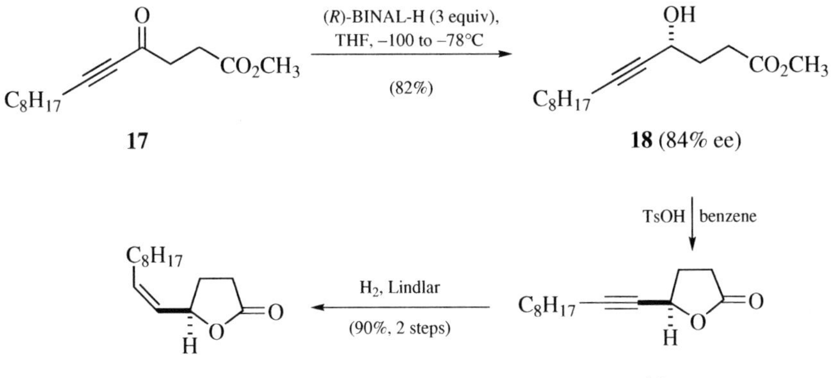

Scheme 4.3f

of **18** to *p*-toluenesulfonic acid in refluxing benzene afforded the γ-lactone **19**, which upon hydrogenation with Lindlar catalyst yielded the target molecule (**16**) in 75% ee.

The BINAL-H enantioselective reduction of α,β-conjugated ketones was used in the total synthesis of (−)-lepadiformine (**20**), which exhibits moderate cytotoxic activities against various tumor cell lines[8] (Scheme 4.3g). A solution of **21** in toluene

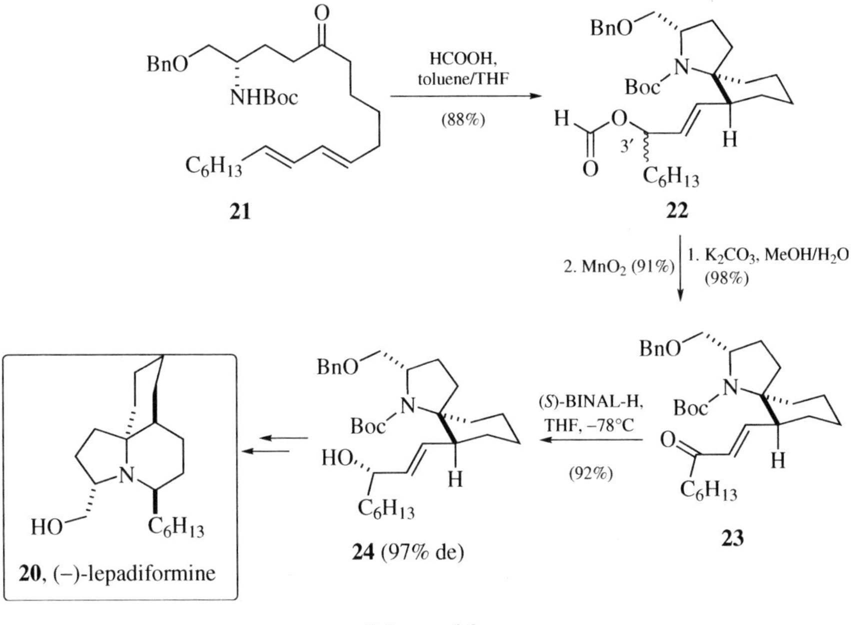

Scheme 4.3g

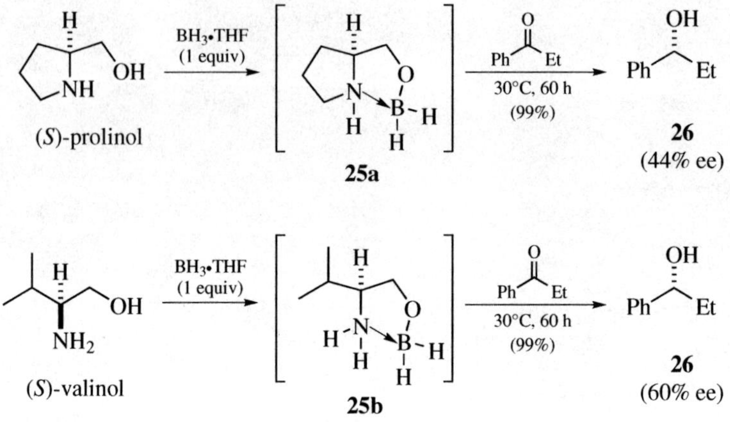

Scheme 4.3h

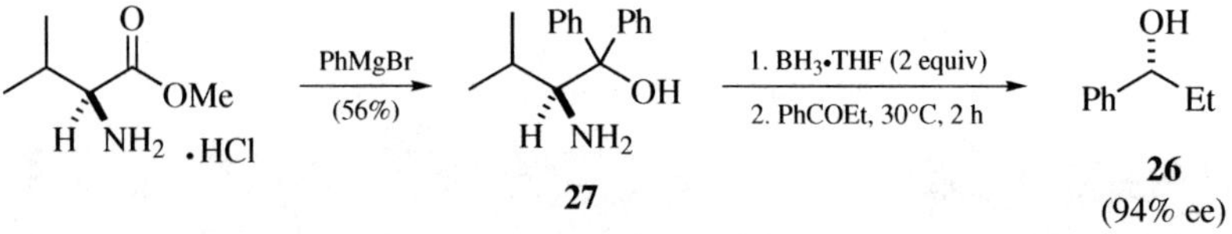

(S)-prolinol + BH₃·THF (1 equiv) → **25a** + PhCOEt, 30°C, 60 h (99%) → Ph-CH(OH)-Et **26** (44% ee)

(S)-valinol + BH₃·THF (1 equiv) → **25b** + PhCOEt, 30°C, 60 h (99%) → Ph-CH(OH)-Et **26** (60% ee)

Scheme 4.3i

Valine methyl ester·HCl + PhMgBr (56%) → **27** → 1. BH₃·THF (2 equiv); 2. PhCOEt, 30°C, 2 h → Ph-CH(OH)-Et **26** (94% ee)

Scheme 4.3j

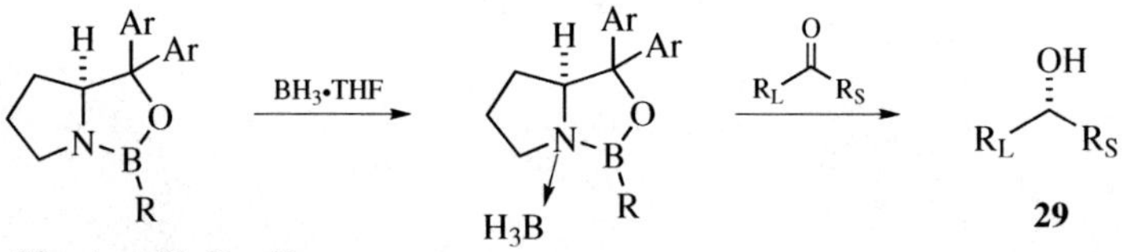

28a, Ar = Ph, R = H
28b, Ar = Ph, R = CH₃
28c, Ar = 2-Naph, R = CH₃

Product **29**: R_LCH(OH)R_S

| | ee, % | | |
Ketone ($R_L COR_S$)[a]	w/**28a**	w/**28b**	w/**28c**
C₆H₅COCH₃	97	96.5	97.8
C₆H₅COC₂H₅	90	96.7	97.4
α-tetralone	89	86.0	94.5
t-BuCOCH₃	92	97.3	92.7

[a] Reaction temperatures: 25°C with **28a**, −10°C with **28b**, 23°C with **28c**. R_L, aromatic substituent for aromatic ketones and tertiary alkyl group for aliphatic ketones.

and THF was treated with formic acid to generate in situ an N-acyliminium species which underwent a spirocyclization reaction to provide the 1-azaspirocyclic formate ester **22** as a 1.6 : 1 mixture favoring the 3′ β-formate isomer. Compound **22** was subjected to basic hydrolysis conditions to give a mixture of diastereomeric allylic alcohols, which were oxidized with MnO_2 to furnish the α β-unsaturated ketone **23**. Asymmetric reduction of **23** with (S)-BINAL-H then afforded the alcohol **24** in 92% yield with excellent diastereoselectivity (97% de).

In 1981, Hirao and others reported that the chiral borane–amine complex **25a**, derived from (S)-prolinol and 1 equivalent of $BH_3 \cdot THF$, enantioselectively reduced propiophenone to afford (R)-1-phenyl-1-propanol (**26**) in 44% ee[9] (Scheme 4.3h). The chiral complex **25b** was even better than **25a**, affording the same secondary alcohol in 60% ee. Two years after the initial disclosure, Hirao et al. uncovered a new catalyst system that improved the previous experimental conditions dramatically[10] (Scheme 4.3i). When the chiral aminoalcohol **27**, prepared from (S)-valine methyl ester hydrochloride and phenylmagnesium bromide, was used along with 2 equivalents of $BH_3 \cdot THF$, the enantioselectivity of the alcohol **26** jumped to 94% ee. In addition, the reaction time was shortened to 2 hours.

In 1987, Corey and co-workers proved that highly enantioselective reduction of ketones could be achieved by using stoichiometric borane in the presence of catalytic amounts of the oxazaborolidine **28a**[11] (Scheme 4.3j). Compound **28a**, synthesized by heating (S)-(−)-2-(diphenylhydroxymethyl)pyrrolidine at reflux in THF with 3 equivalents of $BH_3 \cdot THF$, shows excellent catalytic activity for the asymmetric reduction of acetophenone and other ketones. The B-methylated analog **28b** was later synthesized to improve the air and moisture sensitivity associated with **28a**. The third analog, **28c**, with a 2-naphthyl substituent on the oxazaborolidine ring, has proven to be the best to afford the alcohol **29** with superb levels of enantioselectivity.

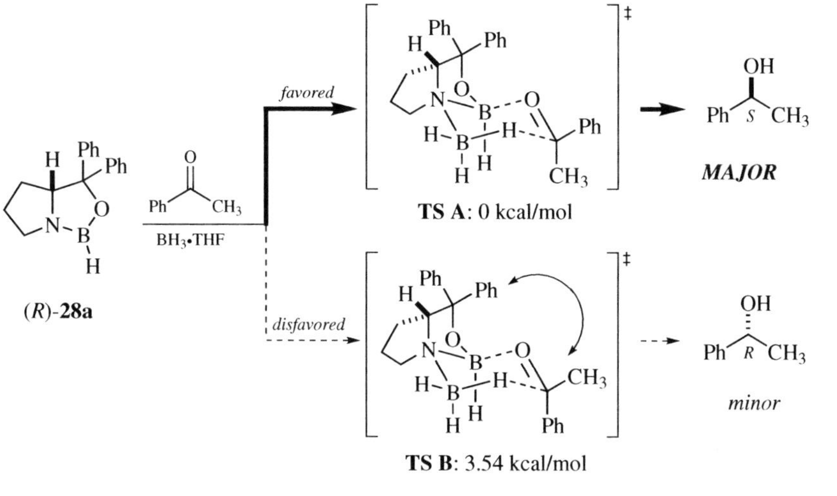

Scheme 4.3k

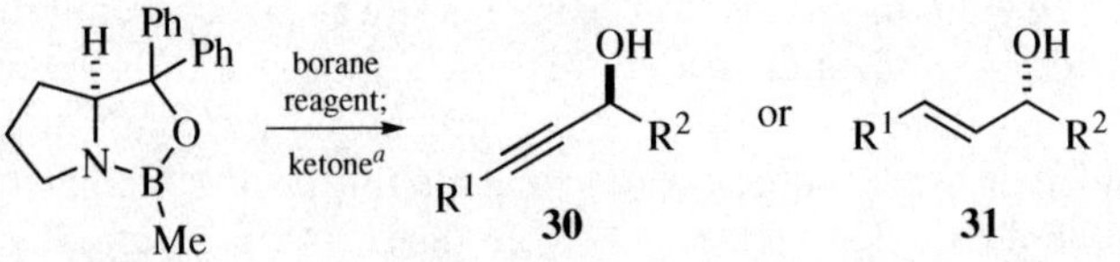

Ketone	Alcohol	Borane	ee, %	Yield, %
PhC≡C-C(O)-CH₃	PhC≡C-CH(OH)-CH₃	BH₃·Me₂S	71	80
(c-C₆H₁₁)C≡C-C(O)-	(c-C₆H₁₁)C≡C-CH(OH)-	BH₃·Me₂S	98	81
(n-C₇H₁₅)C≡C-C(O)-	(n-C₇H₁₅)C≡C-CH(OH)-	BH₃·Me₂S	95	54
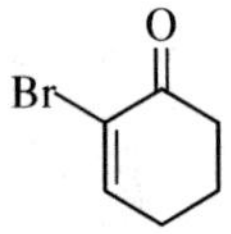 Br-cyclohexenone	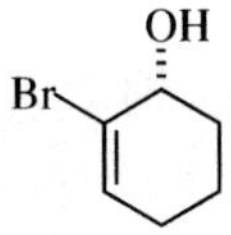 Br-cyclohexenol	BH₃·THF	90	90
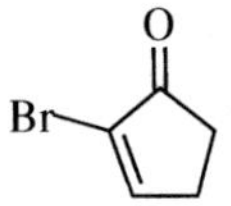 Br-cyclopentenone	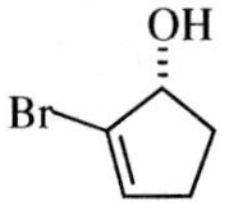 Br-cyclopentenol	BH₃·THF	90	89
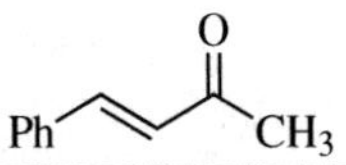 Ph-CH=CH-C(O)-CH₃	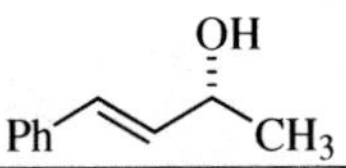 Ph-CH=CH-CH(OH)-CH₃	Catechol–borane	97[b]	90[b]

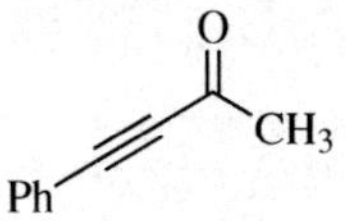

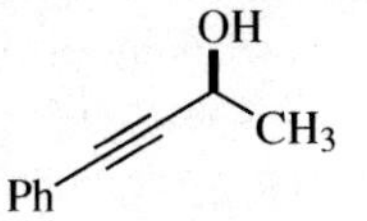

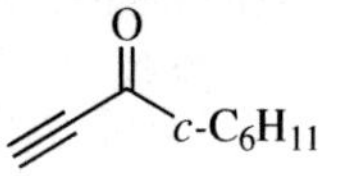

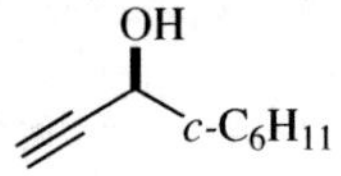

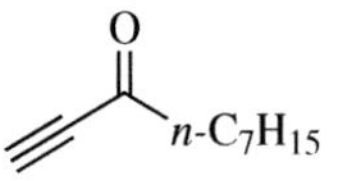

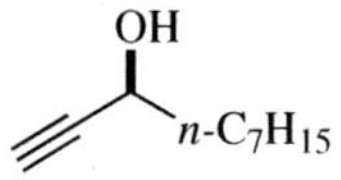

[a] For alkynyl ketones, the reaction was run with 2 equivalents of **28b** and 5 equivalents of BH₃·Me₂S at −30°C.

[b] B-n-Bu oxazaborolidine was used.

Scheme 4.31

Numerous theoretical treatments have been carried out to understand the mode of asymmetric induction of the Corey–Bakshi–Shibata (CBS) reduction, more thoroughly.[12] Liotta et al. carried out computational studies to identify the transition states for CBS reductions of various ketones[13] (Scheme 4.3k). In the asymmetric reduction of acetophenone with the catalyst (R)-**28a**, four transition states were found. Of the lowest energy is chairlike transition state **A**, which would lead to formation of the major enantiomer. In transition state **A**, the phenyl group of acetophenone occupies an equatorial position that is free from any steric interaction, as it is 5.5 Å away from one of the two phenyl groups of the diphenylprolinol ring. On the other hand, transition state **B**, leading to the formation of the minor enantiomer, is less stable than **A** by 3.54 kcal/mol. Two energetically unfavorable interactions are responsible for the high-energy state of **B**: The methyl group of acetophenone experiences a steric interaction with the phenyl substituent of the catalyst, and the phenyl group of acetophenone is engaged in additional interactions with the axial B–H bonds in transition state **B**. The calculations predict 98% ee in the asymmetric reduction of acetophenone, with the catalyst (*R*)-**28a** favoring an (*S*)-enantiomer, which is in excellent agreement with the experimental value of 97% ee.

The CBS reduction has also proven to be an efficient method for asymmetric reduction of α,β-unsaturated enones[14] and ynones[15] (Scheme 4.3l). The asymmetric reduction of alkynyl ketones affords propargylic alcohols **30** with high levels of enantioselectivity and in moderate to good yields. Optimized reaction conditions for the reduction are the use of THF at −30 C, 2 equivalents of chiral oxazaborolidine **28b**, and 5 equivalents of borane methyl sulfide complex.

The CBS reduction has been employed in numerous synthetic applications.[16] The cetirizine hydrochloride **32** (Zyrtec) is an effective treatment as a second-generation histamine H1 antagonist for a range of allergic diseases. Zyrtec is one of the leading antihistamine drugs, with sales of $1.3 billion in 2004 in the United States alone. Corey and Helal prepared a chiral benzylic alcohol intermediate **34** en route to their enantioselective synthesis of Zyrtec[17] (Scheme 4.3m). The asymmetric reduction of the ketone **33** in toluene with catecholborane in

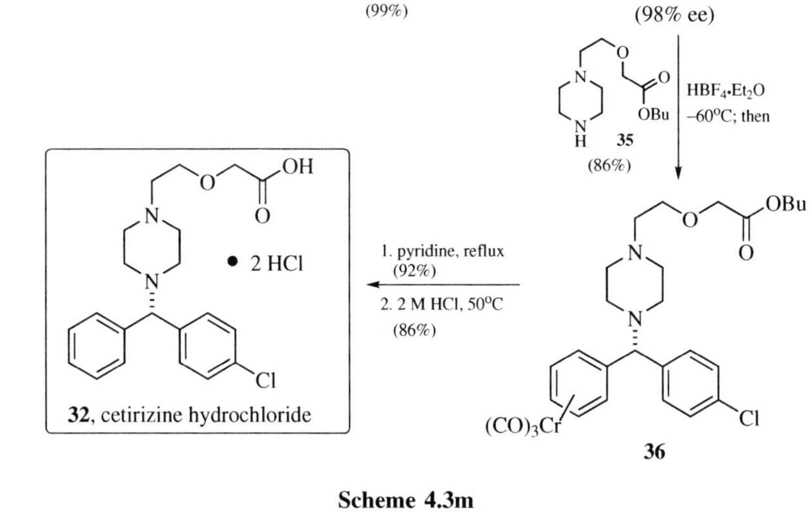

Scheme 4.3m

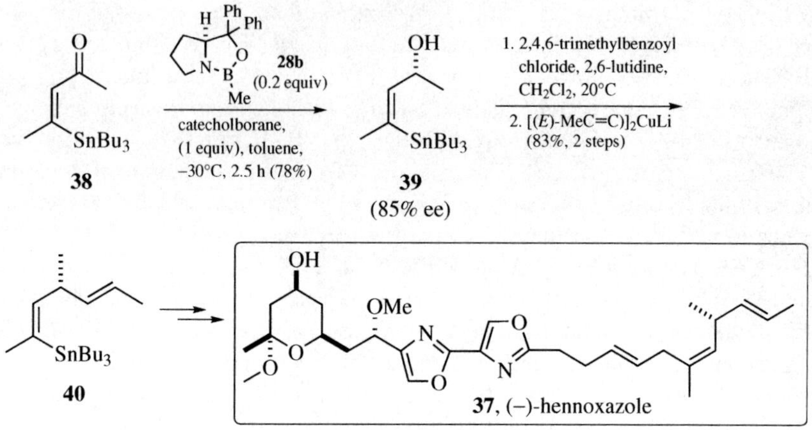

Scheme 4.3n

the presence of a catalytic amount of **28 d** afforded the alcohol **34** in 99% yield and 98% ee. Treatment of the secondary alcohol **34** with tetrafluoroboric acid at $-60\,°C$ followed by addition of the amine **35** yielded compound **36**. Removal of the chromium substituent and subsequent acidic hydrolysis produced **32** in 98% ee, demonstrating a practical application of the CBS oxazaborolidine-catalyzed reduction.

In the synthesis of (−)-hennoxazole (**37**), Wipf and Lim used a CBS reagent to prepare the chiral allylic alcohol **39**[18] (Scheme 4.3n). The enantioselective reduction of the enone **38** using a catalytic amount of the oxazaborolidine **28b**

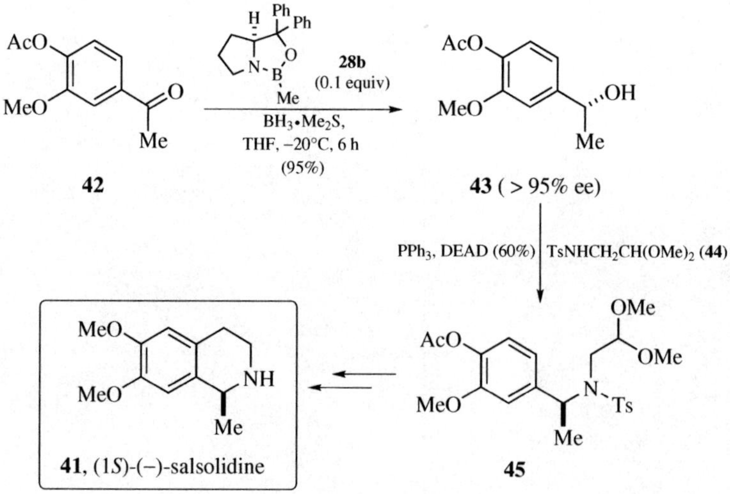

Scheme 4.3o

along with stoichiometric amount of catecholborane delivered the allylic alcohol **39** in 85% ee. Stereoselective addition of the *E*-propenyl cuprate reagent to the 2,4,6-trimethylbenzoate of **39** provided the 1,4-diene **40**.

The oxazaborolidine-catalyzed enantioselective reduction of aryl alkyl ketones was used in the asymmetric synthesis of the naturally occurring molecule (1*S*)-(−)-salsolidine **41**[19] (Scheme 4.3o). The ketone **42** underwent oxazaborolidine-mediated reduction to furnish the alcohol **43** in excellent yield and greater than 95% ee. The alcohol **43** was then coupled with the reagent **44** under Mitsunobu conditions to produce the aminoacetal **45**.

Corey and Roberts reported a total synthesis of the dysidiolide **46**, a marine sponge metabolite with biological activities against A-549 human lung carcinoma and P388 murine leukemia cancer cell lines[20] (Scheme 4.3p). The unwanted alcohol (**47**) was converted to the ketone **48** via Dess–Martin periodinane oxidation. The asymmetric reduction of **48** with the CBS catalyst **28b** efficiently gave the alcohol **49**, which was transformed into the dysidiolide **46** via photochemical oxidation.

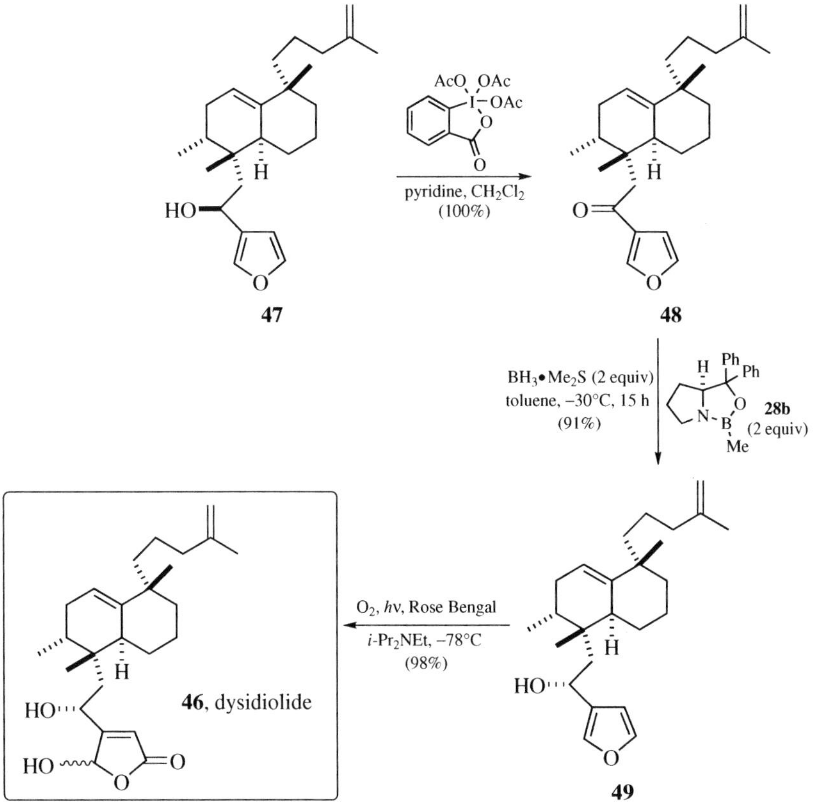

Scheme 4.3p

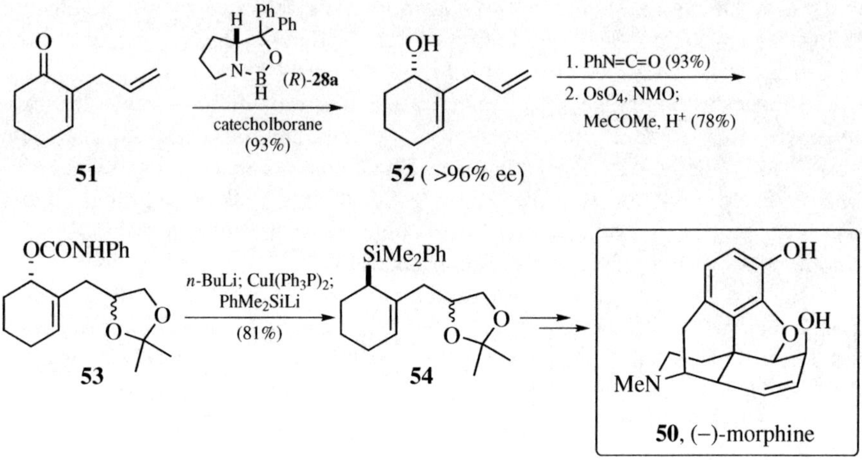

Scheme 4.3q

Overman et al. exercised the CBS reduction strategy during synthesis of the natural opium alkaloid (−)-morphine (**50**)[21] (Scheme 4.3q). Enantioselective reduction of 2-allylcyclohex-2-en-1-one (**51**) with catecholborane in the presence of the (R)-oxazaborolidine catalyst (R)-**28a** provided the corresponding (S)-cyclohexenol **52** in greater than 96% ee. Condensation of this intermediate with phenyl isocyanate, regioselective catalytic dihydroxylation of the terminal double bond, and protection of the resulting diol afforded **53** in 68% overall yield from **51**. The allylic silane **54** for the upcoming iminium ion–allylsilane cyclization step was obtained in 81% yield by a stereoselective S_N2' displacement of allylic carbamate.

(−)-Erinacine B (**55**) is the first xylose-conjugated terpenoid possessing a cyathane core and exhibits significant activity in stimulating nerve growth factor synthesis. In an enantioselective synthesis of **55**, the stereoselective CBS reduction was utilized with excellent stereochemical outcome[22] (Scheme 4.3r). Treatment of the β,γ-epoxy ketone **56** with DBU promoted a clean β-elimination reaction to provide the γ-hydroxy-α,β-unsaturated ketone, which was then converted to the benzoate **57**. CBS reduction of **57** yielded the alcohol desired (**58**) as a single isomer.

Brown et al. disclosed an asymmetric reduction protocol by using (−)-diisopinocampheylchloroborane [**59**; (−)-Ipc$_2$BCl, ^{d}Ipc$_2$BCl, (−)-DIP-Chloride], readily prepared from commercially available (+)-α-pinene in high optical purity of 99% ee[23] (Scheme 4.3s). This reagent reduces aryl alkyl ketones to afford the alcohol **60** with excellent levels of asymmetric induction. Although the asymmetric reduction of less sterically hindered ketones, such as 3-methyl-2-butanone, gives the product in only 32% ee, α-tertiary aliphatic ketones smoothly undergo reduction to afford **60** with high enantioselectivity.[24] Brown et al. proposed that

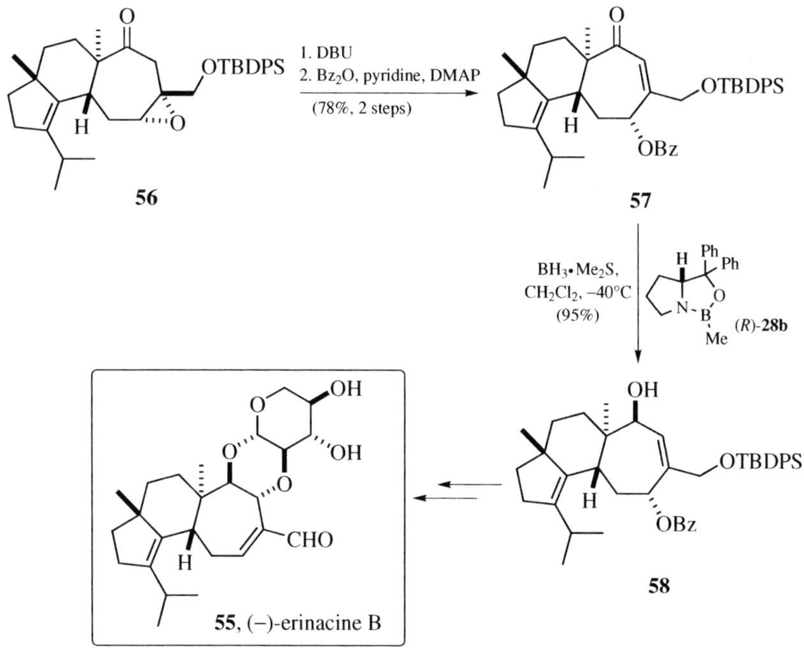

Scheme 4.3r

the reduction would involve a six-membered boatlike transition state[23] (Scheme 4.3t). In the preferred transition state **A**, the smaller group (R_S) has to face an unfavorable 1,3-diaxial interaction with the methyl group, while the larger alkyl group (R_L) assumes a pseudoequatorial position. This explains the formation of the S-enantiomer as the major product.

In collaboration with Brown and others, Rogic undertook a computational study to better understand the transition states of the asymmetric reduction[25] (Scheme 4.3u). In transition state **A**, the six-membered ring is nearly planar, with the B–Cl bond being *syn* to the C_1–H bond. Transition state **B**, on the other hand, is half-chair, the B–Cl bond being *anti* to the C_1–H bond. Calculations show that transition state **A** is more stable than **B** by 2.46 kcal/mol, predicting an (S)-enantiomer of 99% enantiomeric purity.

Brown et al. achieved a highly enantioselective synthesis of (S)-fluoxetine hydrochloride (**61**) (Prozac), an antidepressant medicine[26] (Scheme 4.3v). Reduction of β-chloropropiophenone (**62**) with (+)-Ipc$_2$BCl provided the secondary alcohol **63** in 97% ee and greater than 99% ee after single recrystallization. Mitsunobu reaction of the alcohol with *p*-trifluoromethylphenol afforded the ether **64**, which was then converted to (S)-(+)-fluoxetine hydrochloride (**61**) after subsequent treatment with excess methylamine and a solution of hydrogen chloride in ether.

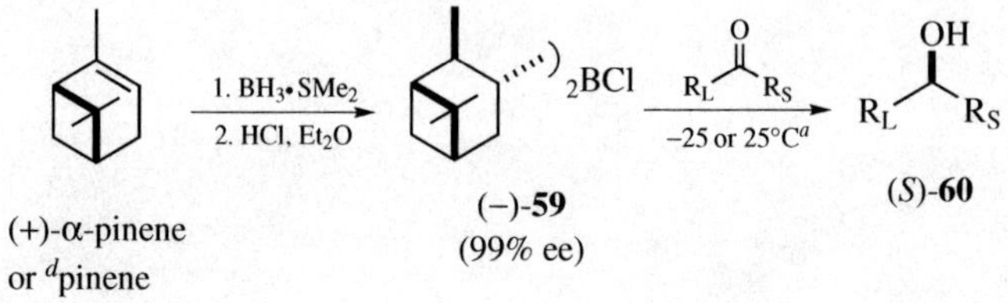

STEREOSELECTIVE REDUCTIONS

(+)-α-pinene or ^{d}pinene → 1. BH$_3$·SMe$_2$; 2. HCl, Et$_2$O → (−)-59 (99% ee) ·2BCl + R$_L$C(O)R$_S$ → (−25 or 25°C^a) → (S)-60 [R$_L$CH(OH)R$_S$]

Ketone (Aromatic)	ee, % (Config.)	Yield, %	Ketone (Aliphatic)	ee, % (Config.)	Yield, %
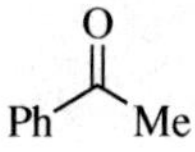 Ph-CO-Me	98 (S)	72	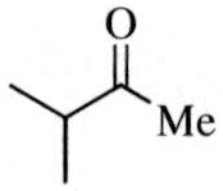 iPr-CO-Me	32 (S)	—
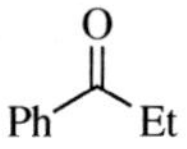 Ph-CO-Et	98 (S)	62	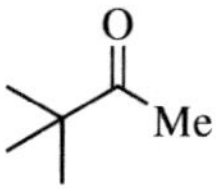 tBu-CO-Me	95 (S)	50
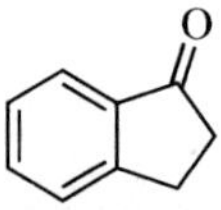 indanone	97 (S)	62	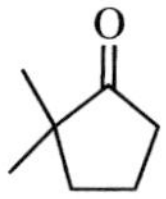 tBu-cyclopentanone	98 (S)	71
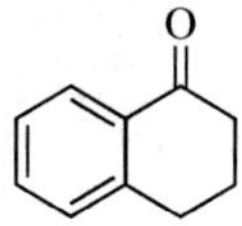 tetralone	86 (S)	70	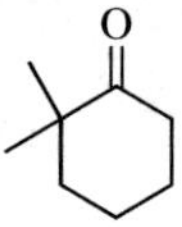 tBu-cyclohexanone	91 (S)	60

^{a} Aromatic ketones: −25°C, 5 h, THF; aliphatic ketones: 25°C, 12 h, neat.
R$_L$, aromatic substituent for aromatic ketones and tertiary alkyl group for aliphatic ketones.

Scheme 4.3s

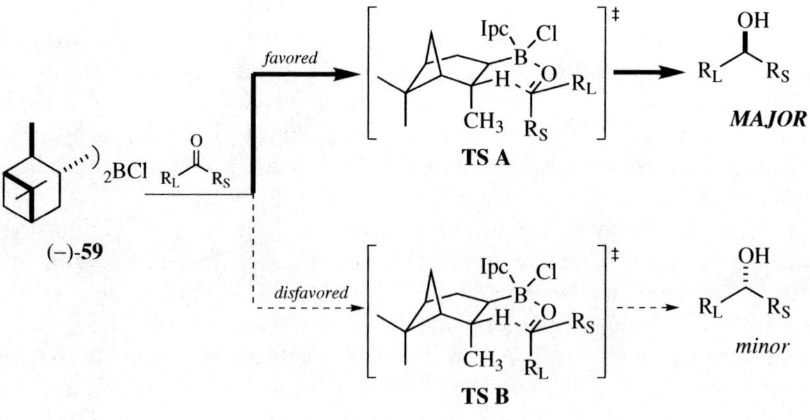

(−)-59 + R$_L$C(O)R$_S$ →
- *favored* → TS A [Ipc, Cl, B, O-R$_L$, H, CH$_3$, R$_S$]‡ → R$_L$CH(OH)R$_S$ **MAJOR**
- *disfavored* → TS B [Ipc, Cl, B, O-R$_S$, H, CH$_3$, R$_L$]‡ → R$_L$CH(OH)R$_S$ *minor*

Scheme 4.3t

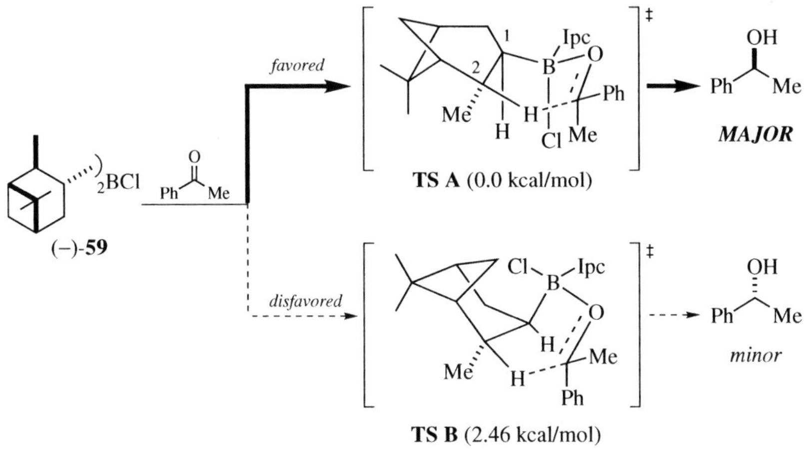

Scheme 4.3u

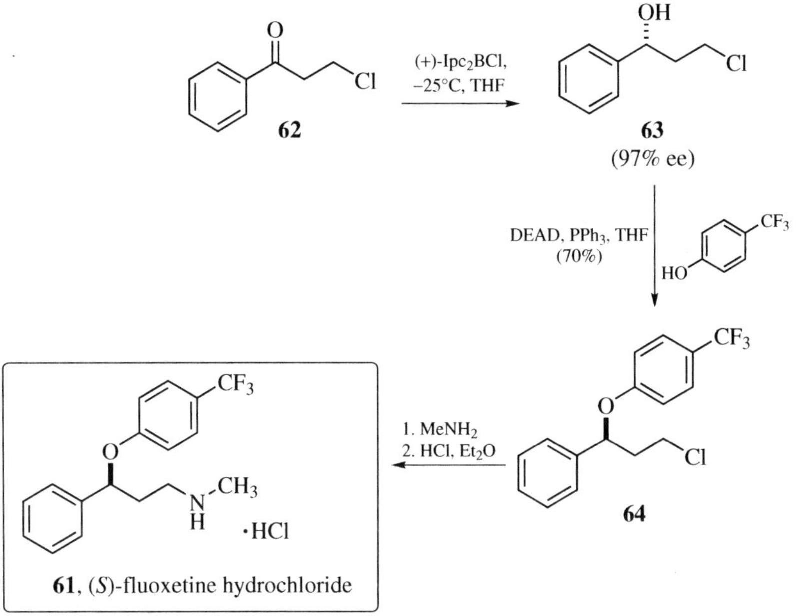

Scheme 4.3v

Nicolaou and Hepworth used the Ipc$_2$BCl reduction method in an efficient synthesis of naphthoquinone alkannin (**65**), which exhibits many interesting biological properties, such as antibacterial, antifungal, anti-inflammatory, and antitumor activities[27] (Scheme 4.3w). The lithium anion of the bromonaphthalene **66** underwent a coupling reaction with the Weinreb amide **67** to provide the

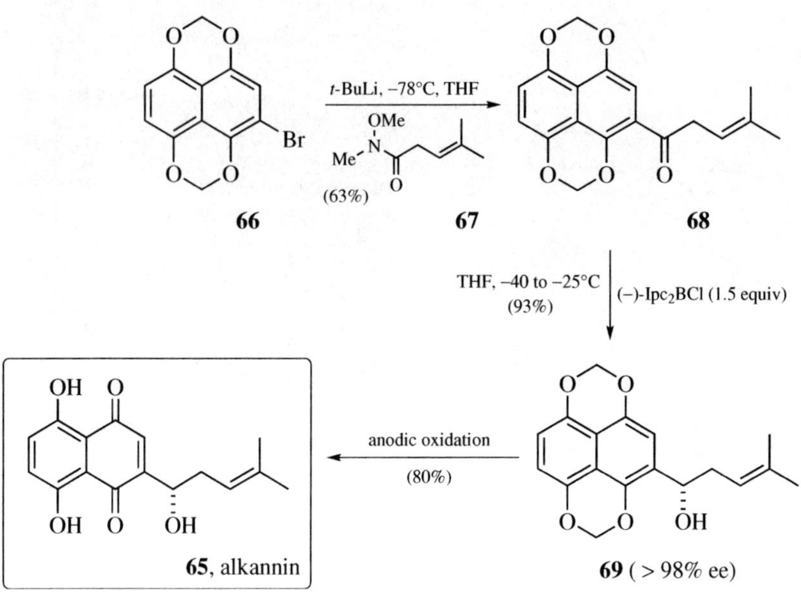

188 STEREOSELECTIVE REDUCTIONS

Scheme 4.3w

ketone **68** in good yield. The asymmetric reduction of **68** was carried out with (−)-Ipc$_2$BCl to form the alcohol **69** in greater than 98% ee and high yield. The subsequent mild anodic oxidation of the free hydroxy derivative **69** afforded the target molecule, alkannin (**65**).

Taber and Zhang synthesized the enediol isofuran **70**, which belongs to a new class of isofurans that show significant biological activities[28] (Scheme 4.3x). The Horner–Wadsworth–Emmons condensation of the aldehyde **71** with the phosphonate **72** gave the α,β-unsaturated ketone **73** in high yield. Asymmetric reduction with (−)-Ipc$_2$BCl proceeded well to afford the allylic alcohol **74** as a single diastereomer.

Trauner et al. completed the total synthesis of (−)-heptemerone B (**75**), a diterpene natural product that strongly inhibits fungal germination of the plant pathogen *Magnaporthe grisea*[29] (Scheme 4.3y). Monolithiation of 3,4-diiodofuran at low temperature followed by addition of the resulting organolithium species to (E) 4-methyl-4-hexenal produced the racemic alcohol (±)-**76**, which upon oxidation furnished the furyl ketone **77**. The ensuing (+)-Ipc$_2$BCl reduction then afforded (+)-**76** in 94% ee. The vinyl iodide **76** underwent intramolecular Heck coupling in the presence of tetra-*n*-butylammonium bromide to give a 5.1:1 mixture of the desired diastereomer (**78**) and its epimer.

Midland and others reported that *B*-isopinocampheyl-9-borabicyclo[3.3.1]nonane [Alpine-Borane; (*R*)-**79**] is an effective reagent for the highly asymmetric reduction of alkynyl ketones to afford the propargylic alcohol **80**[30] (Scheme 4.3z). The reagent (*R*)-**79** is prepared from (+)-α-pinene and 9-borabicyclo[3.3.1]nonane (9-BBN) and often represented as **79***banana*. The levels of asymmetric

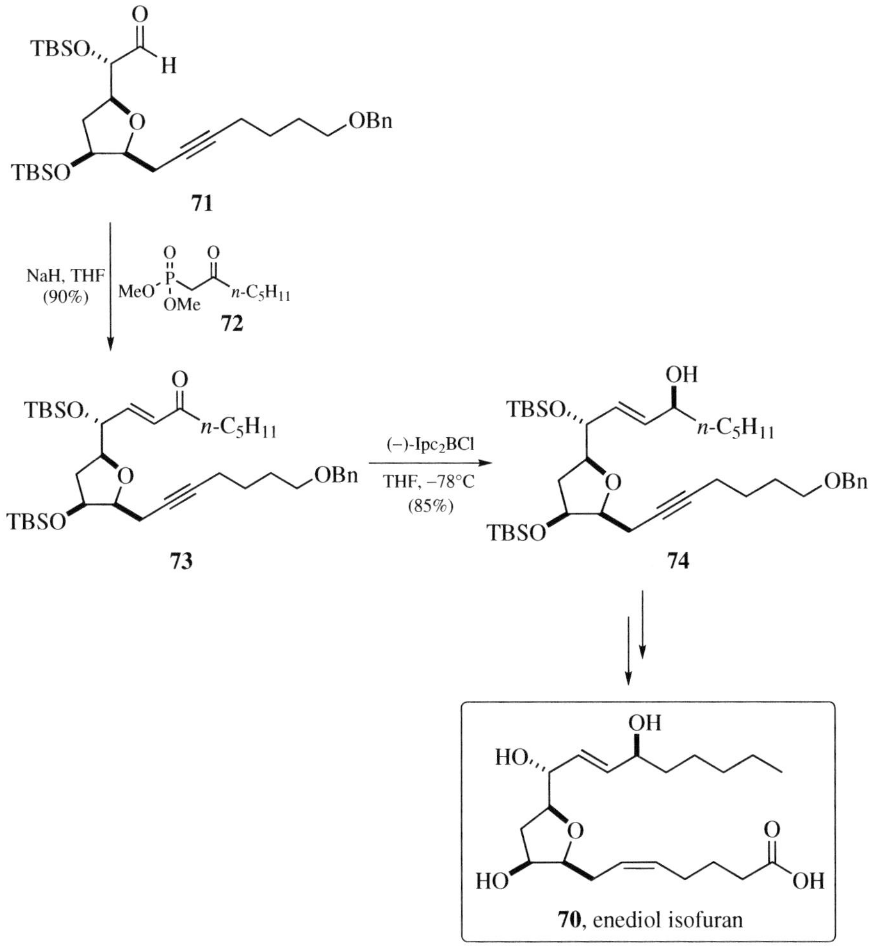

Scheme 4.3x

induction realized with **79** are good to excellent. Particularly noteworthy is the reduction of benzoylacetylenic ketoester, which undergoes reduction with perfect enantiofacial selectivity.[31]

The reduction is considered to proceed through a six-membered transition state[30] (Scheme 4.3aa). In the hydride-bridged six-membered transition state **A**, the acetylenic unit positions itself away from the isopinocampheyl skeleton. The hydrogen β to the boron is then transferred to the carbonyl group from the bottom face of the ketone. Computational analysis on the proposed transition state of the Midland reduction has not, however, been reported.[32]

Midland and Graham completed a total synthesis of (−)-pestalotin (**81**)[33] (Scheme 4.3bb). The asymmetric reduction of the ketone **82** gave the propargylic alcohol **83** with high enantioselectivity. Partial reduction of the alkyne,

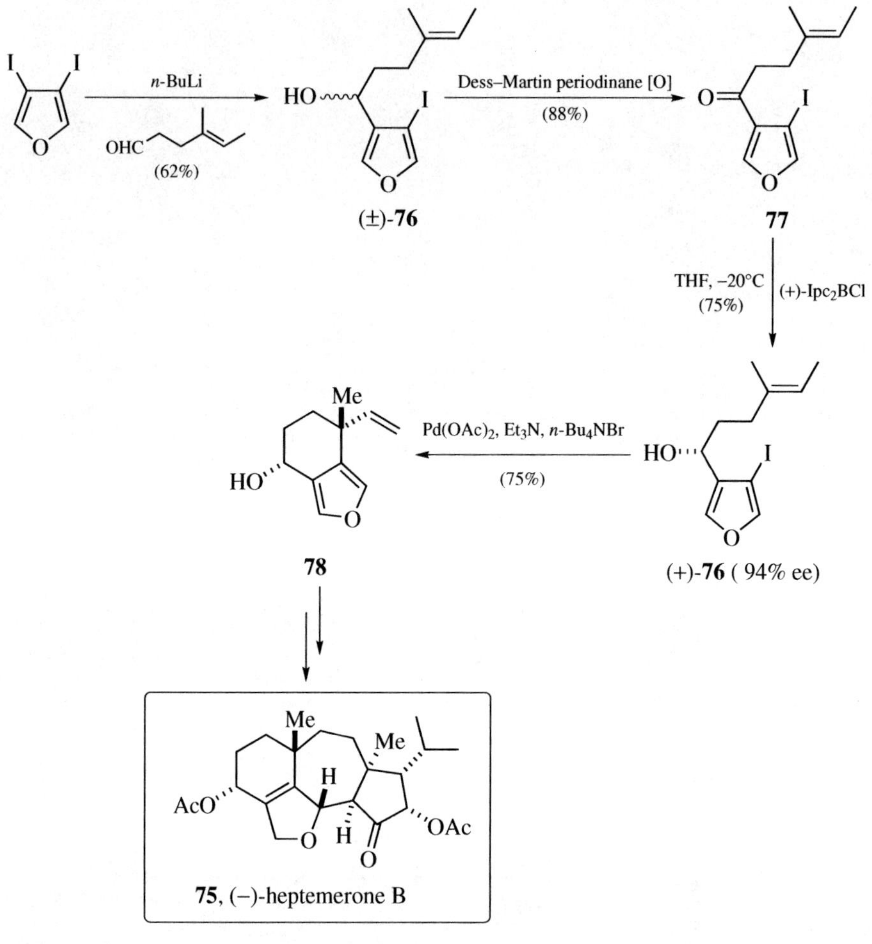

Scheme 4.3y

protection of alcohol as a MEM ether, and ozonolysis afforded the aldehyde **84**. A hetero Diels–Alder reaction of **84** with Brassard's diene (**85**) followed by deprotection provided **81**. In the total synthesis of (−)-chlorothricolide (**86**), Roush and Sciotti used the Midland reduction technology[34] (Scheme 4.3cc). Asymmetric reduction of the acetylenic ketone **87** with (*S*)-**79** afforded the alcohol **88** in 94% ee. Protection of the hydroxyl group as a MOM ether, DIBAL reduction, and subsequent protection of the resulting aldehyde provided **89**.

Mulzer and Berger used the Midland reduction en route to the total synthesis of the boron-containing macrodiolide antibiotic tartrolon B (**90**), which acts

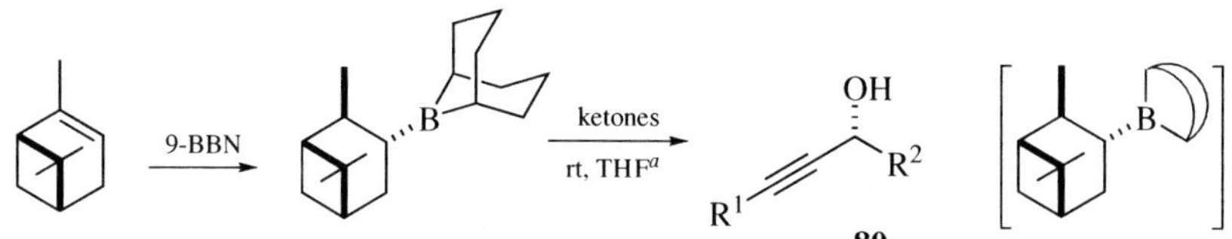

Ketone	Alcohol	ee, %	Yield, %
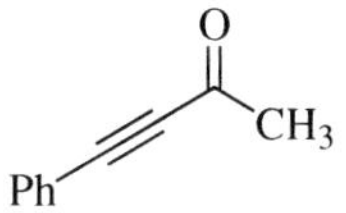 Ph–C≡C–C(O)–CH₃	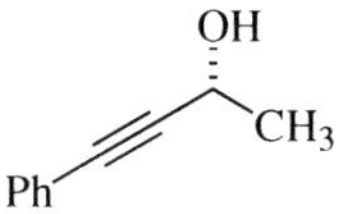 Ph–C≡C–CH(OH)–CH₃	78	98
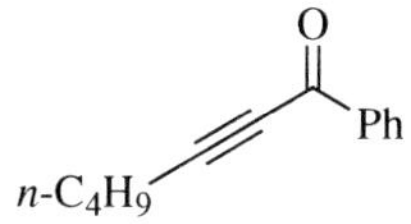 n-C₄H₉–C≡C–C(O)–Ph	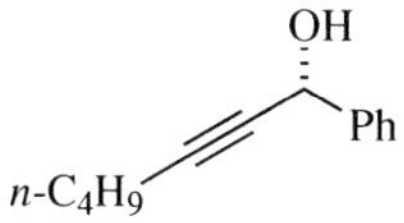 n-C₄H₉–C≡C–CH(OH)–Ph	89	72
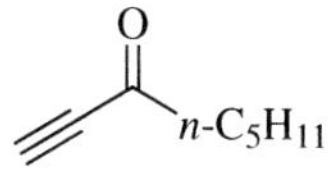 HC≡C–C(O)–n-C₅H₁₁	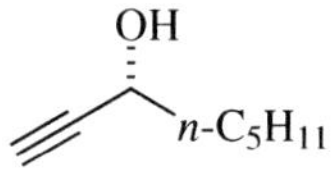 HC≡C–CH(OH)–n-C₅H₁₁	92	65
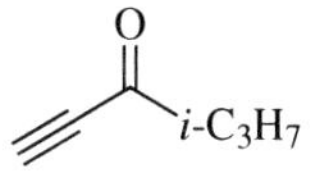 HC≡C–C(O)–i-C₃H₇	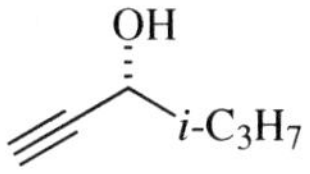 HC≡C–CH(OH)–i-C₃H₇	99	78
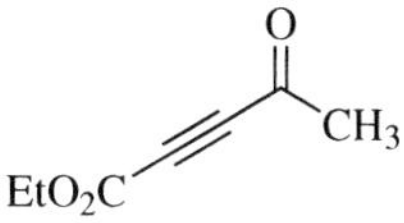 EtO₂C–C≡C–C(O)–CH₃	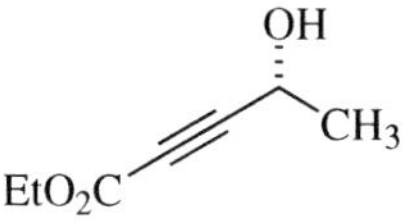 EtO₂C–C≡C–CH(OH)–CH₃	77	59
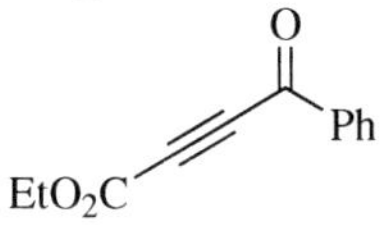 EtO₂C–C≡C–C(O)–Ph	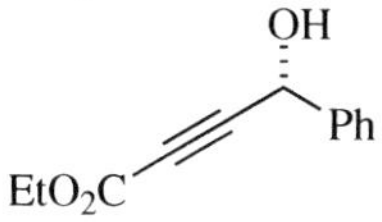 EtO₂C–C≡C–CH(OH)–Ph	100	64

[a] Reaction time: 8 h for terminal ketones and acetylenic ketoesters, 1 to 4 days for internal acetylenic ketones.

Scheme 4.3z

as an active ion carrier against gram-positive bacteria[35] (Scheme 4.3dd). The chiral aldehyde **91** was converted to the corresponding lithium acetylide by a Corey–Fuchs protocol, and subsequent reaction of the anion with the Weinreb amide **92** resulted in the formation of the alkynone **93**. Asymmetric reduction with the boron reagent (R)-**79** afforded the alcohol **94** with 80% de. The alkynol **94** was then subjected to hydrogenation to yield the 1,3-diol derivative **95**.

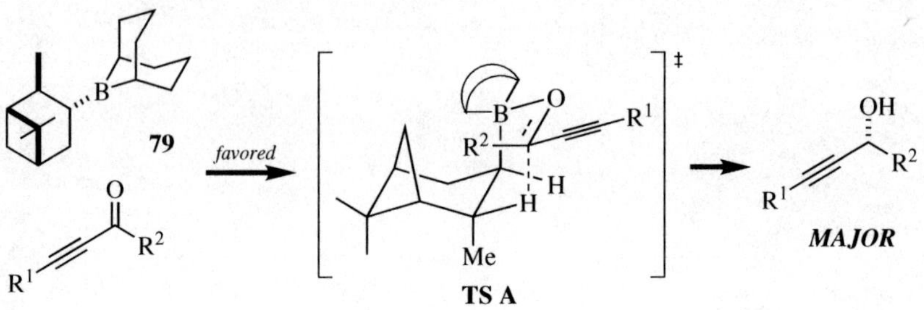

Scheme 4.3aa

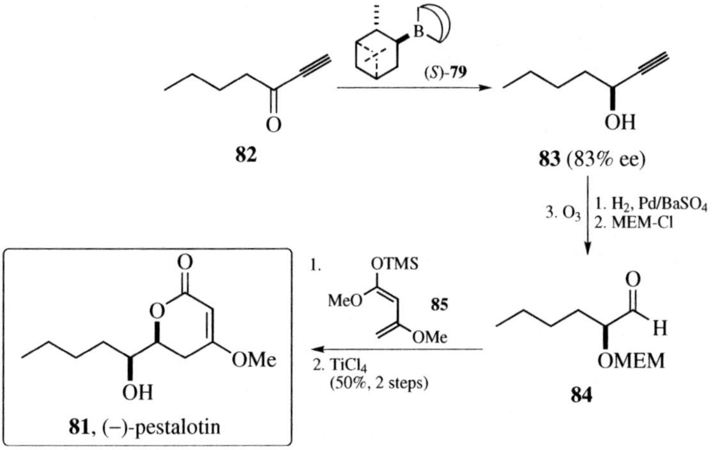

Scheme 4.3bb

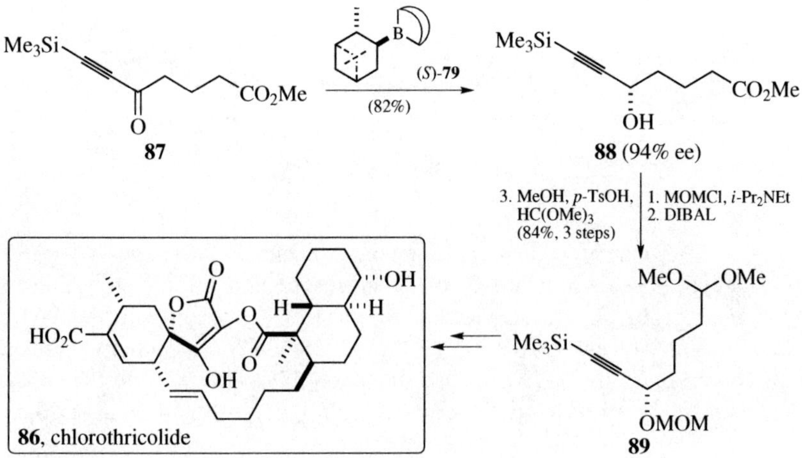

81, (−)-pestalotin

86, chlorothricolide

Scheme 4.3cc

Scheme 4.3dd

Nakada et al. utilized the Midland asymmetric alkynone reduction as a key step in the total synthesis of (+)-phomopsidin (**96**), which shows strong inhibitory activities against the assembly of the microtubule proteins purified from porcine brain[36] (Scheme 4.3ee). The conversion of the δ-lactone **97** to its corresponding Weinreb amide followed by protection of the resulting free alcohol as an ethoxyethyl ether provided compound **98** in excellent yield. The amide **98** was treated with lithium trimethylsilylacetylide to afford the alkynone **99**. Diastereoselective reduction with (*S*)-**79** followed by removal of the TMS group gave the alcohol **100** as a single diastereomer in 88% yield over three steps.

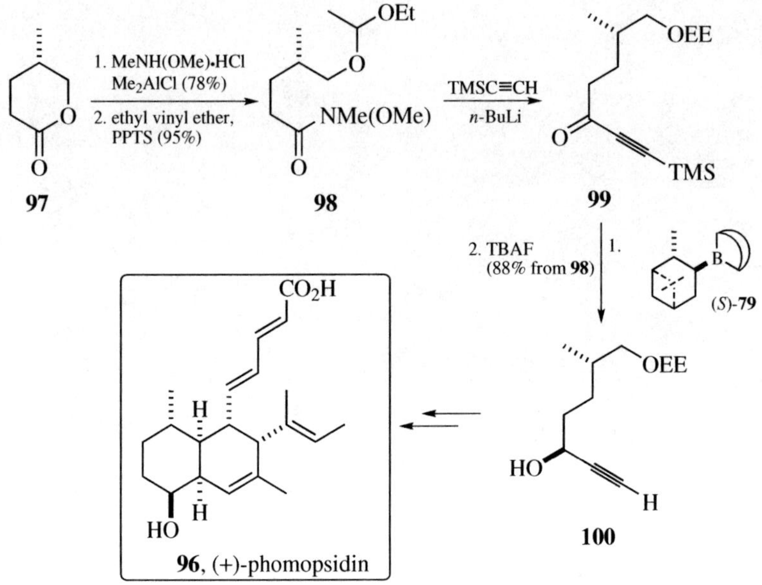

Scheme 4.3ee

REFERENCES

1. Noyori, R.; Tomino, I.; Tanimoto, Y. *J. Am. Chem. Soc.* **1979**, *101*, 3129.
2. (a) Nishizawa, M.; Yamada, M.; Noyori, R. *Tetrahedron Lett.* **1981**, *22*, 247.
   (b) Noyori, R.; Tomino, I.; Yamada, M.; Nishizawa, M. *J. Am. Chem. Soc.* **1984**, *106*, 6717.
3. (a) Noyori, R.; Tanimoto, T. Y.; Nishizawa, M. *J. Am. Chem. Soc.* **1984**, *106*, 6709;
   (b) Noyori, R. *Pure Appl. Chem.* **1981**, *53*, 2315.
4. Noyori, R. *Asymmetric Catalysis in Organic Synthesis*; Wiley: New York, **1994**; pp. 311–319.
5. Suzuki, M.; Yanagisawa, A.; Noyori, R. *J. Am. Chem. Soc.* **1985**, *107*, 3348.
6. van Innis, L.; Plancher, J. M.; Marko, I. E. *Org. Lett.* **2006**, *8*, 6111.
7. Nishizawa, M.; Yamada, M.; Noyori, R. *Tetrahedron Lett.* **1981**, *22*, 247.
8. Abe, H.; Aoyagi, S.; Kibayashi, C. *Angew. Chem. Int. Ed.* **2002**, *41*, 3017.
9. Hirao, A.; Itsuno, S.; Nakahama, S.; Yamazaki, N. *J. Chem. Soc. Chem. Commun.* **1981**, 315.
10. Itsuno, S.; Ito, K.; Hirao, A.; Nakahama, S. *J. Chem. Soc. Chem. Commun.* **1983**, 469.
11. (a) Corey, E. J.; Bakshi, R. K.; Shibata, S. *J. Am. Chem. Soc.* **1987**, *109*, 5551;
    (b) Corey, E. J.; Bakshi, R. K.; Shibata, S.; Chen, C.-P.; Singh, V. K. *J. Am. Chem. Soc.* **1987**, *109*, 7925; (c) Corey, E. J.; Link, J. O. *Tetrahedron Lett.* **1989**, *30*, 6275.
12. (a) Evans, D. A.; *Science* **1988**, *240*, 420; (b) Corey, E. J.; Link, J. O.; Bakshi, R. K. *Tetrahedron Lett.* **1992**, *33*, 7107; (c) Quallich, G. J.; Blake, J. F.; Woodall, T. M.

J. Am. Chem. Soc. **1994**, *116*, 8516; (d) Alagona, G.; Ghio, C.; Persico, M.; Tomasi, S. *J. Am. Chem. Soc.* **2003**, *125*, 10027.
13. Jones, D. K.; Liotta, D. C.; Shinkai, I.; Mathre, D. J. *J. Org. Chem.* **1993**, *58*, 799.
14. Corey, E. J.; Rao, K. S. *Tetrahedron Lett.* **1991**, *32*, 4623.
15. Parker, K. A.; Ledeboer, M. W. *J. Org. Chem.* **1996**, *61*, 3214.
16. (a) Corey, E. J.; Helal, C. J. *Angew. Chem. Int. Ed.* **1998**, *37*, 1986; (b) Farina, V.; Reeves, J. T.; Senanayake, C. H.; Song, J. J. *Chem. Rev.* **2006**, *106*, 2734.
17. Corey, E. J.; Helal, C. J. *Tetrahedron Lett.* **1996**, *37*, 4837.
18. Wipf, P.; Lim, S. *J. Am. Chem. Soc.* **1995**, *117*, 558.
19. Ponzo, V. L.; Kaufman, T. S. *Tetrahedron Lett.* **1995**, *50*, 9105.
20. Corey, E. J.; Roberts, B. E. *J. Am. Chem. Soc.* **1997**, *119*, 12425.
21. Hong, C. Y.; Kado, N.; Overman, L. E. *J. Am. Chem. Soc.* **1993**, *115*, 11028.
22. Watanabe, H.; Takano, M.; Umino, A.; Ito, T.; Ishikawa, H.; Nakada, M. *Org. Lett.* **2007**, *9*, 359.
23. Brown, H. C.; Chandrasekharan, J.; Ramachandran, P. V. *J. Am. Chem. Soc.* **1988**, *110*, 1539.
24. For reviews, see (a) Brown, H. C.; Ramachandran, P. V. *Acc. Chem. Res.* **1992**, *25*, 16; (b) Brown, H. C.; Ramachandran, P. V. *J. Organomet. Chem.* **1995**, *500*, 1.
25. Rogic, M. M.; Ramachandran, P. V.; Zinnen, H.; Brown, L. D.; Zheng, M. *Tetrahedron: Asymmetry* **1997**, *8*, 1287.
26. Srebnik, M.; Ramachandran, P. V.; Brown, H. C. *J. Org. Chem.* **1988**, *53*, 2916.
27. Nicolaou, K. C.; Hepworth, D. *Angew. Chem. Int. Ed.* **1998**, *37*, 839.
28. Taber, D. F.; Zhang, Z. *J. Org. Chem.* **2006**, *71*, 926.
29. Miller, A. K.; Hughes, C. C.; Kennedy-Smith, J. J.; Gradl, S. N.; Trauner, D. *J. Am. Chem. Soc.* **2006**, *128*, 17057.
30. (a) Midland, M. M.; McDowell, D. C.; Hatch, R. L.; Tramontano, A. *J. Am. Chem. Soc.* **1980**, *102*, 867; (b) Midland, M. M.; McLoughlin, J. J. *J. Org. Chem.* **1984**, *49*, 1316.
31. Midland, M. M. *Chem. Rev.* **1989**, *89*, 1553.
32. Rogic, M. M. *J. Org. Chem.* **2000**, *65*, 6868.
33. Midland, M. M.; Graham, R. S. *J. Am. Chem. Soc.* **1984**, *106*, 4294.
34. Roush, W. R.; Sciotti, R. J. *J. Am. Chem. Soc.* **1998**, *120*, 7411.
35. Mulzer, J.; Berger, M. *J. Org. Chem.* **2004**, *69*, 891.
36. Suzuki, T.; Usui, K.; Miyake, Y.; Namikoshi, M.; Nakada, M. *Org. Lett.* **2004**, *6*, 553.

LIST OF COPYRIGHTED MATERIALS

The following transition states were redrawn with permission from the American Chemical Society and Elsevier. The American Chemical Society owns the copyright for material in the *Journal of the American Chemical Society* (JACS) and the *Journal of Organic Chemistry* (JOC). The copyright for articles in *Tetrahedron Letters* (TL) belongs to Elsevier.

Transition State in Scheme:	Redrawn from:
II	Figures on p. 1922 in: Zimmerman and Traxler, *JACS*, **1957**, *79*, 1920.
1.1g	Figure 1 in: Boeckman et al., *JACS*, **2002**, *124*, 190.
1.4d	Scheme 7 in: Fox et al., *JACS*, **1999**, *121*, 5467.
1.4g	Scheme 1 in: Davies and Jin, *JACS*, **2004**, *126*, 10862.
1.5j	Scheme X in: Lee et al., *JACS*, **1990**, *112*, 260.
1.5l	Scheme 3 in: Chen et al., *JACS*, **2000**, *122*, 7424.
1.6e	Equation 2 in: Overman et al., *JACS*, **1983**, *105*, 6629.
2.1q	Scheme IV in: Oppolzer et al., *JACS*, **1990**, *112*, 2767.
2.1v	Scheme 7 in: Crimmins et al., *JOC*, **2001**, *66*, 894.
2.1aa	Figure 1 in: Boeckman and Connell, *JACS*, **1995**, *117*, 12368.
2.2b	Figure I in: Masamune et al., *JACS*, **1986**, *108*, 8279.
2.2r	Figure 1 in: Ghosh and onishi, *JACS*, **1996**, *118*, 2527.
2.3a, 2.3b	Table 2 and Scheme 1 in: List et al., *JACS*, **2000**, *122*, 2395.
2.3e	Figure 3 in: Sakthivel et al., *JACS*, **2001**, *123*, 5260.
3.IV	Figures of 17 and 18 in: Hoffmann and Zeib, *JOC*, **1981**, *46*, 1309.
3.VIII	Figures of 5 and 6 in: Hoffmann and Landmann, *TL*, **1983**, *24*, 3209.
3.1r	Figure of A and B in: Roush et al., *JACS*, **1985**, *107*, 8186.

Six-Membered Transition States in Organic Synthesis, By Jaemoon Yang
Copyright © 2008 John Wiley & Sons, Inc.

Transition State in Scheme:	Redrawn from:
3.1u	Scheme I in: Roush and Palkowitz, *JACS*, **1987**, *109*, 953.
3.2b	Scheme I in: Kira et al., *JACS*, **1988**, *110*, 4599.
3.2i	Figure on p. 6430 in: Sato et al., *JACS*, **1989**, *111*, 6429.
3.2l	Figure 7 in: Denmark et al., *JOC*, **2006**, *71*, 1523.
3.2p	Scheme 2 in: Matsumoto et al., *JOC*, **1994**, *59*, 7152.
4.2b	Scheme IV in: Evans et al., *JACS*, **1988**, *110*, 3560.
4.3c	Figures of 11 and 12 in: Noyon et al., *JACS*, **1984**, *106*, 6709.

ABBREVIATIONS

AD-mix-β	Reagent for Sharpless asymmetric dihydroxylation
9-BBN	9-Borabicyclo[3.3.1]nonyl
Bn	Benzyl
Boc	*t*-Butoxycarbonyl
Bz	Benzoyl
BOM	Benzyloxymethyl
CDI	Carbonyldiimidazole
m-CPBA	*m*-Chloroperoxybenzoic acid
CSA	Camphorsulfonic acid
Cy	Cyclohexyl
DBU	1,8-Diazabicyclo[5.4.0]undec-7-ene
DDQ	2,3-Dichloro-5,6-dicyano-*p*-benzoquinone
DEAD	Diethyl azodicarboxylate
DIAD	Diisopropyl azodicarboxylate
DIBAL-H	Diisobutylaluminum hydride
DIPT	Diisopropyl tartrate
DME	Dimethoxyethane
DMF	*N*,*N*-Dimethylformamide
DMAP	4-Dimethylaminopyridine
DMSO	Dimethyl sulfoxide
EDC	*N*-(3-Dimethylaminopropyl)-*N*'-ethylcarbodiimide
HMPA	Hexamethylphosphoramide
HOBT	1-Hydroxybenzotriazole
KHMDS	Potassium hexamethyldisilazane
LDA	Lithium diisopropylamide
MEM	Methoxyethoxymethyl
MOM	Methoxymethyl
MoOPH	Oxidodiperoxymolybdenum(pyridine)(hexamethylphophoramide)
NaHMDS	Sodium hexamethyldisilazane
NBS	*N*-Bromosuccinimide
NMM	*N*-Methylmorpholine
NMO	*N*-Methylmorpholine *N*-oxide
Piv	Pivaloyl
PMB	*p*-Methoxybenzyl

Six-Membered Transition States in Organic Synthesis, By Jaemoon Yang
Copyright © 2008 John Wiley & Sons, Inc.

PPTS	Pyridinium *p*-toluenesulfonate
Py	Pyridine
TBAF	Tetra-*n*-butylammonium fluoride
TBDPS	*t*-Butyldiphenylsilyl
TBS	*t*-Butyldimethylsilyl
TES	Triethylsilyl
TFA	Trifluoroacetic acid
THF	Tetrahydrofuran
TIPS	Triisopropylsilyl
TMEDA	N,N,N',N'-Tetrmethylethylenediamine
TMS	Trimethylsilyl
Tf	Trifluoromethanesulfonyl
Tol	Toluene
TPAP	Tetrapropylammonium perruthenate
Tr	Trityl
Ts	*p*-Toluenesulfonyl

SUBJECT INDEX

Ab initio calculations, 39
Abbreviations, listing of, 199–200
Acetaldehyde, 49, 62, 98, 133, 142
Acetic acid, 153, 161
Acetone, 91, 93
Acetonide, 62, 136, 156, 165–167
Acetonitrile, 75, 161
Acetophenone, 148, 181
Acetoxy groups, 162
Acetylaldehyde, 54, 133
Acyclic substrates, 9
Acylation, 118
Acyl chloride, 30
Acyloxazolidinone, 81
3-Acylpyrrolidines, 45
Acylthiazolidinethiones, 82
Adda, 31–32
(−)-Aflastatin A, 87
AIDS, 77. *See also* HIV replication
Alcohol(s), 26, 106–107, 110–112, 115, 117–118, 120–121, 124–126, 129, 131, 133, 139–141, 173–176, 181, 183, 188, 190–191, 193
Aldehydes, 13, 16–17, 26, 49, 61–62, 66–67, 70, 79, 84, 91–93, 95–98, 102–103, 107–108, 110–111, 113, 116–117, 123–124, 132, 134–136, 139–141, 170–171, 173, 188, 190–191
Aldol reactions:
 asymmetric *anti*-, 78–90
 asymmetric *syn*,- 57–77
 defined, 49
 diastereoselective, 49–52, 56, 73, 81
 1,3-diaxial interactions, 52
 proline-catalyzed asymmetric, 91–96
Aliphatics, 5, 79, 81, 91, 93–94, 123, 141, 178, 184, 186
Alkannin, 187–188
Alkenes, 9, 25, 121
Alkenyl groups, 175
Alkoxy groups, 95

Alkyl groups, 49, 53, 95, 173, 178, 185
3-Alkylidene tetrahydropyran, 80
Alkynone, 191, 193
Allyboration, Roush method, 118–119
Allylation, 105
2-Allylcyclohex-2-en-1-one, 184
Allyl groups, 9, 13, 123
Allyl iodide, 15
Allylboranates, 116
Allylborane, 115, 125
Allylboration, 106, 115, 122
Allylboronates, 118, 120
Allylboronic acid, 98, 100
Allyldimethylphenylsilane, 140
Allylic alcohol, 24, 26, 40
Allylmagnesium bromide, 122, 170
1-Allyl-2-naphthol, 5
1-Allyl-1-phenylsilacyclobutane, 140
Allylsilane, 140, 184
Allylsilicates, 127–128, 130
Allyltributyltin, 123
Allyltrifluorosilanes, 132–135
Allyltrimethylsilane, 127–128, 139
Allyl vinyl, 13–14
α-Acetoxy enolsilane, 96
α-Alkoxyaldehyde, 95
α-D-glucosaccharinic acid lactone, 30
(+)-α-Pinene, 104, 106, 184, 188, 191
α-Silyloxy, 96
Aluminum hydride, 147
Alzheimer's Disease, 86
Amides, 193
Aminoacetal, 183
Amino acids, 31, 82, 91
Amino alcohols, 46, 77, 179
(2S, 3S, 8S, 9S, 4E, 6E)-3-Amino-9-methoxy-2,6, 8-trimethyl-10- phenyldeca-4,6y-dienoic acid, 31
Ammonium chloride, 5
Amphidinolide T3, 124–125
Anabaena cylindrical, 156

Six-Membered Transition States in Organic Synthesis, By Jaemoon Yang
Copyright © 2008 John Wiley & Sons, Inc.

201

Anachelin H, 156
Anionic Oxy-Cope rearrangement, 36–42
Anodic oxidation, 188
Ansamycin antibiotic group, 117
Anti-aldol products, 58
Antibiotics, 29, 64–65, 121, 158, 162, 190
Anticancer drugs, 25
Antidepressants, 185
Anti-1,3-diol monoester, 26
Anti-isomer, 1, 50
Apoptolidin, 113
Apoptosis, 121
Aromatics, 2, 5, 12, 79, 91–93, 123, 127, 141, 178, 186
(−)-Asteriscanolide, 35
Aza-Cope-Mannich rearrangement, 43, 45–48
Azide, 25
Azumamide A, 109

B-allyldiisopinocampheylborane, 106
Baeyer-Villiger oxidation, 48
Bafilomycin A_1, 117, 121
1BBu$_2$, 151
B-chloropropiophenone, 185
(+)-Bengamide E, 75
Benzaldehyde, 1, 36, 43, 49, 52, 63, 81, 83, 89, 121, 130, 138, 140–141, 159, 167
Benzene, 12, 47, 177
Benzoate, 184
Benzoic acid, 157
Benzoin, 134
Benzoylacetylenic ketoester, 189
Benzyl ester, 31
Benzyloxyacetaldehyde, 77, 165
4-Benzyloxy-3-methoxybenzaldehyde, 87
3-Benzyloxypropanal, 70, 113
β-hydroxy ketones, diastereoselective reduction:
  anti-, 161–173
  syn, 151–160
β-pyrrolidinopropionate ester, 31
B-H bonds, 181
B-isopinocampheyl-9-borabicyclo[3.3.1]nonane, 188
Bi-2-naphthol (BINOL), 173
Binaphthol-modified lithium aluminum hydride (BINAL-H), 173–177, 179
Bisphosphonamide, 137
B-methoxydiisopinocampheylborane, 106
Boatlike transition structure, 1–2, 5, 9–10, 12, 22, 39, 83, 95, 147–148, 185
Boltzmann distribution, 54
9-Borabicyclo[3.3.1]nonane (9-BBN), 51, 143, 188

Borane, 106, 179–180
Borolanes, 122
Boron:
  allylation, 125–126
  enolates, 61–63, 69, 76, 79, 97
  -oxygen bonds, 53
  trifluoride etherate, 111
Bromoborane, 123
Bromonaphthalene, 187
Brown allylation, 107–108
Bryostatin 2, 170
2-Butenes, 111
3-Butenylmagnesium bromide, 66
Butyl groups, 57–58

Calcimycin, 30–31
(−)-Calicheamicinone, 110, 115
(−)-Callystatin A, 72
(+)-Calopin dimethyl ether, 17, 19
(+)-Camphor, 102–103
Camphorsulfonic acid (CSA), 48, 159
Cancer:
  anticancer drugs, 25
  carcinogenesis, 65
  gastric, 26
  leukemia, 183
  tumor cell lines, 170, 177
Carbamate, allylic, 184
Carbohydrates, 96
Carbon-carbon bond, 49, 97, 106
Carbon-hydrogen bond, 39
Carbonyl:
  carbon, 107
  compounds, 51
  groups, 49, 117, 189
  oxygen, 147
Carbonyldiimidazole (CDI), 62
Carbon-oxygen bond, 24, 40
Carboxylic, generally:
  acid, 40, 70
  group, 73
Carcinogenesis, 65
(+)-2-Carene, 104, 107
Catalysis, 162
Catecholborane, 181–183
Cetirizine hydrochloride, 181
Chairlike transition state, 1–2, 5, 8–10, 15, 21–22, 26, 32, 38, 50, 53, 55, 57, 62, 78, 83, 89, 95, 98–99, 102–103, 106, 122–123, 130, 132–134, 136–137, 140, 142, 147–148, 152, 175, 162, 180
Chelation, 49, 71–72, 158–159
Chiral auxiliary, 57, 60–61, 63, 65, 67, 70, 73, 75–76, 81, 83–84, 88–90, 123, 139

Chirality transfer:
  Claisen rearrangements, 13, 15
  Johnson-Claissen rearrangement, 24
  S-Z, 15
Chlorine, 142
Chloroperbenzoic acid, 159
(−)-Chlorothricolide, 190, 192
Chlorotitanium, 70–72
Chlorotrimethylsilane, 81, 87
Chorsimate, 3
Chromatographic analysis:
  column, 124
  high-performance liquid, 63
  thin-layer, 63
Chromophores, 63
Ciguatoxin (CTX3 C), 118
Cinnamaldehyde, 16, 81, 84, 88, 90
Cinnamyl alcohol, 20
Cis-octahydroindoles, 46
Cis-3-methyl-6-phenyl-1,5-heptadiene, 32
Cis-2-octalone, 37
Claisen, Ludwig, 5
Claisen rearrangement, 3–7, 9–10, 13–19, 49
(+)-Clavosolide A, 166
Cleavage, 70, 81, 83, 90, 118, 121, 136, 139, 157, 159, 168, 171
Collins agent, 22
Colony-stimulating factor, 66
Compactin, 153
(+)-Conagenin, 159–160
Cope, Arthur C., 9
Cope-Claisen rearrangement, 34
Cope rearrangement, 5, 9–10, 32–36, 49
Copyrighted materials, list of, 197–198
Corey-Bakshi-Shibata (CBS) reduction, 180–181, 183–184
Corey-Fuchs protocol, 191
Coulombic repulsion, 112
Coulomb interaction, 117
Countercation, 147
(+)-CP-263, 114, 42
Crimmins aldol condensation, 73
Crotylation, 111–113, 128, 130, 132, 136, 141, 143–144
Crotylboration, 113, 117, 121
Crotylboronates, 97–98, 105–106, 117
Crotyldiisopinocampheylborane, 112
Crotylpotassiums, 111
Crotyl propenyl ethers, 7
Crotylsilanes, 133–134, 141, 143–144
Crotylsilicates, 128–130
Crotyltrichlorosilanes, 127, 131–132
Crotyltrifluorosilane, 135–136
Cryptophycin 1, 170

Crystallization, 63, 68
(+)-Curacin A, 105, 108
CTX3 C, 118
Cyclobutene, 35
Cycloheptenone, 34
Cyclohexanes, 3–4, 33, 93
Cyclohexenyl silylketeneacetals, 9
Cyclononadiene, 42
Cyclooctadiene, 36
Cyclopentanols, 46
Cyclopentene, 40
Cyclopentenone, 42, 175
Cyclopentylidene ketal, 166
Cyclopropyl ketone, 134
Cytotoxicity, 72, 142–143, 156, 177

Dean-Stark trap, 47
Decanal, 47
Deep vein mycoses, 171
(+)-9(11)-Dehydroesterone methyl ether, 14, 16
(−)-Denticulatin A, 68–69
6-Deoxyerythronolide, 59
6-Deoxy-L-glucose, 30
Depression-related disorders, 138
Deprotection, 14, 25, 156, 159, 190
Deprotonation, 49
Deprotonation, 75
Dermostatin A, 171–172
Desilylation, 62, 69, 171, 175
Dess-Martin periodinane, 183, 190
D-glyceraldehyde acetonide, 117
Dialdehydes, 22, 142
Dialkenyl cyclobutane, 36
Dialkylboron, 51–52
Diallyl ethers, 15–17
Diaminocyclohexane, 138
Diarylbutane, 143
Diastereofacial selectivity, 94
Diastereomer, 81, 90
Diastereoselectivity, 38, 45, 49, 52, 56, 73, 75–76, 81, 89, 97, 99, 113, 124, 127, 130, 132, 134–135, 140–141, 143, 158. See also β-hydroxy ketones, diastereoselective reduction
Diasterofacial selectivity, 79
Diasteromer, 25, 46
Diaxial interactions, 13, 15, 21, 52
Diazomethane, 31, 156
DIBAL-H, 62, 70
DIBAL reduction, 190
Dibenzocyclooctadiene, 143
Dibutylborinate, 153
Dibutylboron trifluoromethanesulfonate, 52
1,2-Dichlorobenzene, 15, 18

Dichloromethane, 80
Dicyclohexylboron, 78, 80, 171
Dienes, 10, 12, 29, 36, 42, 165, 190
Dienols, 38
Diethylzinc-aldehyde, 16
Diglyme, 38
Dihydrocinnamaldehyde, 126
(+)-Dihdyrocostunolide, 33
(9S)-Dihydroerythronolide A, 73
(+)-Dihydromayurone, 40–41
Dihydroxylation, 136, 168, 184
3,4-Diiodofuran, 188
(−)-Diisopinocampheylchloroborane, 184
1,3-Diketone, 15
Dilithium catecholate, 127
Dimerization, 95–96
2,3-Dimethoxy-4-methylbenzaldehyde, 17
1-Dimethylamino-3-butene, 43
Dimethylborinate, 62
(2S, 5S)-Dimethylborolane
 trifloromethanesulfonate, 78
2,2-Dimethylbutane (DMB), 35
3,5-Dimethylphenol, 149
Dimethylsulfoxide (DMSO), 92
Diols, 22, 25–26, 31, 62, 95, 113, 120, 126, 134, 151–153, 156–159, 166–167, 184, 191
Dioxatetracyclic, 125
(−)-DIP-chloride, 184
2,3,-Diphenyl-3-hydroxypropionic acid, 1
Diphenylprolinol, 180
Dipole-dipole interactions, 99
Dipropionate, 135
Dirhodium
 tetrakis[(S)-N-(dodecylbenzenesulfonyl)
 prolinate, 35
(+)-Discodermolide, 15, 25, 66–67, 119, 121
Disuccinyl carbonate, 75
Dithiane, 156
Dithiolane, 69
1,2-Divinylcyclobutanoxide, 41
Dolabelide D, 143
Dysentery, 40
Dysidiolide, 183

(−)-Ebelactone A, 164–165
(E)-Crotyl propanoate, 29
(E)-4-Decenoic acid, 27
E-Enolates, 49–50, 53–54
Electron repulsion, 142, 175
Electrostatic:
 interactions, 83
 repulsion, 75
(E)-4-Methyl-4-hexenal, 188
(E)-2-Methyl-3-phenyl-2-propen-1-ol, 21

Enantiofacial selectivity, 93, 189
Enantiomer, 142
Enantiomeric ratio, 112
Enantioselectivity, 16, 35–36, 40, 92, 96, 103, 105, 107–108, 121, 123, 126, 137, 141, 143, 148–149, 173, 175–176, 179, 181, 184
Enediol isofuran, 188–189
Enolates, 166
Enol borinates, 51–52
Enone, 159
Epothilone A, 93, 153–154
Epoxidation, 40
Epoxide, 40
(−)-Erinacine B, 184–185
Erythromycins, 59
Erythronolide A, 159–160
Esterases, 73, 165
Esterification, 30–31, 34
Estrogens, 14
1,2-Ethanedithiol, 68
Ethanol, 20, 173
Ethyl:
 acetate, 81
 acetoacetate, 5
 orthoacetate, 20, 22
Ethylene, 35, 18
3-Ethyl-3-pentanethiol group, 78
Evans-Chapman-Carreira reduction, 73, 161–162, 166
Evans-Tishchenko reduction, 25, 167–168, 170–171, 173
(−)-3-Exo-morpholinoisoborneol (MIB), 16–17

FD-891, 70
FKBP12, 168
FK-506, 124
Fluoride, 132
Fluorine, 100
Formaldehyde, 49, 53, 98, 147
Formalin, 45
(±)-Frullanolide, 30–31
Furan, 66

(±)-Garsubellin A., 15, 18
Glucal, 115
Glucolipsin A, 63
Glucose, 95–96
Glycal, 30
Gram-positive bacteria, 191
Grignard reagent, 42

Half-chair transition state, 185

Hardy, Elizabeth M., 9
Hemiacetal, 168
(−)-Hennoxazole, 182
Hepatotoxic cyclic peptides, 31
(−)-Heptemerone B, 188, 190
1,5-Hexadiene derivatives, 10
Hexamethylphosphoramide (HMPA), 29, 31, 34, 156
(+)-Hippospongic acid A, 26
HIV replication, 162
Homoallylic alcohols, 97
Hydroboration, 121, 143
Hydrocinnamaldehyde, 81
Hydrogen:
  characterized, 53, 103
  chloride, 185
  fluoride, 60
Hydrogenation, 121, 136, 177, 191
Hydrogenolysis, 47
Hydrolysis, 30–31, 34, 43, 70, 179, 182
Hydroxy groups, 162
3-Hydroxy-1,5-hexadienes, 37
Hydroxyl group, 45, 88, 110, 159, 190
3-Hydroxy-3-methylglutaryl coenzyme A (HMG-CoA), 153
Hyodeoxycholic acid methyl ester, 118, 121
Hyroindolone, 47

Imidazole, 110
Iminium ion, 43, 45–46, 184
Indanolyloxy oxygen, 89
Indolizidine alkaloid, 24–25
Interiotherin A, 142–143
Iodolactone, 31
Iodolactonization, 31
Ireland-Claisen rearrrangement, 27–32
Iridium, 16
Isobutyraldehyde, 58, 92, 123, 171
Isomerization, 15–17
Isomutyl group, 14
Isopinylboryl enolate, 170
Isopropyl groups, 168
2,3-Isopropylidene-L-erythrose, 24
Isopropylidenetriphenylphosphorane, 24
Isoxazoline, 153
Ivanov reaction, 1

Japanese beetle pheromone, 176–177
Johnson-Claisen rearrangement, 20–26

Ketoacid, 93
Ketoaldehyde, 170

Ketones, 26, 40–41, 47, 52, 97, 121, 124, 131, 134, 147–1513, 156–159, 161–162, 166–168, 173–181, 183–184, 186, 188–190
KOMe, 39

Lactam, bicyclic, 73
Lactols, 68, 143
Lactone, 25–26, 30–31, 177, 193
Lactonization, 25, 31
Lasalocid A (X537A), 29–30
(+)-Lasonolide A, 69, 116, 120
L-deoxythreose ketal, 117
Leighton's silicon reagent, 142
(−)-Lepadiformine, 177
Leucascandrolide A, 171–172
Leukemia cancer cell lines, 183
Leustroducsin B, 66
Ligands, 58, 78, 126–127, 134
Lipases, 73, 165
Lithiation, 34
Lithium:
  acetylide, 191
  alkoxyaluminum hydride, 147
  aluminum hydride, 22, 147–149
  characterized, 187
  diisopropylamide (LDA), 27, 34, 49, 51
  enolates, 52
  hydroperoxide, 62
  peroxide, 64
  trimethylsilylacetylide, 193
L-proline, 91
Lung carcinoma cells, 120, 183
Lutidine, 109, 119, 122, 141–142, 157, 182
LY426965, 138

Macrolactin A, 162–163
Magnaporthe grisea, 188
Magnesium:
  bromide, 84
  chloride, 81
  enolate, 1
Mannose, 96
Mannosidases, 25
Mesityl phenyl ketone, 147
Meso-dialdehyde, 68
Mesylation, 25
Metal allylation reactions:
  boron allylation reaction, 102–126
  characterized, 97–101
  silicon allylation reaction, 127–144
Metal-centered steric effects, 52
Metal enolates, 91
Metallocycle, 89

Metal-oxygen bonds, 53
Methacrolein, 144
Methanol, 39, 153
Methanolysis, 26, 106, 110
Methoxyallyl, 113
Methoxydiethylborane, 153
Methoxyethoxymethyl (MEM), 80
Methoxy ketone, 37
Methoxypropene, 15, 114, 133
Methyl:
 acetals, 9
 groups, 8, 21, 35, 133, 158, 185
 iodide, 41
 monate C, 175–176
 propanoate, 50
Methylamine, 185
Methylation, 41
3-Methyl-2-butanone, 184
Methylenation, 121
(3R,5E)-3-Methyl-3-phenyl-1,5-heptadiene, 32
2-Methyltetrahydropyran, 13
Microcystins, 31
Midland reduction, 189–190, 193
Migrastatin, 87
Mitsunobu conditions/reaction, 115–116, 183
Miyakolide, 79
Monoazide, 25
Monoesters, 171
Monolithiation, 188
Monosilyation, 31, 142
(−)-Morphine, 184
(+)-Murisolin, 111, 115–116

Naphthalenes, meso, 10, 12
2-Naphthyl, 179
Narasaka reduction, 153
N-bromosuccinimide (NBS), 122, 124, 156
Neurodegenerative diseases, 86
N-formylmethionine aminopeptidases, 165
Nikkomycin B, 108, 111
Nitrile oxide, 115
4-Nitrobenzaldehyde, 92
Nitrogen, 142
(−)-N-methylephedrine, 149
N,N-dimethylformamide (DMF), 131–132
Nonactin, 156
Norephedrine, 61, 78, 80
Norpseudoephedrine, 136
N-oxazolidinethione propionates, 71
Noyori reduction, 175
Nozaki-Hiyama-Kishi reaction, 26
N-propionylbornanesultam, 68–69
N-propionyloxazolidinone, 63, 66, 83, 87, 165, 170

N-propionyloxazolidinethione, 72
N-propionylsultam, 67
Nuclear magnetic resonance (NMR), 67, 132
Nucleophilicity, 132
(2E,6E)-Octa-2,6-diene, 12
(2Z,6Z)-Octa-2,6-diene, 12

Olefination reaction, 165, 168
Olefins, 17, 136
Organolithium, 188
Orostanal, 121
Oxazaborolidine, 179–181, 183
Oxazolidine, 47
Oxazolidinones, 61–63, 75
Oxidation, 22, 48, 73, 80, 112, 158, 165, 183, 188
Oxidative lactonization, 25
Oxirane, 41
Oxygen, 106, 116, 158
Ozonolysis, 108, 110, 112–113, 118, 142, 190

(±)-Pancracine, 45–46
Parkinson's Disease, 86
3-Pentanone, 50
(−)-Pestalotin, 189, 192
Pharmacotherapy, 138
Phenylacetic acid, 1
1-Phenyl-3-buten-1-ol, 139
Phenyl groups, 79, 130, 134, 175, 180–181
Phenylmagnesium bromide, 179
Phenyls, 103, 137
(+)-Phomopsidin, 193–194
Phorbol, 65
Phorboxazole A, 106, 108, 123–124
Phosphonate, 188
Phosphonium bromide, 31
Pictet-Spengler cyclization, 47
Pinacolboronates, 100
Pinacolone, 148
Pivalic acid, 156–157
4-(Pivaloyloxy)benzaldehyde, 112
(±)-Pleuromutilin, 40, 46
p-methoxybenzyl (PMB), 65, 144, 170
p-methoxybenzyl ether, 83
p-nitrobenzoate, 170
Polyols, 83, 124
Potassium:
 hydride, 37, 40
 hydroxide, 125
 methoxide, 39
Prelog-Djerassi lactonic acid, 60
(+)-Preussin, 47–48
Propanal dihydroborinate, 53
2-Propenyllithium, 22

2-Propenylmagnesium bromide, 166
Propionamide, 65
Propionic acid, 20, 22, 25
Propionyl chloride, 31, 73, 75
Prostaglandin A_2, 24
Prostaglandin E_2 (PGE2), 175–176
Proteases, 89
Prozac, 185
Pseudoenantiomerics, 71
Pseudoephedrine, 138, 141
Pseudomonic acid C^6, 175
Psymberin, 141–143
P388 murine leukemia cancer cell lines, 183
p-toluenesulfonic acid, 70, 177
p-trifluoromethylphenol, 185
Pyridine, 31, 109–110, 114, 176, 183, 185
Pyridinium *p*-toluenesulfonate (PPTS), 15, 156, 166
Pyrrolidines, 31, 45–46, 48, 93, 137

Quantum mechanics, 5
Quenching, 153

Racemic 3,4-dimethylhexa-1,5-diene, 12
(−)-Rapamycin, 168–169
Ras farnesyltransferase, 42
Rat glia cells, 113
(*R*)-4-Benzyl-3-propionyloxazolidin-2-one, 73
Recrystallization, 185
R group, 13, 27, 49, 52
Rhizoxin D, 80, 84
Rhodium, 36, 40
Rifamycin S, 114, 117
RK-397, 165–166
(+)-Roxaticin, 164, 166
(*R*)-Phenylglycine, 84
(*R*)-1-Phenyl-1-propanol, 179
(*R*,*R*)-Bromoborane, 124
(*R*, *R*)-Dimethyl tartrate allylboronate, 117
R^2, 162
Rutamycin B, 64–65
Ruthenium, 15, 35

(−)-Salsolene oxide, 40–41
(1*S*)-(−)-Salsolidine, 182–183
Samarium, 168, 170–171
Saponification, 31–32, 170
Saquinavir, 76–77
(+)-Saudin, 15, 17
Scandium, 103, 106
(−)-Scopadulcic acid A, 34
(*S*)-Cyclohexenol, 184
(*S*)-(−)-2-(Diphenylhydroxymethyl)pyrrolidine, 179

Selenoxide, 26
Serotonin, 138
Sesquiterpenes, 40
S-3-(3-Ethyl)pentylpropanethioate, 78
(*S*)-Fluoxetine hydrochloride, 185, 187
Sharpless dihydroxylation, 25
[3,3]-Sigmatropic rearrangements:
  anionic Oxy-Cope, 36–42
  aza-Cope-Mannich, 43, 45–48
  Claisen, 3–7, 9–10, 13–19, 49
  Cope, 5, 9–10, 32–36, 49
  Ireland-Claisen, 27, 29–32
  Johnson-Claisen, 20–26
Signal transduction, 66
Silacyclobutanes, allylic, 140
Silanes, 131, 136–138, 171
Silicon, 140
Silylation, 70, 81, 121, 124, 170
Silyl enol, 14
Silylketene acetals, 27, 31
Silyloxy cycloheptadiene, 34
Silyls, 25, 31
(*S*)-mandelic acid, 57, 59
Smoking cessation, 138
Sodium:
  borohydride, 147, 153
  hypochlorite, 115
  metaperiodate, 60
(−)-Sparteine, 71–73
(*S*)-Phenylalanine, 63
Spirocyclization reaction, 179
(*S*)-Prolinol, 178–179
Squalene, 22–23, 42
Stannane, 124
(−)-Stemoamide, 83–84, 88
Stemona, 83
Stereochemistry, 116
Stereoselective reactions:
  asymmetric reduction, 173–194
  characterized, 147–150
  diastereoselective *anti*-reduction of β-hydroxy ketones, 161–173
  diastereoselective *syn*-reduction of β-hydroxy ketones, 151–160
Stereoselectivity, 14, 20, 26, 78, 93, 134
Steric repulsion, 9, 57
Stien, 123
Stilbenediamine, 123
Stoichiometric borane, 179
Streptomyces:
  roseosporus, 159
  toxytricini, 89
(−)-Strychinine, 46–47
Sulfate, 156

Sulfones, 168
Sultam propionate, 70
Sumarium iodide, 153
Suzukii-Miyaura coupling, 143
(S)-Valine methyl ester hydrochloride, 179
(S)-Valinol, 61
(−)-Swainsonine, 24–25
Swern oxidation, 73, 80, 121, 165
Swinholide A, 155–156
Syn-aldol reaction:
  asymmetric, 57–77
  characterized, 52, 57
2,3-Syn-disubstituted 4-heptenol, 17
Syn-isomer, 50–52, 62
Syn\anti, 21, 49

(−)-Talaumidin, 86, 88
Tartrolon B, 190, 193
TBSOTf, 70
Terpenoids, 184
Tert-butyl-dimethylsilyl (TBS):
  characterized, 9, 88, 118
  chloride, 27, 31
  ether, 143
  keteneacetal, 10, 27, 157
Tert-butyldiphenylsilyl (TBDPS) ether, 70, 112
(+)-Testudinariol A, 80, 83
Tetrafluoroboric acid, 182
Tetrahydrofuran (THF), 15, 27–28, 30–31, 51, 125, 179
(−)-Tetrahydrolipstatin, 89–90
Tetrahydropyran, 176
Tetramethylammonium triacetoxyborohydride, 166
Tetraol, 171
Tetrodotoxin, 15, 18
Tetronic acid, 115
Thermal rearrangement, 32
Thianes, 73
Thiazolidinethione, 81, 84
Three-carbon system, 10
Threo-isomer, 1
Tirandamycin, 158–159
Titanium, 71–72, 88–90
TMSCl, 34
Toluene, 30–31, 42, 114, 117–119, 138, 141, 181–182
Transamidation, 66
Transition state. See specific types of transition states
Transmetalation, 124

Trans-3-methyl-6-phenyl-1,5-heptadiene, 32
Triacetoxyborohydride reduction, 167
Trialkylsilanes, 127
Trichloroacetimidate, 63
(+)-Trienomycinol TBS ether, 62–63
(+)-Trienomycins, 62
Triethylamine, 24, 76, 78, 80–81, 133
Triethyl borane/sodium borohydride, 153
Triethylsilylketene acetal, 31
Trifluoroacetic acid, 121
Trifluoroperoxyacetic acid, 48
Triisopropylsilyl triflate, 170
Trimethylbenzoyl chloride, 182
Trimethyl orthoacetate, 24
Trimethylsilyl, 123, 139, 159
4-Trimethylsilyloxyl-1,2-dihydronaphthalene, 35
Trisallylborane, 126
Tris(triphenylphosphine)ruthenium(II) dichloride, 15
Tumor cell lines, 170, 177
Twist boat-like transition, 53, 55, 82, 98, 152

U.S. Food and Drug Administration (FDA), 77

Valilactone, 73–74
Valinol, 178
Vinyl bromide, 26
Vinyldiazoacetate, 36
Vinyl iodide, 165, 184
Vinyllithium, 41

Weinreb amide, 62, 66, 83, 170, 187, 191, 193
Wilkinson's catalyst, 121
Wittig olefination, 165
Wittig reaction, 32

Xenical, 90
X537A, 29–30
X-14994, 166
Xylene, 25

Z-boronate, 97
Z-enolates, 49, 52–54, 58
Zimmerman-Traxler transition state, 2–3, 49, 68, 79, 81, 89
Zinc:
  alkoxide, 16
  borohydride, 157–158
Zincophorin, 136
Zyrtec, 181

SCHEME INDEX OF NATURAL PRODUCTS

adda, 1.3 h
(−)-aflastatin A, 2.2n
alkannin, 4.3w
amphidinolide T3, 3.1gg
anachelin H, 4.1i
apoptolidin, 3.1l
(−)-asteriscanolide, 1.4e
azumamide A, 3.1 m

(−)-bafilomycin A$_1$, 3.1x
(+)-bengamide E, 2.1bb
bryostatin 2, 4.2 l

calcimycin, 1.3 g
(−)-calicheamicinone, 3.1o
(−)-callystatin A, 2.1w
(+)-calopin dimethyl ether, 1.1l
cetirizine hydrochloride (Zyrtec), 4.3 m
(−)-chlorothricolide, 4.3cc
(+)-clavosolide A, 4.2 g
compactin, 4.1e
(+)-conagenin, 4.1n
(+)-CP-263,114, 1.5 l
cryptophycin 1, 4.2k
CTX3 C, 3.1v
(+)-curacin A, 3.1 h

(+)-9(11)-dehydroestrone methyl ether, 1.1f
(−)-denticulatin A, 2.1r
6-deoxyerythronolide B, 2.1e
dermostatin A, 4.2 m
(+)-dihydrocostunolide, 1.4c
(9S)-dihydroerythronolide A, 2.1x
(+)-dihydromayurone, 1.5j
(+)-discodermolide, 1.2i, 2.1o, 3.1z
dolabelide D, 3.2w
dysidiolide, 4.3p

(−)-ebelactone A, 4.2 d

enediol isofuran, 4.3x
epothilone A, 2.3c, 4.1f
(−)-erinacine B, 4.3r
erythronolide A, 4.1 m

FD-891, 2.1t
FK-506, 3.1ee
(S)-(+)-fluoxetine hydrochloride (Prozac), 4.3v
(±)-frullanolide, 1.3f

(±)-garsubellin A, 1.1i
glucolipsin A, 2.1j
glucose, 2.3 g

(−)-hennoxazole, 4.3n
(−)-heptemerone B, 4.3y
(+)-hippospongic acid A, 1.2j

interiotherin A, 3.2v

Japanese beetle pheromone, 4.3f

lasalocid A (X537A), 1.3e
(+)-lasonolide A, 2.1s, 3.1w
(−)-lepadiformine, 4.3 g
leucascandrolide A, 4.2n
leustroducsin B, 2.1n
LY426965, 3.2m

macrolactin A, 4.2c
mannose, 2.3 g
methyl monate C, 4.3e
migrastatin, 2.2 m
miyakolide, 2.2e
(−)-morphine, 4.3q
(+)-murisolin, 3.1p

nikkomycin B, 3.1k
nonactin, 4.1 h

Six-Membered Transition States in Organic Synthesis, By Jaemoon Yang
Copyright © 2008 John Wiley & Sons, Inc.

orostanal, 3.1y

(±)-pancracine, 1.6f
(−)-pestalotin, 4.3bb
(+)-phomopsidin, 4.3ee
phorbol, 2.1m
phorboxazole A, 3.1i, 3.1ff
(±)-pleuromutilin, 1.5i
(+)-preussin, 1.6h
prostaglandin A_2, 1.2g
prostaglandin E_2, 4.3d
Prozac ((S)-(+)-fluoxetine hydrochloride), 4.3v
psymberin, 3.2u

(−)-rapamycin, 4.2j
rhizoxin D, 2.2g
rifamycin S, 3.1u
RK-397, 4.2f
(+)-roxaticin, 4.2e
rutamycin B, 2.1l

(−)-salsolene oxide, 1.5k
(1S)-(−)-salsolidine, 4.3o

saquinavir, 2.1dd
(+)-saudin, 1.1g
(−)-scopadulcic acid A, 1.4d
squalene, 1.2f
(−)-stemoamide, 2.2o
(−)-strychnine, 1.6g
(−)-swainsonine, 1.2h
swinholide A, 4.1g

(−)-talaumidin, 2.2p
tartrolon B, 4.3dd
(+)-testudinariol A, 2.2f
(−)-tetrahydrolipstatin, 2.2t
tetrodotoxin, 1.1h
tirandamycin, 4.1l
(+)-trienomycinol TBS ether, 2.1i

valilactone, 2.1y

X537A (lasalocid A), 1.3e

zincophorin, 3.2j
Zyrtec (cetirizine hydrochloride), 4.3m